MAKING FACES

MAKING FACES
The Evolutionary Origins of the Human Face

ADAM S. WILKINS

Illustrated by Sarah Kennedy

The Belknap Press of Harvard University Press
Cambridge, Massachusetts
London, England
2017

First printing

Library of Congress Cataloging-in-Publication Data

Names: Wilkins, A. S. (Adam S.), 1945– author.
Title: Making faces : the evolutionary origins of the human face / Adam S. Wilkins;
illustrated by Sarah Kennedy.
Description: Cambridge, Massachusetts : The Belknap Press of Harvard University Press, 2017. |
Includes bibliographical references and index.
Identifiers: LCCN 2016017279 | ISBN 9780674725522 (hard cover : alk. paper)
Subjects: LCSH: Face—Evolution. | Face—Differentiation. | Human evolution.
Classification: LCC GN281.4 .W53 2017 | DDC 599.93/8—dc23
LC record available at https://lccn.loc.gov/2016017279

In memory of my father, Alvin Meyer (1916–1963); my mother, Sophie Wilkins (1915–2003); and my stepfather, Thurman Wilkins (1916–1997)

CONTENTS

PREFACE

Familiarity need not breed contempt, but it often induces blindness. That blindness can obscure either the flaws of the familiar entity or, more surprisingly, its distinctive or unique properties. Take human language, for example. Precisely because we use it all the time without thinking, we rarely pause to reflect on just what an extraordinary ability it is. Another example of something whose familiarity dulls the sense of wonder is the subject of this book: the face. The human face is highly distinctive not only in its set of physical features compared to our animal cousins, but also in being the most expressive face of any creature on Earth, and as such it plays a crucial part in our social existence. Of course, everyone is aware of the faces of others, as well as her or his own, and senses that human faces are "interesting," yet for the most part we simply do not see the face's exceptionally special qualities; those are simply part of our daily existence.

To appreciate fully what makes something unusual, however, we must understand its origins. If the object of interest is biological, and if one does not believe divine creation to be its source, that awareness involves understanding its evolution. In this book, I will attempt to explain the human face in terms of its evolutionary history. As we shall see, that history is connected to that other distinctive human attribute mentioned above: language.

Although the past ten to fifteen years have seen a plethora of new scholarly and popular books on the human face, most writers neglect its evolutionary history, giving it relatively brief attention at most. This book will attempt to correct that omission. While neither a definitive nor a final account—no evolutionary history can aspire to be either—it will describe that history, beginning with the first vertebrate faces half a billion years ago and closing with the final fashioning of the face in our recent human ancestors. In particular, we will see how the physical shaping of the human face was rooted in our high sociability, a property bequeathed to us by our anthropoid primate ancestors. To the extent that human faces and those of our anthropoid relatives were shaped by the evolution of social interactions, their history diverges from that of all other higher animals.

This book is intended primarily for general readers, especially those interested in human biology, though I believe it will also be of interest to my fellow biologists. Accordingly, the writing style is comparable to that of a series of public lectures. As such, I occasionally employ personal pronouns. When I write *we,* I am referring to you and me—the readers and the author—though in some places I am referring to humanity in general; the context will make clear the intended sense. In addition, I refer to myself occasionally. After all, this book is the product not of some disembodied spirit but of an individual with his own viewpoints, judgments, and, inevitably, blind spots.

Indeed, every book is to some extent a personal statement of the author and has its origins in personal sources. Let me state mine. First, like all children, I found the faces of my parents and others around me interesting and, undoubtedly from earliest infancy, inspected them for clues to what they were feeling. At a certain point, I cannot remember when, I became conscious that I was doing so—again, this may be common to all or most children—and from then on paid even more attention to their faces. Second, I have always been good at recognizing and remembering faces in general and frequently look at faces in crowds as I walk. I have always found faces extremely interesting, though whether this was simply an extension of my being a keen observer of my parents' faces, I do not know. Third, since my first visit to the American Museum of Natural History in New York City at age five or six, shepherded by my father, I have been fascinated by the phenomenon of evolution, which he explained to me in simple terms on one of the early visits. I remember being especially intrigued in the old Hall of Mammals by the variety of extinct elephants, rhinoceroses, and horses—in particular, their different sizes and bodily proportions and shapes of head as seen directly in the skeletons and in the wonderfully evocative paintings by Charles R. Knight that graced the upper walls. I felt that the source of those differences among obviously related animals must be significant, although I was nearly a decade away from learning in school about that source, namely, biological development. In retrospect, the distance from my childhood fascination with different but similar mammals to a strong interest in how the human face evolved does not seem large. Perhaps it was almost inevitable that a lifetime's fascination with faces would combine with nearly as long an interest in evolution to yield a desire to write about the evolutionary origins of the human face.

As highly visual, sentient animals, we habitually look out on the world from our faces, continually assessing the scene before us. In this book, we will instead direct our gaze toward ourselves, specifically toward our shared human face, to understand how it came to be the way it is and how we as a species came to be what we are.

MAKING FACES

The history of the human face is a book we don't tire of, if we can get its grand truths, and learn them by heart.

Julia Margaret Cameron (1815–1879), photographer

— one —

THINKING ABOUT THE HUMAN
FACE AS A PRODUCT OF EVOLUTION

Introduction: The Remarkable Human Face

In discussing anything, it often helps to start by defining it. What precisely is meant by the term "the face"? If we leave aside metaphorical usages (such as the "face" of a watch or a cliff), the word denotes the front part of an animal's head that features a pair of eyes above a mouth. That definition, however, gives scant clue to what makes the face interesting. For that, one needs to know what it does, what its function is.

The clue to the face's function is that it houses three (of five) of the key sensory organs—those for vision, olfaction, and taste. It thus constitutes a compact site for receiving vital information about food, possible mates, and potential threats. In effect, the face serves as the sensory headquarters of the animal. Of course, it also contains the mouth, which not only houses the organs of taste but also is crucial for ingesting food items. Indeed, the original function of the sensory apparatus of the face was almost certainly to help direct the animal toward potential food for ingestion.[1]

Given the importance of an animal's face for guiding its owner through the world—to seek its delights and avoid its dangers—it is not surprising that the face is a universal feature of the complex animals we designate as **vertebrates**. From sharks to frogs, from snakes and birds to mice and men, all vertebrates have faces, in contrast with such features as lungs, limbs or tails, each of which is missing in some vertebrate groups.

Nevertheless, the variety of animal faces is truly striking. Even if we leave aside the intriguing and diverse faces of fish, frogs, snakes, and birds and consider only mammals, the range of facial types is still impressive. The faces of porpoises, warthogs, vampire bats, rabbits, elephants, anteaters, and gorillas, for example, are all highly distinctive. It may be surprising to realize, therefore, that the human face, which we unconsciously take as the standard for judging other animals' faces in terms of their strangeness or comicality, is actually one of the oddest of all.

That oddness is captured in a statement by Donald Enlow, one of the twentieth century's great experts on craniofacial structure, namely, the structure of the head and face:

> The human face is different. By ordinary mammalian standards, our facial features are unusual, specialized, and perhaps even grotesque. The long functional muzzle that marks the face of most other mammalian forms is all but lacking in man. The associated snout is reduced to a curious overhanging vestige. The face is flat, wide, and vertically disposed. Instead of a graceful facial contour sloping back to the skull roof, the human face possesses a unique, bulbous, upright forehead in front of an enormous braincase. The flattened face is diminutive in character relative to the remainder of an enlarged head. The eyes are close together and they face straight forward. The human dental arches are disproportionately small relative to the size of the whole body.[2]

Some of these distinctive differences between human faces and those of other **mammals** are illustrated in Figure 1.1. The face of a fox on the left has a far more typical mammalian face than the human on the right. Apart from the features mentioned by Enlow, another obvious difference is that the fox's face is fur covered while the human face is bare. In addition, foxes, like most mammals, possess "wet" noses while humans possess dry ones. Other unusual features of the human face include a bone-buttressed nose, a chin, everted lips, and eyes in which the colored portion, the iris, is conspicuously surrounded by a white area (the sclera). Now look at the animal face shown in the middle of the figure, that of a chimpanzee: it is intermediate between fox and human in its mix of features. The instantaneous impression, however, is that its face is more similar to the human face than that of the fox. Indeed, on several grounds, we know that chimpanzees are more closely related to humans than to foxes, a fact their faces ratify.

The human face differs strongly from other mammals not just physically, however, but in its "behavior"—that is, in its mobility and expressiveness. If you were to observe human beings, chimpanzees, and foxes interacting with their fellows (their **conspecifics**), you would notice that all three creatures exhibit facial expressions but that chimpanzees and

FIGURE 1.1 Comparison of fox, chimpanzee, and human faces. The fox has a rather generic mammalian face, while the human face is a true oddity among mammals (see text). In its mix of features, the chimpanzee face is in between the generic mammalian face and the human face but closer to the latter, befitting its evolutionary relationship to humans.

humans faces have far more expressive faces than foxes and that the human face is still more expressive than the chimp's. Human facial expressivity is particularly evident in conversations: when two humans converse directly, literally "face to face," there is a constant interplay of facial expressions, something without parallel in the worlds of the fox and the chimpanzee. (The latter can make a range of vocalizations, but these are not speech.) Human facial expressions are an important adjunct to understanding the meaning of the spoken words; made quickly and automatically, they either reinforce or undermine the sense of the spoken statements. The human face is indeed an exquisitely sensitive communications device capable of expressing an enormous range of emotional states, a great many of them highly subtle and involving miniscule displacements of features. For example, a slightly furrowed brow, with eyes slightly narrowed, denotes puzzlement; that same expression when accompanied by a slightly downturned mouth displays skepticism; slightly upturned corners of a mouth with parted lips signal happiness or amusement. while tightly closed lips can signify distrust. We make these expressions unconsciously, reflecting our inner psychic states, and also instantly "read" them in others. Thus, the vibrant play of facial expressions that accompanies talk constitutes a form of shadow dialogue behind the verbal exchange that we regard as the actual dialogue. They provide the emotional subtext that is often as important as the information conveyed by the spoken words. Given our dependence on language and speaking, we use our faces for communicating with our conspecifics far more frequently than any other animal species, even our nearest primate relatives, who can be highly facially expressive.[3]

Evidently, the human face does more than act as the individual's sensory headquarters, important as that general function remains; the face is integral to so much of our social interactions. We can see the special significance of the face in human existence in light of this fact: in the natural scheme of things, a human being's life begins and (often) ends with the sight of another person's face. Indeed, one might say that a normal human life is bookended by images of human faces. The first thing a newborn infant sees at all clearly, within a minute or so of birth, is a face. It is the mother's, and the baby sees it just after having been placed in her arms. The image is undoubtedly a bit blurry, and the infant has, at best, only the dimmest sense of what she or he is looking at—the concept of "who" as distinct from "what" will only develop later—but it is a face, and this particular face will be one of the most important objects in the universe of that infant. Many decades later, if that same individual—no longer an infant but an elderly person—has had the good fortune to have experienced close human relationships and is approaching the end in a conscious state, the last image seen is likely also to be a face, this time not a mother's but that of a beloved spouse or a grown-up child or grandchild or, perhaps, a close friend. In Westernized societies of the twenty-first century, this kind of peaceful leave-taking of life is less and less common but throughout most of human history, the sight of key faces at the beginning and the end of a life was far more typical.

In between birth and death, of course, everyone experiences the sight and associations of numerous other human faces. The actual number, of course, is a matter of chance but also a function of the kind of society in which one lives. In prehistorical times or the old peasant societies of medieval Europe or even today in geographically remote tribal groups in Africa, Asia and South America, the number of faces encountered in a lifetime might be only a few dozen, perhaps 200 at most. In modern industrial societies, in contrast, everyone is exposed to innumerable faces, both directly and indirectly through the various forms of electronic media. Most of the faces seen in crowds or as images on the media are too transient to register but, on average, in the course of a lifetime, one probably recognizes and remembers several thousand different faces, each having distinct associations. Blind individuals lack this element of experience, yet even for those blind from birth there is often a compensatory way of knowing someone via the face—namely, touching that person's face. For most people, faces are the identity tags we present to one another when we meet.

Furthermore, an individual's face matters to its possessor, as a synecdoche for oneself: the importance we attach to presenting our "best face" to the world is closely connected to the feeling that our face represents who we are as individuals. Such facial self-consciousness is probably more extensive today than at most earlier times in human history, but it is a feature of all societies in which the concept of the individual exists and where reflecting surfaces exist, permitting a self-identification with one's face.

How did the face come to have such importance in human life? Ultimately, that is a question about deep time and our species, a question of our evolution (that is, if one does not subscribe to creationist views). That evolutionary history is the subject of this book, which will attempt to answer the question. Yet before entering into the evolutionary arena, we should take a brief look at the phenomenon just mentioned, namely, awareness of the human face as an embodiment of individuality. That is a matter of cultural, not biological, history, but it is relevant to the beginnings of the scientific inquiry into the evolution of the face. Cultural beliefs and practices are also relevant to the biological story because they have played a real, if indirect, role in the evolutionary shaping of the human face in human history, as we shall see.

Face Consciousness in History and the First Scientific Explorations of the Face

We do not know when humans first began to be interested in their own faces or became fully conscious of the idea that the face embodies individuality. Reacting to the faces of one's fellows is a basic characteristic not just of ourselves but also of our primate cousins, hence it is an ancient one. Conscious awareness of faces, however, and what they do is another matter. The only certainty is that the first strong indications of interest in the human face date to the beginnings of the different major civilizations, 10,000 to 5,000 years ago, as judged by the artifacts they left behind. There had previously been comparatively rare artistic representations of human heads and faces, some as early as 25,000 to 30,000 years ago, in the form of sculpted rock heads, but the artistic representation of individual faces seems to have begun with the rise of human civilization. It was within those large social structures that a stronger and more general interest in human faces and their variety seems to have bloomed, as judged from the pictorial art of the past 5,000 years, and especially during the last 2,500 years, with the face becoming a more frequent motif in art. This turn toward the depiction of the face in art was paralleled by attempts to decipher the deeper meaning of faces, especially what they reveal about character. Such efforts are collectively denoted as those of *physiognomy*, which began in ancient Greece approximately 2,500 years ago and has been in different forms a feature of Western civilization ever since. Though physiognomy is not a science and exhibited its last robust expression in the late nineteenth and early twentieth centuries, most people are intuitive physiognomists, feeling that a person's face is, in its habitual expressions if not its physical features, an index to his or her character.[4]

Genuine scientific investigation of the human face, however, has had a much shorter history. Its origins can be dated to the publication of a book in 1806, titled *Essays on the Anatomy of Expression in Painting,* by the prominent British physiologist, Sir Charles Bell (1774–1842). It was Bell who first described and noted the special importance of the muscles of the face, the **mimetic muscles**, concentrated around the mouth and eyes and making the wide range of human facial expressions possible. In identifying these muscles as the immediate source of human facial expression, Bell brought true science to the study of the face. Impressed as he was by the great facial expressive capacity of human beings, however, he believed these muscles were unique to humans and concluded they had been bestowed on humanity by God for the special purpose of expressing human feelings. In this belief, he represented his intellectual traditions, those of eighteenth-century natural philosophy.

Yet by the third and final version of Bell's book, published posthumously in 1844, that philosophical tradition was on the wane. The winds of conceptual change were blowing with ever greater force in the mid-nineteenth century, and it was none other than Charles Darwin who set out to rebut Bell's thesis. He did so in a book titled *The Expression of the Emotions in Man and the Animals,* published in 1872. There Darwin documented the wide range of expressive capacity in animals, both in body language and facial expressions specifically. In particular, he pointedly described similar expressive facial capacity to that of humans in our nearest animal cousins, the monkeys and apes. As Darwin made clear in his introductory remarks, the principal purpose of his book was to refute Bell's claim of a uniquely human facial expressive capacity.

Although the matter of expressions in animals might seem too specialized a subject to have warranted book-length treatment by Darwin, he clearly did not think so. After all, Bell had been a major scientific figure, and his views could not be ignored. Had he been right, Darwin's central claim—namely, that all biological features arise by evolutionary descent with modification—would have been fatally wounded. Indeed, if Bell were right about the mimetic muscles, why could not all special organismal properties be attributed to divine creation, as claimed by traditional religious belief? It was important to Darwin that Bell's argument be rejected, and *The Expression of the Emotions* set out to do so.

After Darwin's book, however, the subject of facial expressions received relatively little scientific attention for many decades. It only reemerged strongly in the early 1970s, with the work of Paul Ekman and his colleagues. That work focused on the communicative functions of human expressions and characterized them in exquisite detail, showing their power to display feelings. Today, this subject receives much attention, yet the resulting impressive body of work on human facial expressions largely gives lip service to its evolutionary dimensions but mainly neglects it, though this aspect was central to Darwin's interest. Indeed, some of the current literature is so focused on the distinctive aspects of human expressiveness that it is effectively, if not intentionally, antievolutionary in flavor.[5]

I believe the last book-length treatment of the evolution of the face was written in the late 1920s by a scientist at the American Museum of Natural History in New York City, William K. Gregory, the museum's curator of fishes. *Our Face from Fish to Man,* appearing in 1929, was written with erudition, insight, and wit. Gregory's focus was different from Darwin's, however. His book dealt not with facial expressions but with the physical or morphological evolution of the vertebrate face. The book is a delightful tour of the manifold variations of the vertebrate face that evolution has wrought, but Gregory's essential point was that the basic superstructure of the vertebrate face created by the bones of the skull has remained much the same from early fishes to modern humans, thus providing strong evidence that human beings are a product of evolution. While Gregory drew extensively on both fossil evidence and the comparative anatomy of living species, he recognized that the subject needed information drawn from other fields—in particular, developmental biology, genetics, neurobiology, and animal behavior studies. At the time, these disciplines furnished relatively little relevant information, but his belief in their potential was prescient and, as we will see, all these fields now contribute substantially.[6]

The starting point of any scientific inquiry into the face, however, remains Darwin's key insight: that the human face and its expressive capabilities have an evolutionary history as shown by our nearest animal relatives. The faces of monkeys and apes not only bear clear physical similarities to ours but also exhibit many facial expressions similar to ours. One has no difficulty in detecting fear, surprise, happiness, or anger, for example, in the facial expressions of a chimpanzee, and scientific work shows that this is not mere anthropomorphizing. In contrast, while nonprimate mammals such as cats and dogs have facial expressiveness, especially for showing fear or threat but also pleasure, their range of expressions is considerably more limited, as noted earlier.

That evolutionary history explains shared features of expression between human faces and our mammalian cousins, yet the employment of facial expressions, particularly in conjunction with speech, shows unique human aspects. Evidently, some novel evolutionary events linking the face and the brain must have taken place in the line of descent or **evolutionary lineage** that led from ape-like prehuman ancestors to humans. That line of descent is termed the **hominin lineage** and denotes the branch of the primates that diverged from that of the chimpanzees and gave rise to modern humans. It follows that any account that tries to do justice to the full story of human evolution must consider both the elements that are widely shared with other primates, nonprimate mammals, and even vertebrates generally and the elements in the human face that are novel. This book will sketch the origins of both sets of features, although its main focus will be on the events that generated the unique properties of the human face.[7]

Interrogating the Evolutionary History of the Face via Five Questions

It is frequently remarked that the key to success in science is to ask the right questions. That seemingly straightforward statement, however, im-

Table 1.1. Five General Questions about Evolutionary Change as Applied to the Face		
Question Designation	**Specifics of the Question**	**Kinds of Evidence Needed to Answer**
1. The "What happened?" question	What were the actual morphological changes in animal faces in vertebrate history (from fish to humans)?	Fossil evidence
2. The "When?" question	When did this sequence of key changes occur?	Fossil evidence with radioisotopic dating, faunal association, molecular clocks
3. The "Where?" question	Where did the these events in the evolutionary shaping of the face first occur?	Fossil evidence in conjunction with maps of changing continental and regional positions
4. The "Foundations?" question	What were the genetic and developmental changes that underlay the evolution of the human face?	Comparative studies of developing embryos of birds and mammals
5. The "Why?" question	How did the changing physical—and social—environments select for the changes that drove the evolution of the human face?	Clues from skulls and teeth Faunal association and physical evidence on environments Comparative studies of development, behavior, and brain structure using the methods of comparative neurobiology, psychology, and animal behavior

mediately prompts a question itself: how does one know when a scientific question is "right"? The rough answer is that a good scientific question is one that provides a clue as to how its subject might be investigated. By that criterion, such queries as "How did life first arise on Earth?" or "How does the brain work?" are not good scientific questions; they provide no hint to where to begin. In any scientific inquiry, therefore, one must pose narrower questions, and frequently answering them involves measurement. If one can measure something, it is then possible to make comparisons between different entities or situations and compare the relevant measurements, which in turn often generate new insights.[8]

Keeping such thoughts in mind, we can see that the question this book addresses, "How did the human face arise through evolution?" is not a good one. It, too, needs to be broken down into smaller questions, which must then be applied to specific events in evolution and indeed repeatedly to sequential steps in that history. For evolution, generally, there are five such component questions, and these are listed here in Table 1.1. Let us look at them and see how they apply to the story of the face.

The "What Happened?" Question: What Were the Specific Visible Changes in the Evolutionary Sequence?

This question focuses on the events that created visible changes in shape, detailed form, and size in evolution. These events are in the domain of **morphology**, the scientific study of the form of organisms. Those differences in form are the dramatic manifestation and embodiment of evolutionary change. If, for example, you are looking at the dinosaur skeletons in a natural history museum and want to trace the evolutionary route by which giant creatures such as *Apatosaurus* or *Tyranosaurus* evolved from earlier, smaller dinosaurs, you are asking the "What happened?" question. Good fossil evidence is essential to providing information relevant to answering the question, including facts not only about the morphology but also about the physiology and behavior of the animal that left the remains. For animals and plants that left little or no fossil record, whether purely by chance or as a function of their tissues being unable to fossilize, the changes in morphology can never be ascertained. Nevertheless, even for species that left an abundant fossil record, vital pieces of evidence are often missing; in such cases, investigators must use creative inference to reconstruct the missing events.

Yet apart from such technical difficulties in answering the "What happened?" question, the question itself generates a problem for anyone trying to write about evolutionary history: where does one begin the story? (Indeed, this is a problem of historical accounts generally since, for any chosen starting point, there must have been preceding events that led to it.) For the human face, for example, one could concentrate solely on the hominin lineage, beginning with the chimpanzee-like earliest hominins, whose first species appeared about 6 million to 7 million years ago. This, however, would be analogous to writing a history of the building of the Eiffel Tower without discussing how its foundations were laid. In the case of the human face today, its biological foundations are shared not just with our hominin ancestors but also, as noted earlier, with our nearest animal cousins, the great apes and the monkeys, who make up the large group of animals known as the **anthropoid primates** (literally, "manlike primates").[9]

The faces of some representative species of this group are shown in **Plates 1–4**. Despite the obvious differences between them, their "family" facial resemblance, including their resemblances to humans, is obvious. Clearly, an understanding of the origins of the human face needs to take account of how those shared anthropoid primate facial features arose in evolution.

Nevertheless, the story also does not begin with the anthropoid primates because they too had evolutionary precursors. Those ancestral primates possessed more generic mammalian facial features, namely, a projecting muzzle, more typical mammalian teeth (either larger canines or larger incisors and less developed and probably fewer premolars or molars), more widely spaced eyes, faces covered with fur—and, almost certainly, lesser facial expressiveness. In turn, those early primates had evolved from a succession of earlier mammalian and premammalian animal species on a trajectory that can be traced in reverse far back in time to the first vertebrates.

Hence, to return to the question, where should one begin the story of the evolution of the face? I decided to begin with the very origins of the vertebrate face, namely, with the first vertebrates—tiny fishes lacking jaws—that appeared on Earth more than half a billion years ago in the geological interval of Earth's history designated as the **Cambrian period**. Doing so has the advantage of being the real start of the history since there was no ancestral face as such before the vertebrates arose, although of course their ancestors possessed heads and mouths. There is, however, an additional reason: those beginnings involve a central element of the whole story, the appearance of a new kind of cell, one not seen in any of the ancestors of the vertebrates: **neural crest cells**. These comparatively little known cells, only rarely mentioned in the popular science literature, appear early in embryonic development in vertebrates but then move to other locations in the embryo. There they multiply and are replaced by the various kinds of specialized cells they give rise to, each characteristic of its location. Those descendant cells contribute to a wealth of different tissues and structures in the body, including the bones of the face and the mimetic muscles. Thus, the neural crest cells are essential for the development of vertebrate faces—as well as much else. Correspondingly, much of the evolutionary differences between different kinds of faces are due to evolved differences in the activities of neural crest cells.

To begin the story of the human face with the first vertebrates, however, risks overwhelming the story—not to mention the reader—with a great deal of material. To make this vast sweep of time and change manageable, I have therefore divided it, quasi-arbitrarily, into seven stages, as

Table 1.2. Seven Stages in the Evolution of the Face in the Long Hominocentric Lineage Path

Number	Biological Stage	Approximate Dates*	Salient Facial Feature Changes
1.	Origins of the vertebrate face	520–500 mya	Formation of first faces, in jawless fishes, from primitive chordate ancestors
2.	The first gnathostomes	420–410 mya	Evolution of jaws
3.	Origins of amphibians, then amniotes from first amphibians, then the mammal-like reptiles (synapsids)	370–290 mya	Alterations in skull bones, jaws Beginning of simplification and strengthening of jaws Diversification of teeth
4.	From early synapsids to Mammaliaformes to first true mammals	200–125 mya	Further changes in jaws and dentition Evolution of fur Evolution of true mammalian muzzle Improvements of olfaction and vision
5	Placental mammals to first primates	80–64 mya	?
6.	First primates to anthropoid primates	63–55 mya	Further dentition changes Reduction of muzzle More convergent (binocular) vision and better acute, close-up vision Improved color vision Loss of facial fur in anthropoid primates
7.	Apelike to hominins to humans	6 mya–200 kya	Increasing brain size and development of foreheads Further reduction of muzzle Development of chins and bone-buttressed noses

*mya = millions of years ago; kya = thousands of years ago

shown in Table 1.2. Each stage contributed vital elements in the evolutionary shaping of what became, in the evolutionary lineage that led to human beings, the human face.

Given the plethora of different vertebrates that have populated the Earth since the first fishes appeared in the Cambrian period, there have been many, indeed innumerable, different vertebrate faces, each the product of its own long evolutionary lineage, the ensemble creating an immense branching "bush" of vertebrate evolutionary lineages. Much of that **phylogeny** will have to be ignored in order to concentrate on tracing the specific lineage of animals that led to human beings. To do so does not imply that it was intrinsically more special than any other, let alone foreordained, but simply to maintain focus on the sequence of evolutionary events that led to humans specifically. I will call this evolutionary pathway the **long hominocentric lineage path** (LHLP). It denotes the direct line of descent—the sequence of species—from the first fishes to the first humans, so it begins long before there were primates, let alone hominins. This is a new and admittedly somewhat awkward term, but it will serve our purposes, namely, to focus on the evolutionary sequence of changes that in retrospect can be seen to have led to humans and the human face specifically. In a crude, highly simplified and condensed form, this particular path is diagrammed in Figure 1.2. In contrast to the entire LHLP, the term **hominin lineage** only designates the late and comparatively brief phase of the LHLP in which hominins branched off from the chimpanzee lineage.

The "When" Question: Establishing When Key Events in the Evolutionary Shaping of the Face Took Place

Ascertaining when specific things happened in evolution or the durations of certain phases in that history is often of vital importance to understanding evolutionary events. Biologists who deal with the distant past always want to know such things as: When did change A (creating a new structure) take place? Did it occur before, concomitant with, or after change B? If before, did it potentiate change B? Over how long a period did a particular evolutionary transformation take place? When did specific extinction events take place? All evolutionary histories require such information.

By the late eighteenth century, geologists had shown that Earth had a far longer history than that deduced from the Bible but had to settle for the relative timing of different events; absolute ages could not be assessed. Such comparative dating was made from the evidence of fossils in sedimentary rocks and the relative positions of the different strata in those rocks that contained those fossils. Since sedimentary rock formations are produced by the laying down of silt in water to form rocky strata, and these necessarily build up from bottom to top, layers closer to the top were

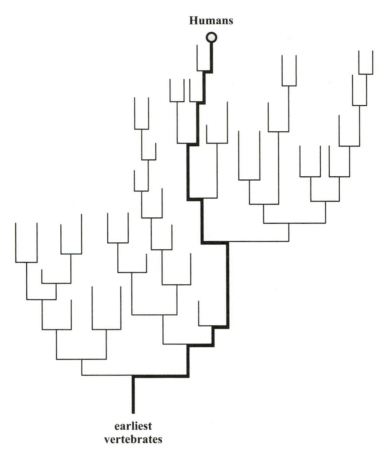

FIGURE 1.2 The *long hominocentric lineage path* (LHLP) shown in crude schematic form within a highly simplified evolutionary or phylogenetic "tree" of the vertebrates. All of the details have been omitted, and the tree is indicated in highly schematic form. The point of the picture is to illustrate the concept of a lineage path from its beginnings to a particular terminus—in this case, our species.

formed more recently than those lower down. Correspondingly, fossils found in upper layers must be younger than fossils found in deeper layers.[10]

Such relative dating provided the essential temporal context of fossil (paleontological) studies throughout the nineteenth century and into the twentieth, but actual specific dates could only be approximated by indirect and often hugely inaccurate methods. Nevertheless, in the late nineteenth century, questions of time were considered crucial to weighing the validity of Darwin's theory: had there really been enough time in Earth's history to support the extent of evolutionary change demanded by the theory? While Earth was clearly far older than the biblically allotted

6,000 years, the initial estimate by Lord Kelvin, one of the most respected physicists of his time, of 75 million years was still far too short to accommodate the span required by Darwinian evolution and correspondingly caused Darwin distress. (Kelvin's estimate was based on assumptions about the rate of cooling of Earth from its original igneous state in ignorance of the fact of heating of Earth's core by radioactive decay.)

The advent of **radiometric dating** in the first half of the twentieth century provided the first reliable estimates of Earth's age at 4.5 billion years, a result that would have deeply reassured Darwin had he lived to see it. This method, based on the invariant rates of decay of different radioisotopes, also made it possible to obtain absolute dates for fossil specimens, hence the time of existence of the organisms from which those fossils were the remaining vestiges. These methods helped transform both geology and paleontology from largely qualitative sciences to quantitative ones (Box 1.1).

Then, in the 1960s, an ingenious method was developed using biological material to date evolutionary events. Its special gift was that it could be used to estimate the times of divergence of evolutionary lineages, namely, the times in the distant past at which related lineages, call them B and C, arose from an earlier common ancestral lineage, A. This involved measuring the extents of difference between the DNA sequences of living species of the descendant lineages. Since those differences accumulate as a function of elapsed time from the temporal point of divergence from the common ancestor, they constitute a kind of timing device. For this reason, the method is termed the **molecular clock**. The basic methodology is discussed in Box 1.2.

Molecular clock analyses have major potential problems and corresponding uncertainties, but they can be highly informative when the various necessary corrections are made. Thus, for example, it was molecular clock data that provided a critical date in the study of human evolution: the estimated time of splitting of the chimpanzee and hominin lineages from a chimpanzee-like common ancestor. The derived estimate for this divergence is about 6 million to 7 million years ago, a much shorter interval than had been estimated by paleontologists earlier from the rather patchy fossil record. Thus, all of the features of the human face that differentiate our face from those of our nearest primate cousins, chimpanzees, took place within the relatively short time span of 6 million to 7 million

BOX 1.1 Radiometric Dating Methods

Radiometric (or "radioactive") dating methods provide estimates of absolute ages of samples. They rely on the fact that certain isotopes of certain elements—termed *nuclides*—undergo spontaneous decay, either with the emission of electrons from neutrons to yield an atom with one more proton in the nucleus and hence a new atomic number (that of another element) or by fission into two new kinds of atoms. In the latter case, one or both of the products then often further fissions into stable products. The basic property of radioactive decay is that it occurs at a fixed rate, irrespective of local conditions (temperature, pressure, different features of the chemical environment) and amount; it is said to be a *zero-order decay process*. The decay rate is given in terms of the *half-life* of that element, namely, the time at which half of the original amount has decayed. Thus, for a hypothetical nuclide with a half-life of 100,000 years, a sample of the substance will have only half the amount of original nuclide after 100,000 years and only one-quarter after 200,000 years. If there is a unique product of the decay, as there usually is, and no contamination of the sample with other material, then measuring the amount of product relative to the amount of the original nuclide can permit calculation the age of the sample using the known half-life. Since each nuclide has a characteristic decay rate, different nuclides of different elements will be useful for measuring different extents of elapsed time from the time of the sample's origin.

Therefore, for dating fossils, one needs to do the measurements using a nuclide whose unique products are present in the sample at measurable amounts and whose half-life is such that informative ratios of the product to the original nuclide can be obtained. Some of the most useful and frequently used methods are two uranium–lead conversions, each with a different initial uranium nuclide and its own half-life, and the potassium–argon decay system. Most of the approximate evolutionary datings given in this book come from these methods. Since animal bodies normally do not contain the elements or nuclides of interest, the measurements are typically of the surrounding mineral material in which the fossils are embedded.

For a comparatively few nuclides, what is measured is their loss, not the accumulation of a unique product. Carbon-14 (^{14}C), which decays to carbon-12 (^{12}C), is an example. It forms a constant proportion of the carbon in the carbon dioxide in the atmosphere due to bombardment with cosmic rays, becomes incorporated first in plant material through photosynthesis and then in animal bodies after those plants have been ingested, and subsequently decays with a half-life of 5,730 years. This is one of the few methods where the fossil material itself can be tested—given that carbon is a major constituent of living things—but the short half-life means that, in practice, ^{14}C dating cannot be done accurately on samples much older than 40,000 years. That time period, however, covers much of the recent history of *Homo sapiens*, including a large part of the time elapsed since the exodus from Africa. This method has thus been of great value in assessing the dates of particular archeological artifacts.[1]

---

1   For a simple, general and short discussion of the premises, methods and varieties of radiometric datings, the Wikipedia entry is good. A more detailed and authoritative account is in Faure (1998).

BOX 1.2 The Molecular Clock

The term *molecular clock* denotes both a hypothesis about the way genetic material changes over time and a set of techniques for estimating the times in the distant past at which different lines of organisms diverged from a common ancestor. The basic premise is that in all organismal lineages, from distant ancestors to modern-day descendants, heritable changes steadily accumulate in the genetic material, deoxyribonucleic acid (DNA), as it is passed from generation to generation. Because DNA consists of linear strings of chemical units that constitute a chemical four letter alphabet (see Chapter 3), with each gene or DNA segment having a characteristic sequence of those "letters," genetic changes are simply alterations in that sequence, often a single letter at a time. Hence, the farther back in time that two lineages of organisms diverged from a common ancestor, the greater the number of differences that will have accumulated in their DNA sequences.

By measuring the amount of DNA sequence difference between comparable DNA sequences—for example, the same gene—of two different organisms, one can obtain a relative measure of the length of time since they diverged from their common ancestor. Such relative rates can then be used to calculate the actual, if approximate, time at which that divergence took place. This is possible if there is good fossil evidence, dated radiometrically, for one of the early points in the evolutionary sequence. This permits "calibration" and hence conversion of relative rates to actual times. For example, we know that the modern cat family evolved long after the earliest true mammals had appeared on Earth. Hence, house cats and lions, which evolved from a common feline ancestor, will show more similarities in their DNA sequences than either will to the comparable sequences in kangaroos or platypuses, whose lineages in the mammalian evolutionary tree are much older. With appropriate fossils to "calibrate" the molecular clock, one can estimate the age of the cat family and of the kangaroo and platypus lineages from the number of differences in their DNA sequences.

The idea of the molecular clock first arose from empirical observations of proteins, which are the products of genes and whose sequences reflect those of their genes (Chapter 3). Its first detailed exposition was given by Linus Pauling and Emilio Zuckerkandl in 1965. Since that time, the idea has been directly and amply verified with DNA sequences. The

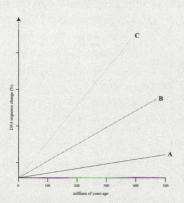

FIGURE BOX 1.2 This diagram of idealized molecular clock data shows rates of change over long evolutionary time spans for three genes, A, B, and C. A changes at the slowest rate—it has the slowest molecular clock—and C at the greatest rate, with B showing an intermediate rate of "ticking" of the molecular clock.

(continued)

molecular clock reflects the existence of constant mutation rates and the fact that not all of the DNA sequence "letters" play vital roles in the function of the genetic material—they can be replaced by other such letters without loss. Replaceable letters of this kind are said to be "neutral." Since the number and proportion of those replaceable letters is a function of the particular gene, different genes or DNA sequences will change over time at different relative rates. In effect, each gene has its own molecular clock.

Even though the idea of the molecular clock has been of inestimable value in evolutionary studies, the method is not without major complications and difficulties. We will look at an example of the problems and their solutions in discussing some critical events and their dates in primate evolution in Chapter 6.[1]

---

1   The first glimmerings of the idea of the molecular clock were obtained in the late nineteenth century (see Bearn 1993, pp. 64–65). The classic formulation is that of Zuckerkandl and Pauling (1965), but the contribution of Margoliash (1963) is also important. Pulquerio and Nichols (2007) and Drummond et al. (2006) provide some useful qualifications and critiques.

years ago, a period making up only the last 10 percent or so of the history of primates.

Finally, there is the method of **faunal association**. It relies on the presence of other animal fossils from species ("index species") in the same rock stratum where the fossil of interest had been found. If the two fossils come from the same fossil bed, then the respective animals must have lived at the same time; if the index species fossils have been dated radiometrically, one obtains an approximate date for the new species. Faunal association patterns also often provide clues to the nature of the environment—whether savannah, marsh, jungle, forest—in which the species of interest lived, and hence to the kinds of food that that environment was most likely to provide. Changes of diet that are forced by changes of climate and environment were crucial in the sequence of events leading to the evolution of the human face at various points, as we shall see.

The "Where?" Question
Establishing the Locations of Certain Species and Anatomical Changes
This third question concerns the places and conditions in which key evolutionary changes were initiated. There is, of course, a long series of "where" questions for the evolution of the face, each corresponding to a specific evolutionary change in the LHLP. For the very beginnings of the

vertebrate face, involving the origins of the first fishlike animals from their invertebrate ancestors (stage 1 in Table 1.2), the site of those origins is unknown; the only (near) certainty is that these events took place in the geological period designated the late Cambrian in shallow seas somewhere on the planet more than 500 million years ago.

The same geographical vagueness envelops the second stage of vertebrate evolution in which vertebrates acquired movable jaws (stage 2, Table 1.2), a step that led to vertebrates becoming the top predators and thus dominating food chains in both the sea and on land. It took place somewhere in the world's oceanic waters, probably along a continental shelf; again, nothing more specific can be said.

With respect to many of the subsequent evolutionary events that took place after vertebrates began to evolve on land, the answers are better but still hedged with uncertainty. The absence of informative fossils is a recurring problem, but beyond that there is a more general difficulty: the impossibility of knowing whether particular fossils that reveal new features were the first of their type or whether still earlier progenitors had originated elsewhere but had migrated into the area where the later fossils were formed. A complicating factor concerns visualization of the global position of relevant sites in terms of contemporary geography, given that the continents have been moving, changing shape and size, through mergers, separations, and fragmentations for hundreds of millions of years. Yet these patterns of continental movement and shape change have been well reconstructed, permitting a good picture of earlier locations on the planet of present-day land sites. For example, the broad region in which mammals first arose and in which the primates first appeared can be provisionally identified. It was probably in the Southern Hemisphere in the regions that correspond to parts of present-day southern Asia, Africa, and South America, which were joined together in the former supercontinent of **Gondwanaland**. (The rest of the modern continental masses—namely, Europe, northern Asia, and North America—were part of the other early supercontinent **Laurasia**.)

As to the place of origin of the primates (stage 5 in Table 1.2) and specifically the anthropoid primates (stage 6), who probably originated about 60 million to 55 million years ago, this was probably either in Southeast Asia or Africa. Only for the much later arising hominins, and *Homo sapiens* specifically (stage 7 in Table 1.2), however, can a definite site of origin be given: Africa. Correspondingly, the first modern human face appeared in Africa. Darwin had argued that Africa was the probable place of origin of humanity in *The Descent of Man,* a conclusion he based on the visible anatomical resemblances that exist between the African apes and humans. His inference, however, was rejected by many others—in particular, his chief European disciple, Ernst Haeckel (1834–1919), who favored Asia and whose writings influenced other prominent scientists in that belief until the early 1930s. Today, however, the near universal consensus is that Darwin was right: modern humans originated in Africa and migrated out of that continent, eventually populating every continent except Antarctica.

The *where* question, however, concerns more than geography. It also refers to the nature of the local environment in which a particular evolutionary change occurred, whether, for example, it was a jungle, swamp, forest, or desert. Frequently, such information can be gleaned from chemical analysis of the sites in which a particular fossil is embedded and from the kinds of associated fossils. In addition, the well-preserved teeth of fossils provide clues to the probable main items of diet and hence to climate and environment. In the case of human evolution, that information about the changing environments of the different hominin species has provided vital clues to the external factors helping to drive their evolution.

The "Foundations" Question
Genetic Material and Developmental Processes: Morphological Changes and the Fossil Record

The fourth general question concerns the hidden biological foundations from which the visible morphology of an organism springs: the genetic material of the organisms and their developmental processes, ultimately derived from that information, which create the living forms. The importance of genetic changes for evolutionary change is a basic tenet of Darwinian evolutionary theory. Its core proposition is the occurrence of heritable (genetic) changes that are propagated and spread through populations by **natural selection**. That spread of hereditary variations, each affecting the form or physiology of the animals, transforming the population, takes place because those changes convey an adaptive benefit, allowing the individuals that exhibit those alterations to leave more offspring than those

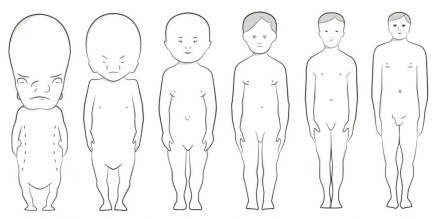

FIGURE 1.3 The developmental stages of humans, showing the changing proportions of head and body during development as the head becomes relatively smaller with respect to the body because of prolonged body growth. The human face, however, is one of the most slowly maturing parts of the human body, only reaching its final mature state in late teens, along with the termination of body growth. The six stages of human development, from left to right, are: (1) embryo, (2) fetus, (3) newborn, (4) juvenile, (5) teenager, and (6) adult.

not so favored. Such hereditary changes are said to increase the organism's **fitness**.

While the importance of genetic changes in animal evolution is obvious, when seen through the Darwinian lens, the need to understand their associated developmental changes may be less so. For complex animals, the word *development* refers principally to the whole set of transformational processes that take place between the fertilized egg and the newly emerged young and on through later and subtler changes in juvenile stages. (It usually excludes, however, still later changes due to aging.) For humans, six distinct stages of development are conventionally recognized. The first begins with the fertilization of an egg and continues with the formation and development of the embryo (zero to ten weeks) and with human identity of the embryo clear by the end of week ten. The second is the period of fetal development, which runs from the end of week ten until birth at thirty-eight weeks of gestation. The third starts with birth and runs through infancy. The final three stages are those of the juvenile (ages three to twelve years), the teenager (thirteen to nineteen years), and the adult (twenty years onward). This sequence is diagrammed in Figure 1.3. The most dramatic changes are those taking place in the first two stages while the most obvious general change over the whole sequence, seen in this picture, is the relative reduction of head to total body size. All of the first five stages, however, make distinctive contributions to the shaping of the human face, as we shall see.

The reason for delving into development is that, ultimately, the distinctive shape of the adults, their morphologies, within an evolving lineage—whether dinosaurs, whales, mice, or humans—are the final products of those changes in developmental processes. It is those processes that ultimately lead to the species-characteristic form of the adult, and it is within those processes that the evolutionary changes actually take place. Thus, if one wants a true understanding of evolutionary changes in shape and form, it is necessary to know what their developmental foundations in development were. This idea only came into clear focus in the 1920s in the writings of a few perspicacious biologists but was then largely lost to view for nearly fifty years.[11]

Yet how does one study the genetic and developmental processes of animals that went extinct millions or even hundreds of millions of years ago? Development is primarily a matter of changes in soft tissues and, unlike bone, those rapidly degrade, leaving no trace except under exceptional conditions. Furthermore, even where those preservation conditions exist, embryos are small and for that reason also are much less likely to have been preserved (or found, even when they left a fossil trace). Only a relative handful of presumed fossilized vertebrate embryos are known.

Is it possible to obtain the information in some other way? The answer is "yes," but the explanation may be surprising: by appropriate comparative study of living species, whose evolutionary relationships are known, one can reconstruct critical features of their common ancestors. When this approach is applied to developing embryos of animal species alive today, such **comparative biology** can be used to infer aspects of the development of long-extinct animals, including those that died hundreds of millions of years ago. From the presence and absence of particular features, and if the evolutionary pattern has been correctly deduced, one can often reconstruct the sequence in which particular features arose. The basic logic of the approach is illustrated in Figure 1.4 and explained in Box 1.3.[12]

Thus, by comparing the developing face in embryos of a few living species, discovering both what is common to and different between them in terms of their developmental processes and the **gene activities** that are central to those processes, it becomes possible to make inferences and

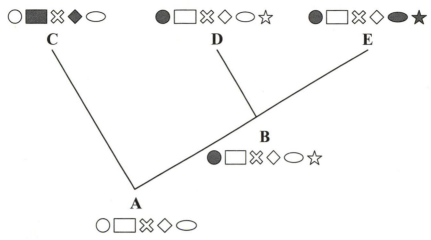

FIGURE 1.4 Using traits of living species to establish their relationships and reconstruct likely ancestral forms. The basic working assumption is that evolutionary change can be visualized as involving dichotomous branchings; forms that are traced back to the same putative ancestor are said to be **sister groups**. In this phylogenetic tree (or *cladogram*), three living species are related via shared traits (synapomorphies). Each trait is designated by a particular geometric shape (a circle, a rectangle, etc.), and alternative forms of a particular trait are indicated by shading, either black or white. By comparing the living forms, it is possible to deduce the probable sequence of changes as they took place in evolution and the characteristics of putative ancestral forms. Thus, in this example, all three living species, C, D, and E share five of the six traits, but C is missing one (the star). In addition, D and E share similar forms of four traits (the circle, rectangle, X and diamond). By these criteria, D and E are probable sister groups whose common ancestor, B, probably had the six traits and the same forms of the four shared traits. C is more distantly related through the putative ancestral form A. (For further discussion of these points, see Box 1.3.)

hypotheses about the evolutionary changes in those genetic elements in their ancestors. This form of analysis is possible in vertebrates because so much of vertebrate development and its underlying genetic basis are shared from fish to human—the main point of W. K. Gregory's book, though he could only judge this from comparative anatomical studies. The revelation produced by many molecular studies in the 1980s and 1990s was that not only are developmental processes in the vertebrates held in common but also that so is much of the underlying genetic machinery for development—specific genes employed in comparable ways. In the jargon of the trade, these shared molecules and processes are said to be highly **conserved** in evolution. The seemingly large differences in adult morphology that we observe—for example, between sharks and shrews and humans—come from often seemingly small "tweaks" in the key genes that regulate the activities of other genes (a matter that will be explained

BOX 1.3   A Brief History and Description of Phylogenetic Systematics or Cladistics

Phylogenetic trees are diagrams that attempt to show evolutionary relationships in terms of splitting events. They have been drawn ever since Darwin's single frontispiece figure in *The Origin of Species*. Earlier ones were based on the presumption that the overall extents of similarity of traits are a measure of the closeness of evolutionary relatedness. One weakness of such trees is that what often looks like the "same" trait had evolved separately and independently in different lineages; this is known as the phenomenon of *homoplasy*. Another difficulty is that others can be lost from one or more otherwise closely related lineages. Thus, overall similarity of traits is not the most reliable guide to evolutionary history. True evolutionary relationships are better shown in terms of reconstructed patterns of acquisition and loss of specific traits. That is the basic principle of *phylogenetic systematics*, which was developed by German entomologist Willi Hennig in his book of the same name, which was published in 1950 in German but later (1966) presented in a shorter version in English . Hennig realized that evolutionary diversification can be drawn in terms of a sequence of bifurcating branchings in which traits appear either before the branch point (in those cases, both branches share the same trait) or afterward but in only one branch. In effect, it is the pattern of trait acquisitions, both shared and unique, relative to the time of branchings that describes the overall pattern of relatedness and, implicitly, the evolutionary sequence. *Clade* is the Latin word for "branch" and, correspondingly, *cladistics* has become the name for the technique (displacing *phylogenetic systematics*). Finally, the name of the kind of diagram showing such relationships is called a *cladogram*.

An abstract example of a cladogram is shown in Figure 1.3 to illustrate the method. This one is based on only three living species (C, D, and E) and involves the reconstruction of two ancestors. With such small numbers and such small numbers of traits (six in this case), errors would be probable. Cladograms based on real comparisons would normally involve more living species and more traits and would correspondingly be more certain. The basic reasoning is explained in the legend of the figure, but a few more points of basic terminology are needed.

Where a trait appears before a node, it will be shared by all or most of those above it; such shared traits are termed **synapomorphies**. An example in the figure would be the star trait shared by living species D and E but also by their common ancestor, B. A trait that is shared by all living members of the group (C, D, and E) was probably present at the origin of the group and is thus regarded as a primitive trait or **plesiomorphy**; the circle, the rectangle, the cross, the diamond, and the oval are such. Traits that are unique to a single branch, not illustrated here, are said to be **autapomorphies**. Note that all of these designations, however, are relative to particular nodes. Thus, the star is a synapomorphy shared by D and E, presumably present in ancestor B, but is an autapomorphy with respect to C (and inferred ancestor A).

Cladistic reasoning is not restricted to morphological traits but can also be used with molecular data (Box 5.1). As with the molecular clock, there is a large and somewhat contentious literature about cladistics, although the differences in point of view have diminished with time.[1]

---

1   A brief, clear account of the principles of phylogenetic systematics is that of Hennig (1966). A history of the revolution in systematics wrought by cladistic thinking and its accompanying controversies is that of Hull (1988).

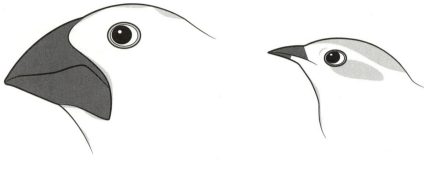

Geospiza magnirostris   Certhidea olivacea

FIGURE 1.5 Two Galapagos finches, illustrating the differences in beak size and shape. The large ground finch, *Geospiza magnirostris*, eats seeds, which requires a strong beak for crushing them. In contrast, the green warbler finch, *Certhidea olivacea*, uses its needle-like beak to probe decomposing vegetation for insect larvae.

more fully.) The processes of development and growth amplify those invisible molecular changes into large visible morphological ones. From comparing what is conserved and what is different among the embryos of species whose evolutionary relationships are known, it is possible to make specific hypotheses about particular gene activities changed in location or duration. These ideas can often then be experimentally tested by creating similar changes in particular special living species (**model systems**) and observing the effects.[13]

An example involving bird faces illustrates the general approach. The fourteen species of finch that inhabit the Galapagos Islands in the eastern Pacific Ocean, more than 1,000 miles off the South American coast, were first described by Darwin as prime evidence for his theory of evolution, which he termed "descent with modification." As hypothesized by Darwin and confirmed by modern DNA analyses, all the living species are descended from a common ancestral species from the South American mainland that migrated to the islands several million years ago. These species differ conspicuously in overall size, head shape, and beak sizes and shapes, which range in size from small and almost needle-like to others that are large and robust. They correlate with the kinds of food that each species eats, and each beak is clearly an adaptation for a specific diet. Two contrasting Galapagos finches, which specialize in different foods, are drawn in Figure 1.5.

The beak develops from a special bone termed the *premaxilla*, and different beak shapes and sizes reflect alterations in its development. Molecular analysis has shown that it is different amounts of two specific molecules during beak formation in the embryos that largely account for these differences in beak dimensions: the more **calmodulin** (CaM) present during beak development, the longer the beak will be, while the more **bone morphogenetic protein-4** (BMP-4) present during the critical period of beak development, the wider and deeper the beak—and the face—will be. By sophisticated experimental manipulation of the amounts of these two kinds of molecules in chick embryos, it was possible to establish these quantitative relationships.[14]

The crucial inference is that, starting from the ancestral mainland species, different Galapagos finches arose that had evolved different size and shaped beaks via genetic changes, **mutations**, that altered the amounts of these two molecules to produce the different beak shapes. Hence, if one knows something about a conserved developmental process and its genetic foundations, one can make hypotheses about how those processes were tweaked in evolution, often millions or tens of millions of years earlier.

The comparative approach not only is useful for morphological changes but also can illuminate behaviors and mentalities. For instance, Darwin showed and others have confirmed that six basic expressions are common across human cultures and ethnic groups: anger, happiness, sadness, curiosity, fear, and disgust. They are even seen in persons blind from birth onwards and hence are not learned. Although these expressions are modulated by different cultural milieus in different ways, the fact that they are shared by all people indicates they are part of our common human genetic evolutionary heritage. These six basic expressions are also shared with our closest primate cousins, chimps.[15]

Since both the generation of these expressions and their "reading" by other individuals involves vital brain activities, understanding the evolutionary roots of the face and facial expressions entails comprehending something of the evolution of the brain. Indeed, a full account of how the face evolved must include analysis of how various brain functions evolved in parallel with the face—in particular, those brain properties involved in making and reading facial expressions. Such joint evolution of two distinct features of an organism is termed **coevolution**, and face–brain coevolution is an integral aspect of the evolutionary history of the face.

The "Why" Question

Changing Physical, Biological, and Social Environments Driving
   the Evolution of the Human Face

The fifth and final broad evolutionary question, the "why" question, involves the most uncertainties because it involves the longest sequences of inference. The question is, what were the forces of natural selection—or other change-promoting processes—that favored and propelled the visible changes in form of interest—in this case, of the face? Answering it requires, in particular, information about the environments, both physical and biological, in which the animals lived and then a synthesis of those findings with what is known about the biology of the animals.

For much of primate history, it is possible to reconstruct the probable nature of the environments in which particular species lived. From those inferences and from detailed morphologies of teeth, it is often possible to deduce what foods the animals relied upon. From those conclusions, other inferences often follow. Thus, for example, there are many findings relating to how the physical and biological environments changed in humankind's birthplace, Africa, as that continent became cooler and drier and as jungle increasingly gave way to forest and, then, forest to the open grasslands known as *savannahs*. One consequence would have been changes in diet. These climatic and environmental shifts also selected for the transition from tree-living, quadrupedal apelike creatures to ground-dwelling, fully bipedal animals requiring further changes in diet. Their forelimbs and hands, already suitable for manipulating various kinds of food, came to be used by some of their descendants to manipulate other objects, some of which could then be used as simple tools. These capabilities, in turn, had selective consequences not just on their bodies but also on their brains and faces. Another example: in earlier stages of anthropoid primate evolution, changes in climate led to changes in available food and a shift in diet away from insects and leaves toward a heavy dependence on tender leaves and fruit. For finding and picking such plant material, high-resolution, acute close-up vision, which anthropoid primates possess, would have been a strong adaptive advantage and almost certainly favored by natural selection.

Selective forces, however, are not limited to physical or dietary environmental factors: for highly social animals such as primates, social forces can be selective ones. In effect, the social environments that highly gregarious species create for themselves are also shapers of evolutionary change in those species. How social environments and pressures affected the evolution of the face will emerge as a major theme of this book.

Organization of the Book

A few words of explanation about this book's organization might be useful. Although its focus is on evolution, I will not begin with evolutionary history. As previously discussed, to understand evolutionary change in animal form, one must first understand the specifics of the animals' development. The following two chapters will provide that background. Chapter 2 first describes the visible developmental events that give rise to the head and face in the embryo and fetus as they can be seen by microscopic examination and then discusses the developmental events that shape them postnatally in childhood and adolescence. A key player in these processes is the neural crest cells, already introduced as a key element in the origins of the vertebrate head. Chapter 3 will examine these same events but from underneath as it were, namely, in terms of their genetic and molecular underpinnings. That material, however, prompts a question: how does the genetics of face formation relate to the extent of human facial diversity? Chapter 4 addresses this question and discusses various genetic hypotheses that might explain the extent of human facial difference.

We will then turn to the evolutionary story. Chapter 5 describes the first five stages of face evolution (Table 1.2) from the first vertebrate faces to the first primate faces. This embraces a lot of evolutionary history, involving a time span of approximately 450 million years, but it will concentrate on the early and foundational events along the LHLP. Chapter 6 will then describe the evolution of faces from the earliest primates to the later-appearing anthropoid primates and then from the first hominins to modern *Homo sapiens*. It was in this period of 55 million to 50 million years—in particular, its final 6 million years—in which the defining features of the human face came into being. This chapter will close with a discussion of the possible origins of the most distinctive of human traits, namely, the capacity for language and its expression through speech.

Turning to those matters, however, shifts the focus of the narrative from the externalities of the face to the internal, neural underpinnings of what faces do as directed by that most remarkable of all structures, the human brain. Chapter 7 deals with this subject directly and the ways in which the face and the brain mutually influenced their coevolution. It will deal specifically with three topics: (1) the brain structures and neural circuitry for recognizing the faces of different individuals; (2) the neural and muscular mechanisms that create facial expressions, including those accompanying speech; and (3) the neural capacity for "reading" those facial expressions (the prelude to reacting to them). The evolution of acute, close-up vision (probably first evolved for foraging for fruit or edible leaves) is essential for two of these properties, face recognition and reading expressions, and it will also be discussed.

Chapter 8 deals with evolutionary changes following the emergence of *Homo sapiens* as a distinct species. It will describe the spreading out of humans from Africa, starting about 72,000 to 60,000 years ago and the partial genetic differentiation that accompanied this huge geographical expansion across the planet. It will also continue discussing the brain–face connection, describing two relatively late events in human evolution. The first concerns properties that are highly visible yet have far less biological importance than traditionally attributed to them: "racial" differences. It will be argued that Darwin's idea that these differences arose largely as a by-product of **sexual selection**, the process of selection of properties that attract mates, can help explain how and why many of those differences arose. (The role of natural selection, however, in promoting different skin colors, however, will also be discussed.) The other set of mental and behavioral properties that may have influenced the human face have the reverse character: they would have created a relatively subtle modulation of facial features—a mild juvenilization—but could have specifically affected our capacity for social and cultural evolution. I will argue that this process existed and stemmed from the phenomenon of *self-domestication*. Just as human beings over the last 15,000 years have domesticated and tamed many other animals for various needs, our own evolutionary history probably involved an element of self-taming, allowing much greater sociality that, in turn, would have made possible the astonishing capacity of our species to construct complex societies. Intriguingly, the story of self-domestication may, like that of the origins of the face and its evolution, also involve neural crest cells.

The penultimate chapter, Chapter 9, discusses the future of the human face with the inevitable uncertainties acknowledged. Three aspects of that future will be treated: culture, evolution, and science. It begins with the idea of face consciousness, defined here as *conscious* awareness of the importance of the face as an individuating element in general and of one's own face. As stated previously, this phase is probably largely concordant with the rise of human civilizations, and the chapter will begin by giving the argument more fully. The roles of culture and globalization, with increasing population mixing, in affecting human faces in the future will then be sketched. Finally, perhaps reflecting the current cultural wave of interest in the face, the old "science" of physiognomy, may be undergoing a mild revival albeit in a different guise among some geneticists; this, too, will be discussed.

Chapter 10 will attempt to knit the whole story and its implications into a coherent whole. The central argument, sketched in the preceding four chapters but brought out more fully here, is that the special and growing demands of social existence played an especially strong role—via selection for more sociable individuals in situations where group survival was promoted by social coherence—in shaping the evolution of the human face. Those changes, in turn, would have fostered still more communication and greater sociality.

The book can be regarded as having two parts. The first, consisting of the first five chapters, introduce and explain the human face and its early evolutionary foundations. The second part, the final five chapters, documents how the demands of sociality began to influence the evolution of the face beginning with the appearance of the anthropoid primates. The face, of course, did not lose its primary role as "sensory headquarters," but it acquired a second major function as a source of information about the individual, making possible an ever-richer and more complex social existence.

I will close this introductory chapter with a few words about the format of this book. For lay readers in particular, I have included explicit background explanatory material, particularly at the beginnings of Chapters 2, 3, 4, and 7, and a glossary of specialist terms and words (appearing first in the text in bold). There are also occasional text boxes providing supplementary material for certain topics. Not least, endnotes contain further comments and references to sources. To deal with the problem of technical terms that are long and repeated many times, I have used acronyms, though sparingly, giving the full term at the point at which it is introduced. A complete list is at the back of the book.

To set the stage for the evolutionary story, we will, as mentioned, begin by looking at the development of the human face in the embryo and fetus, a phenomenon that takes place every time a human being is conceived and grows in the mother's womb.

— two —

HOW THE FACE DEVELOPS: FROM EARLY EMBRYO TO OLDER TEENAGER

Introduction: Rethinking Animal Evolution in Terms of Developmental Processes

The first chapter introduced the idea that evolutionary changes in the forms of mature animals result from alterations in the developmental processes that create them. The idea was illustrated with the beaks of different Darwin's finches, whose various shapes and sizes are now known to reflect changes in the modulation of two biochemical processes during beak development in the embryo. Perhaps, however, the general idea—that changes in adult form are the consequences of developmental change—seems trite. After all, everyone knows that the embryo is the forerunner of the child and the child is the father of the man. Thus, if the form of an adult animal is the end result of its development, then evolutionary changes in that form *must* reflect changes in its development.

The mainstream modern theory of evolution, the *evolutionary synthesis* as it came to be called in the twentieth century, virtually ignored this fact, however. First, too little was known about the processes of **development** for it to be usefully incorporated in the theory. Second, the attention of the founders was focused not on details of morphology but on the general process and dynamics of evolutionary change as mediated by natural selection. In effect, development, in all its seemingly messy complexity, was shoved into a conceptual "black box" and left there. Yet its de facto omission from the theory had consequences. Not least, it affected the popular view of evolution, which came to be visualized (unconsciously) as sequential transformations of *adult* forms of one kind of species into another. The classic depiction of human evolution as a steady transformation of shape changes from (adult) ape to (adult) modern human epitomizes this view (Figure 2.1). This may be termed the *computer graphic* view of evolutionary change.[1]

Development does matter, however. Anyone who seeks to understand the evolutionary origins of the human face, for example, must know how

FIGURE 2.1 The classic depiction of human evolution from ape to human. For decades, it was the iconic visual representation of human evolutionary history, with changes perceived as the slow morphing of one adult form into another.

the face develops. To illustrate how the developmental perspective deepens evolutionary understanding, let us take an example more dramatic than the beaks of finches: the wing of the bat, a structure that allows these animals to do what no other kind of mammal can—fly. In purely functional terms, the bat's wing is an unambiguous **evolutionary novelty,** although in structural terms, it is merely a modified mammalian forelimb, the equivalent of a human arm.

To understand how a striking new functional property, bat flight, could result from relatively modest developmental alterations, the basic facts of how limbs develop are needed. The crucial one is that all vertebrate limbs, both forelimbs and hind limbs, develop from small outgrowths on the embryo termed **limb buds**. These grow out at fixed locations on the embryo trunk and, as they grow outward, generate the regions that will give rise to successively more distal (further from the body) parts of the limb. Thus, in the growing human forelimb bud, the region that eventually gives rise to the upper arm develops before the part that becomes the lower arm while the last part of the limb bud to be formed develops into the most distal part, the hand in humans. This terminal part of a limb bud is designated the **autopod.** The general process of limb bud outgrowth is schematized in Figure 2.2A.

A.

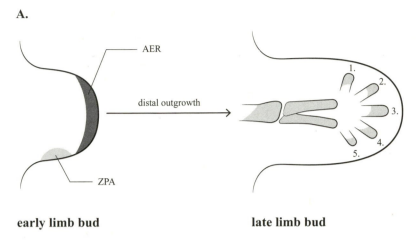

early limb bud   **late limb bud**

AER - apical ectodermal ridge
ZPA - zone of polarizing activity

B.

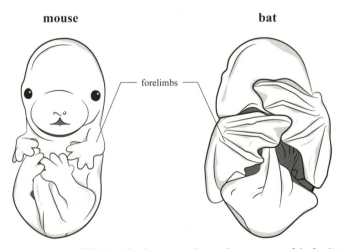

FIGURE 2.2 A. Schematic of limb bud development, indicating future regions of the forelimb that will develop. The numbers on the right indicate the numbers of the digits that will develop from the component bones, the phalanges. The "apical ectodermal ridge" is the zone of active cell proliferation that elongates the limb bud, while the "zone of polarizing activity" creates a chemical gradient that leads to differences in the digits of the autopod. **B.** Mouse and bat embryos compared. The end result of the evolution of the anterior autopods of the bat is a greatly enlarged pair of forelimbs (relative to the more rodentlike ancestor), due to the selective, greatly increased growth of the phalanges in digits two through five, and the absence of cell death in the interdigital membranes, generating a continuous wing-like surface.

Bats evolved from nonflying rodentlike creatures and their wings evolved as a result of two specific developmental alterations of the forelimb buds. Where the ancestors of bats would have developed a typical rodentlike paw from their autopod, with closely packed short digits, bats develop several inordinately long and well-spaced digits in theirs. The bones that constitute the digits are termed **phalanges** and the bat's forelimb is distinguished by exceptionally long phalanges. Differential growth of the digits, however, would not have been sufficient to turn the bat forelimb into a wing; that required the second change. In the developing autopods of all mammals, skin stretches between the digits but, late in the development of the autopod, the extensive death of the cells between the digits eliminates that interdigital webbing. In the developing bat embryo, that process of programmed cell death, **apoptosis**, does not occur, and the flesh extending between the digits survives to become the webbing of the bat's wing. (A similar suppression of apoptosis of the interdigital membranes develops in other mammals that possess webbing in their forelimbs, including seals, whales, and Labrador dogs.) Ultimately, all the differences between the different kinds of mammalian forelimb—from rodent paw to human foot to dolphin flipper to bat wing—lie in modulations of extents of growth and apoptosis during embryogenesis and fetal development. The differences in forelimb size and structure between a rodent and a bat are already evident in the late embryo stage as shown in Figure 2.2B.

The preceding description of limb development may seem a long digression from our subject, the face, but it is relevant in two ways. First, it exemplifies an important general principle—that *quantitative changes* in a developing structure can lead to *major qualitative differences in function* of that structure. We will see several instances of this in the evolution of the face. Second, as we will see in the next chapter, limb development bears important genetic and evolutionary relationships to the development of the face.

Human Development and That of the Face: Some Basic Considerations

Human development, as noted earlier, is conventionally divided into six distinct stages: fertilized egg → embryo → fetus → infant → juvenile / teenager → adult (Figure 1.2). All stages and their transitions are essential to the full development of the face, yet the second and third, the embryo and fetus, are particularly important because they lay the foundations of the face. All changes in later stages of face development build on those foundations and involve modulations of growth and shaping, to ultimately yield the face of the adult.

The face does not develop in isolation, of course, but as part of the head, so the development of the head is an integral part of the story. Per average unit volume, the head is the most complex part of an animal's body and the most diverse in its functions. Indeed, if the head did no more than house the brain, the most complex organ, it would still merit that distinction. The face, however, is also a strong contributor to the complexity of the head. In its intricate, stereotypic arrangement of bone, sense organs, nerves, blood vessels, and skin, the face is architecturally the most complex part of the head. These statements may seem purely subjective judgments, but there is an objective measure of complexity that supports them. It involves the relationship between intricacy of a constructed object and the errors likely to occur in its construction: the more complex it is, the more difficult and the more error prone is its construction. Among live human births in which newborns exhibit visible defects, three-quarters—a high proportion—involve abnormalities of the head or face. These are termed *craniofacial defects* and are the major focus of pediatric surgery.[2]

The face initially forms and takes on its basic shape in the embryo and early fetus, a process occurring roughly between the fourth and tenth weeks of pregnancy. In humans, embryogenesis—when the major organ systems have been formed—is completed during the first eight weeks of development, while fetal development begins in the ninth week and lasts until birth some thirty-two weeks later.

The sketch of human development presented shortly is based in part on technically sophisticated observations of living human embryos inside their mother wombs and, even more, on human embryos and fetuses that did not come to term. Many of the details, however, have been inferred from information initially derived from experiments on animal embryos, animals as diverse as fish and chickens and mice, and then extrapolated to humans. The reason it is valid to do so is that many general features in face and head development are shared (*conserved*) between different vertebrates, as previously discussed. Hence, those found in both fish and mice, for example, are likely to be present in humans as well. In contrast, the differences between the embryos of those animals supply clues to the probable points of divergence in their development that ultimately generate their adult morphological differences.

When reading the somewhat complex material that follows, remember that this is an account of events that everyone—you the reader, I the author, our children, parents, grandparents, all our ancestors, and not to mention (though with different details) our cats and dogs—has undergone. Thus, at the beginning of your life, in your mother's womb, your face took shape via the events that are described in this chapter.

Cells, Tissues, and Tissue Interactions in Development of the Face

Some facts about the essential building blocks of development—cells—are needed at the start. The **cell** is the fundamental unit of the animal body. Although each human cell is tiny, their aggregate numbers in any animal can be huge; the average adult human body contains approximately 10^{13} cells—in other words, 10 trillion. Despite the small size of the average cell, approximately 1/100,000th of a meter in diameter (10 microns), it is an intricate entity.

A schematic diagram of an average animal cell is shown in Figure 2.3A. One element is a centrally located, membrane-bound, densely staining spherical body—the **nucleus**—which houses the genetic material, partitioned among individual entities termed **chromosomes.** In humans, there are forty-six chromosomes per cell. When chemically stained to enhance their visibility, they are usually depicted as tight little rods, or sometimes as little X's or V's, which is how they appear under the light microscope if caught at the midpoint of cell division, or **mitosis.** Each such cell division gives rise to two "daughter cells," each containing one copy of all the chromosomes—again, forty-six. During most of a cell's existence, however, the chromosomes exist as long threads within the nucleus, none individually distinguishable, the stained nucleus looking like a ball of wool. How does the cell fit all these thin threads of that length into a nucleus, itself perhaps no more than one or two microns (1–2 millionths of a meter) in diameter? The answer is by successive stages of coiling, just as one can take a long piece of string, coil it once to make it thicker and then coil that

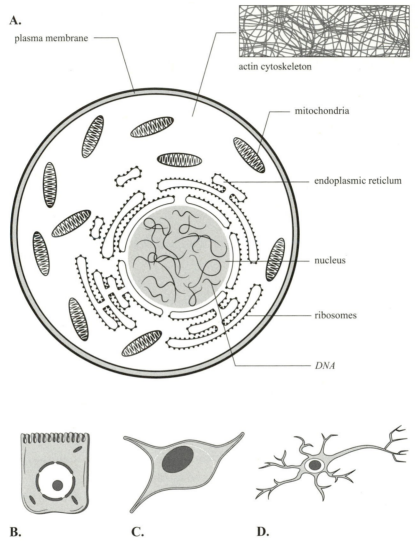

A.

plasma membrane

actin cytoskeleton

mitochondria

endoplasmic reticlum

nucleus

ribosomes

DNA

B.   **C.**   **D.**

FIGURE 2.3 A. Diagram of a generalized animal cell. Some of the important internal structures of the cell are the nucleus, which contains the genetic material; the endoplasmic reticulum, the site of protein synthesis; the small RNA-rich particles; the ribosomes, which bind to the endoplasmic reticulum and on which protein synthesis takes place; and the mitochondria, the small bodies that produce the energy currency (ATP molecules) of the cell. For illustrative purposes, the chromosomes are shown as short threads; in reality, they are long and indistinguishable from one another in the nondividing cell. **B.** An epithelial cell. **C.** A motile mesenchymal cell. **D.** A neuron.

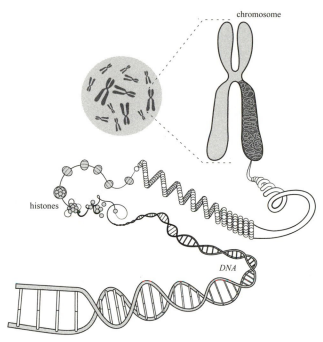

FIGURE 2.4 Levels of coiling within a chromosome, showing how an immensely long DNA molecule of a chromosome (bottom) can be successively coiled after it has bound the proteins known as *histones* into successively tighter, more compact structures. At the top left is a "spread" of condensed chromosomes as they appear in a squashed and stained cell caught in mitosis. One X-shaped chromosome is shown to the right, and its successive layers of coiling are shown in the lower half of the figure.

coil to make it even thicker, and so forth. This process is diagrammed in Figure 2.4.

The nucleus is surrounded by a semifluid space, the **cytoplasm.** Although the cytoplasm was regarded well into the twentieth century as a rather simple homogeneous colloidal gel, in reality it is complex in its molecular composition and highly structured, being criss-crossed by a form of internal skeleton or scaffold, the **cytoskeleton.** The cytoskeleton consists of **proteins** that give it its basic shape, structure, and rigidity and others that give it potential for mobility. Cell mobility (for cells that move) and contractility are made possible by two kinds of proteins specifically, actin and myosin, which form joint actin–myosin filaments, which are the basis of contractility. (Muscle cells, whose main purpose is to enable contraction of the muscles they compose, are greatly enriched in these two proteins.) In contrast, basic cell shape and rigidity of cell structure are made possible by another protein, termed *tubulin*, that aggregates into microscopically visible **microtubules**, and a group of proteins termed **intermediate filament proteins** (named for their intermediate width between actin filaments and microtubules).

The internal contents of the entire cell (cytoplasm and nucleus) are bounded and contained within a thin lipid and protein-rich envelope, the cell membrane. That is not, however, the only membranous structure of the cell: the cytoplasm also contains a set of internal membranes connected to the nucleus that carry small, electron-dense bodies named **ribosomes,** on which the essential synthesis of all of the cell's proteins takes place. It is the proteins of the cell that carry out its activities and give it its special characteristics. In addition, there are the distinct and complex football-shaped entities known as **mitochondria**, which supply energy to the cell, and of which there can be hundreds per cell.[3]

As a result of differences in protein composition, cytoskeletal construction, and consequent other properties, cells come in many different types. The currently estimated number of cell types found in mammalian bodies is 400, based on properties visible with microscopy. Three major cell types, all found in the face, are diagrammed in Figures 2.3B, 2.3C, and 2.3D. They differ not just in shape and structure—reflecting differences in the organizations of their cytoskeletons—but also in their internal biochemistry, which helps make possible their differing functions.[4]

While cells are the fundamental building blocks of the animal body, many cell types form extensive aggregates of cells of the same type: **tissues.** Every tissue is a compact arrangement of many cells of one or a small number of cell types and is found in one or more characteristic locations within the body, developing in a stereotypical fashion in the embryo or fetus, each having a characteristic function. An example is lung tissue, whose cells deliver oxygen to the red blood cells that pass through the lung and deliver the oxygen to the cells throughout the rest of the body. (The lung cells also pick up carbon dioxide waste generated by the cells of the body and carried by red blood cells returning to the lungs via the veins.)

Some kinds of tissues, however, are defined more by their general physical properties rather than by a specific biochemical or physiological activity. Thus, many tissues in the body are said to be **epithelial**, consisting of columnar-shaped cells tightly packed in a single layer, rather like a large number of shoeboxes packed together side by side on their ends, each one in close contact with four others on its four sides. These epithelial layers cover the different major organ systems such as the lungs and the liver. (An epithelial cell is diagrammed in Figure 2.3B.) Underneath the single layer of epithelial tissue and interacting with it and supporting it, are so-called stromal tissues made up of **mesenchymal cells** that provide further physical buttressing to the epithelial layer and other supporting functions.

Yet epithelial cell shape is not always rigidly maintained. In particular, certain embryonic epithelial cells found initially as surrounding layers for newly developing structures and organs can transform to become migratory, amoebalike mesenchymal cells. Such transformations are termed **epithelial-mesenchymal transitions**. This initially involves a loosening of the chemical junctions at the sides that bind these cells together, along with biochemical changes at the two ends of each cell, the "apical" end, which faces outward, and the "basal" end, in contact with the stromal cells. It also involves the activation of molecular "machinery" to provide mobility to the newly liberated epithelial cells. One of the most important of these transitions takes place in the early embryo, on the uppermost side of a structure known as the **neural tube**, which is the precursor of the **central nervous system (CNS)**, namely, the spinal cord and the brain. From this site, a large number of cells undergo the transition and become migrating mesenchymal cells. These are the neural crest cells, mentioned in Chapter 1, that are integral to formation—and evolution—of the face. We will return to them shortly.

In general, embryonic development features transformations of cell and tissue types, taking place in a highly ordered, indeed, stereotypic fashion. Yet the great diversity of cell types derives ultimately from a small set of general founder tissues. Those initial foundational tissues are known as the **primary germ layers**, and they form early in the life of the embryo; their discovery was crucial in the development of modern embryology in the early nineteenth century (Box 2.1). They are distinguished and named for their relative positions in the early embryo, but their significance lies in the specific set of adult cell types and tissues that each gives rise to, those outcomes termed their *prospective fates*.

Figure 2.5A presents a schematic diagram of the placement of the three originally designated germ layers in an early human embryo. The innermost germ layer is designated the **endoderm** (literally "inner layer"), and it gives rise to the digestive apparatus of the animal, the long tract that leads

BOX 2.1 The Idea of the Germ Layers and the Beginnings of Embryology

Although today all scientific disciplines are international enterprises, many have discernible roots in a particular national scientific culture through the work of certain individuals and their immediate colleagues. For developmental biology, the modern descendant of classical embryology, those roots are in German scientific culture and specifically in the idea of *germ layers*. Three scientists from that scientific-cultural milieu were involved. The first was the Berlin-born Caspar Friedrich Wolff (1733–1794), who is best remembered today for his study of mammalian reproductive systems, especially the discovery of the male-specific structures that are essential for sperm delivery; in fact, the Wolffian ducts are named after him. Wolff only sketched the germ layer concept, however, and did not give it much emphasis, but he is undoubtedly the father of the idea. It was the second scientist, Heinz Christian Pander (1794–1869), who fully developed it and coined the term from his study of chick embryos. Pander, like Wolff, was German in descent and culture but was from the city of Riga, now in Latvia. He developed the idea from his meticulous observations of chick embryos, whose development in opened eggs can be directly observed with a dissecting microscope; he published his observations in 1817 when he was only twenty-three.

The crucial figure in the development of the idea, however, was Karl William von Baer (1792–1876), who began by working with Pander. Von Baer, born in what is present-day Estonia, was, like Pander, a member of the large German community in the Baltic region. It was von Baer's further observations on various kinds of vertebrate embryos that established the generality of the germ layer concept. Given the breadth of his discoveries and insights, he is widely regarded as the founding father of embryology. One of those discoveries was the mammalian ovum or egg cell, thus linking mammalian reproduction with that of the rest of the animal kingdom. He also showed that early vertebrate embryos have far more in common in their structures than do their later stages. That observation was later used by Charles Darwin as crucial support from embryology for the shared evolutionary roots of all vertebrates. Ironically, von Baer himself was antagonistic to Darwinian evolution, believing like many nineteenth-century scientists that it was far too haphazard a process to account for the complex and beautiful structures of living things. This was one of von Baer's few failures of insight.

from the lining of the mouth through the stomach and intestines to the anus. The outermost germ layer is known as the **ectoderm** ("outer layer") and it generates both the outer covering of the young animal, namely its skin or epidermis, and the central nervous system via the neural tube. The third germ layer is located between the endoderm and the ectoderm and is termed the **mesoderm** (literally "middle layer"). The mesoderm is the ultimate source of other major tissues of the animal—namely, the muscles,

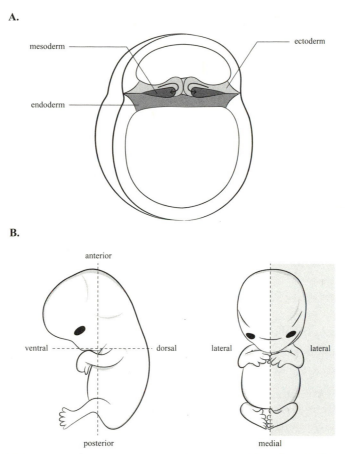

FIGURE 2.5 A. Diagram of a human embryo after the ingression of the cells that will establish the middle germ layer, the mesoderm. Indicated are the ectoderm (outer germ layer), mesoderm, and endoderm (innermost layer). **B.** Schematic diagram of an early human embryo, indicating the basic axes: the antero-posterior or longitudinal axis, the dorso-ventral or top-to-bottom axis, the left-right axis, and the bilaterally symmetric medial-lateral (distal) axis.

connective tissue (such as tendons and ligaments), the blood-generating tissues (and the resulting blood cells), and the bone-generating tissues that will form the skeleton, including the skull. The early embryo is traditionally referred to as the *trilaminar embryo* in reference to its three germ layers. In human embryos, it has formed by early in the third week after conception. The three classic germ layers, however, are not the direct and exclusive source of all tissues in the body. The neural crest cells (not shown here), which were discovered decades after the classic germ layers, are the source of a large number of cell and tissue types. Although derived from the embryonic ectoderm, the neural crest cells are sufficiently distinctive in their properties and developmental fates that they are now frequently regarded as, in effect, a fourth primary germ layer.[5]

Although the traditional concept of the germ layers is still useful, it requires a qualification. Traditionally, it carried the implication that the cells derived from any particular germ layer have their developmental destinies, their "prospective fates," written into them. The term *germ layer* thus connotes autonomy and inevitability in tissue development. In reality, the complete development of most of the differentiated tissues and structures is highly dependent on interactions between cells of different tissues, often between cell and tissue types from different germ layers. The development of the face exemplifies such interactions, as we will see.

The Early Human Embryo: Laying the Foundations of the Face

The formation of the human face takes place within the first ten weeks of embryonic development. By that point, the embryo has what is clearly a human face, though considerable transformation and growth into an individual face is yet to come. The whole process can be divided, somewhat arbitrarily, into two stages. The first takes place in the first four weeks and establishes the basic geometry and organization of the embryo and the rudiments of the face. In the second part, taking place in the following six weeks, those facial rudiments grow and come together, knitting themselves together to form the face. This latter period comprises the last four weeks of embryogenesis and the first two weeks of fetal development.[6]

As part of the head, the face always develops at the most anterior (forward) part of the embryo; indeed, the position of the head defines the anterior end of the embryo. When the face visibly begins to take shape, the embryo has lengthened, filled out, and acquired a distinctive geometry defined by three main axes as shown in Figure 2.5B. The first is the antero-posterior axis, which defines the lengthwise extension of the developing embryo from head to tail. The second is the dorso-ventral axis, with the top (future back) of the embryo designated the *dorsal* side and the bottom (future stomach side) the *ventral* side. The third is the left-right axis, which features basic bilateral symmetry with mirror-image left and right sides. (Each half can be considered to define a fourth axis, which occurs in duplicate, the medial-distal axis running from the center of the embryo to the most distal point, either to the left or the right.) That mirror-image symmetry of the early embryo foreshadows and translates during later development into the bilateral, mirror-image symmetry of the body as a whole, including the face.

The embryo, of course, does not begin with such a longitudinal form. It starts as a single cell, the large fertilized egg, generated by the union of the father's sperm with the mother's egg, a much bigger cell. That fertilized egg undergoes a relatively rapid series of cell divisions, occurring without substantial growth, that converts the initial single cell into a nearly spherical ball of smaller cells termed the *blastocyst*. In human embryos, it forms within the first week of development, and between days eight and ten it has begun to implant itself in the lining, the endometrium, of the mother's uterus. This attachment, in turn, begins the process of formation of the placenta, the complex, thick multitissue, organ rich in blood vessels through which the developing embryo and later the fetus receives nutrients from the mother and through which it expels its metabolic waste products. With implantation, the blastocyst develops further, and the embryo begins to take in nutrients from the mother's blood circulation as the placenta develops. This influx of nutrients will support rapid cell divisions and growth of the embryo.

Inside the implanting blastocyst, there is a special group of internal cells that make up the direct precursor of the embryo. Termed the *embryo disc*, this structure is flattened and somewhat elongated. The rest of the blastocyst will form various membranes that surround the embryo proper or will form part of the placenta. Within the embryo disc, a number of infoldings soon take place that establish the trilaminar embryo and then create the basic longitudinally shaped embryo. The first of these infoldings is termed **gastrulation** and leads to the formation of the internally located mesodermal cells, the third germ layer, located between the external ectoderm of the embryo disc and the internal endodermal cells (Figure 2.5A). The subsequent growth of those cells takes place asymmetrically, extending the embryo in a direction that becomes the future antero-posterior (head-tail) axis. It is this initial extension that forms the basis of the subsequent tubular structure of the whole embryo with its three axes. In human embryos, these events are complete within the first fifteen to seventeen days of development.

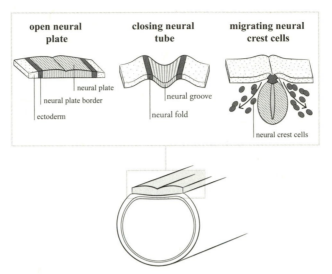

FIGURE 2.6 As shown schematically at the bottom of the figure, the neural plate from which the neural tube will develop is located in the dorsal part of the embryo. (The internal details of the embryo have been omitted.) It undergoes the stages of folding shown at the top from the open neural plate stage, via invagination to create the dorsal neural tube. From the fully formed neural tube, the neural crest cells migrate ventrally from the dorsal edge of the neural tube. The cellular descendants of the neural crest cells will go on to form many structures in the embryo, including the bones of the face.

It is, however, the second infolding of cells that lays the initial foundation for the subsequent development of the face, though many steps intervene between this first step and the actual formation of the face. This event consists of the invagination of a long strip of tissue along the top (dorsal) side of the embryo, which leads to the formation of the neural tube. The anterior or cranial half of the neural tube will give rise to the brain, while the posterior half, the trunk section to the spinal cord. As the embryo develops, the trunk region will become much longer than the cranial region, but the cranial and trunk regions are of nearly equal size at the beginning.

The process of neural tube formation begins in the middle of the third week of human embryonic development, around day eighteen, as a thickened strip; the *neural plate* of the outer ectoderm (the tissue that encases the embryo) forms along the dorsal side of the embryo. As diagrammed in Figure 2.6, the neural plate first folds inward to form a long longitudinal bulge at the dorsal side. As the fold deepens, it bends at the two sides, and the two topmost folds come together and fuse at their edges, turning the infolding bulge into a tubelike structure. At this stage, the embryo is often described as a "tube within a tube," namely, the neural tube within the larger of the embryo itself. It is the upper (dorsal) side of the neural tube from which the neural crest cells originate. Yet there is also a second tubelike structure, the future gut, running the length of the embryo on the ventral side. It is derived from the endoderm at the end of the first four weeks postconception, hence the embryo here is better described as "two tubes within a tube."

The Brain and the Face in Embryonic Development

The story of the face is inextricably connected with that of the brain. Those connections are multiple and diverse and extend from the beginnings of life until its end. Here, however, we will focus on the initial relationships, those that unfold in the embryo and in which the brain plays two distinct roles in the development of the face. The first takes place in embryogenesis; it is quite brief but crucial in initiating the formation of the face. The second takes place during fetal development and involves the shaping of the face during a prolonged phase of mutual interaction between the developing face and the developing brain.

The early steps that lead to initiation of face formation in the embryo involve a series of chemical "signals"—molecules that trigger certain aspects of development—between different cell types. To keep things relatively simple, I will only give their names and roles here and explain the names later. These two molecules are known as **fibroblast growth factor 8** (FGF8) and **sonic hedgehog** (SHH). In the brief account that follows, I will not discuss the evidence behind the conclusions (sources, however, are given in the endnotes).

While the position of the developing head defines the anterior or "cranial" end of the embryo (and thereby its antero-posterior axis), the developing brain is the initial main event in the development of the head. Shortly after the neural tube has formed, the future anterior (cranial) region becomes divided into three separate bulges, each surrounding a fluid-filled space, its *ventricle*. These three bulges are the precursors of what will become the brain's three main divisions. Proceeding from the anterior end, they are, respectively, the **prosencephalon** or future "forebrain"; the

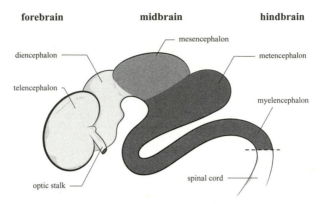

FIGURE 2.7 Sketch of the early mammalian brain, showing its three main divisions: forebrain, midbrain, and hindbrain. The forebrain, at the front (anterior), consists of the telencephalon and the diencephalon; the early midbrain is unsegmented and consists of the mesencephalon. The hindbrain, or rhombencephalon, has two main divisions, the metencephalon and the myelencephalon. The development of the brain and its relationship to face behavior is described in Chapter 7.

compartment just behind (posterior) it, the **mesencephalon** or "midbrain" and; the posterior-most, the **rhombencephalon**, the future "hindbrain." Each of these grows and expands while developing further subdivisions and complexities. A particularly important division, with respect to our story, is that of the forebrain into two parts, the more anterior **telencephalon** and the section just posterior to it, the **diencephalon**. It is part of the telencephalon that will help initiate formation of the face. A diagram of the early embryonic tripartite brain is shown in Figure 2.7.[7]

The development of the brain, however, does not unfold automatically. Rather, it depends on one of the chemical signals mentioned above, FGF8. This protein is secreted by the migrating neural crest cells and interacts with the cells of the newly forming brain. Without FGF8 at this point, the cells of the forebrain and midbrain undergo cell death—apoptosis—with the result that these two brain regions do not develop properly, becoming only vestiges. The consequence of that failure of brain development, however, is equally dramatic for the face: in the absence of the forebrain, the face itself does not develop or even begin to form. Hence, an early step in the sequence of events that leads to forming the face is the neural-crest–stimulated "rescue" of the anterior parts of the brain so that they can grow and develop (and then contribute to formation of the face). The step can be summarized as:

                    Emit chemical signal, FGF8
Neural crest cells   ──────────────────────>   Forebrain development.[8]

How does the forebrain initiate formation of the face? The answer is connected to the way in which the forebrain grows. As it grows along the antero-posterior axis, its most anterior section, the telencephalon, becomes constrained by the membranous **amniotic sac** that surrounds the embryo and is forced to fold over The telencephalon now possesses distinct dorsal (top) and ventral (bottom) surfaces. This second step can be represented as:

                    Telencephalon folds over
Forebrain grows   ──────────────────────>   Telencephalon divided into
                                               dorsal and ventral sides

The newly formed ventral side of the telencephalon now lies opposite the underside (ventral surface) of the outer ectoderm at the front end of the embryo. It is from this ectodermal region that the face will develop.[9]

It is not, however, physical contact between the ventral telencephalon and this ectodermal area that is essential for initiating the face but another chemical signaling event. The ventral telencephalon produces and secretes a second molecule, SHH, that stimulates this area of ectoderm to begin forming the rudiments of the upper portions of the face. That region is now termed the **frontonasal ectodermal zone** (FEZ**).** This third step can be depicted as:

                    Emits chemical signal SHH
Ventral telencephalon   ──────────────────────>   FEZ activated

It is with this step that face development truly begins. The "activation" of the FEZ consists of it switching on its own synthesis of *both* signals, FGF8 and SHH. Migrating neural crest cells, derived from the region covering the diencephalon and midbrain, are now attracted to six specific sites on the inside of this ectoderm, settle there, and begin to proliferate under the influence of these molecules.[10]

The consequence is that these six sites begin to grow and bulge outward. The proliferating cells at these sites, which are descended from the neural crest cells that migrated to these positions, are mesenchymal in nature; they are surrounded by an ectodermal "jacket" derived from the outer ectodermal cells at the original site. These six sites at this early point are designated the **facial primordia** and constitute the rudiments of the "midface"—that part of the face from and including the upper jaw and upper lips to (what will become) the eyebrows. Thus, the fourth of the early events can be summarized as:

As the facial primordia grow outward from the ventral ectodermal surface of the embryo and become prominent bulges, they can now be referred to as **facial prominences.** They will eventually fuse to generate the midface, rather like a self-assembling jigsaw puzzle, as will shortly be described.

In addition to the six facial prominences derived from the FEZ are two other facial primordia, but these have a different origin. These are the *mandibular primordia*, from which the lower jaws will develop. They arise from a pair of sac-like internal structures termed the first **branchial (or pharyngeal) arches**, and they receive their neural crest cells from the region over the hindbrain. Like the other six facial primordia, they develop under the stimulus of SHH, although in their case it comes from the anterior ventral endoderm, not the brain.[11]

Thus, in mammalian embryos, there are initially four pairs or a total of eight facial prominences. From top to bottom, they consist of two medial nasal prominences in close proximity to the midline; two lateral nasal prominences, one on either side of the medial nasal prominences; two maxillary prominences just below the two pairs of nasal prominences and above the mouth; and two mandibular prominences just below the mouth. The medial and nasal prominences will grow and fuse to become a structure termed the **frontonasal process** (FNP). The maxillary facial prominences, which will fuse with the FNP, will become the upper jaw. The lowermost facial primordia, the mandibular prominences, develop into the lower jaw. In contrast to the other six facial primordia, they ini-

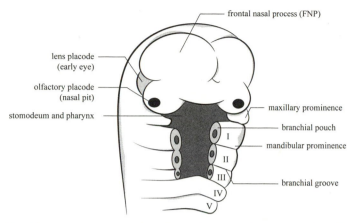

frontal nasal process (FNP)

lens placode
(early eye)

olfactory placode
(nasal pit)

stomodeum and pharynx

maxillary prominence

branchial pouch

mandibular prominence

branchial groove

I

II

III

IV

V

FIGURE 2.8 Diagrammatic representation of an early stage of face formation from the facial pri-mordia in a four-week-old human embryo. The initially separate frontonasal (or medial) prominence gives rise eventually to the central part of the face: the middle part of the nose, the cheeks (when joined with the lateral nasal prominences), the primary palate (internal), and the philtrum, or shallow depres-sion that runs between the bottom of the nose and upper lip. The lateral nasal prominences initially generate the sides of the nose, while the maxillary and mandibular prominences give rise to the upper and lower jaws, respectively. At this stage, the FNP has already formed. The mandibular prominences make up the first pair of the branchial arches, which are labelled I through V. (Modified from Lieberman 2011.)

tially have an outer lining of endodermal cells—reflecting their origins in the anterior endoderm—and later acquire an outer ectodermal covering. The embryonic face of a four-week-old embryo is shown schematically in Figure 2.8; the FNP has formed at this point.

Putting the Facial Prominences Jigsaw Puzzle Together to Form the Face

Bird and mammalian embryos differ somewhat in the details of growth and fusion of their facial prominences, in particular in the formation of the FNP, but here we will discuss only the mammalian pattern. In mam-malian embryos, as we have seen, the two medial (central) nasal promi-nences fuse first and then the lateral nasal prominences fuse on each side to that medial central mass. The latter soon develops two distinct thick-enings on its sides, *nasal placodes*, each of which becomes divided during the fifth week of development in human embryos into two parts by the de-velopment of a single nasal pit, which will become a nostril. These two depressions effectively divide each placode into a medial (central) and lateral half. As the placodes grow, by late in the seventh week, there is clearly a single central entity, the FNP.

During this period, the maxillary prominences, the source of the upper jaw, have also been growing. These now come together at the front, being connected by an internal bridge, the intermaxillary process. Externally, a shallow groove—the **philtrum**—develops from the bottom of the nose to the upper lip. (A glance in the mirror reveals it at once. It is unique to the human face.) Behind the intermaxillary process, the fused maxillary prominences give rise to the primary, bony palate above the mouth. The fleshy *secondary palate* that underlies the primary palate, and which creates a distinct compartment for nasal breathing, is also formed by the maxillary primordia. Starting out as two soft *palatal shelves* initially growing downward at the sides of the oral cavity, they turn horizontally and fuse via loss of the epithelial cells at the fusing edges.

The mandibular prominences, precursors of the lower jaw, are joined at a slight depression, which is subsequently filled in by cell proliferation of the underlying mesenchymal cells. This process will give rise to the lower lip. By late in the seventh week of development, the mandibular and maxillary processes have fused at the sides of the future mouth to produce the future cheek regions. (The true mouth appears earlier, during the fifth week, when the internal membrane separating the foregut from the stomodeal depression ruptures.) The following three weeks of development are largely devoted to further growth and closer integration of the prominences and the creation of a truly integrated face. By the end of the tenth week in the human embryo, there is a discernible human face that becomes more obviously so over the next ten weeks. By week twenty, the face of the fetus shows clear individual identity. (This is a purely physical and biological identity, however; there is no evidence of anything approaching consciousness or individual personality at this point.)

A few additional details of the development of the face are worth noting. In particular, the pair of second branchial arches, which lie just behind the first (the mandibulary prominences), are the source of the progenitor cells of the twenty-one different kinds of facial muscles. These are the mimetic muscles, the muscles of facial expression, which, as we have seen, were the subject of the first scientific studies of the face by Sir Charles Bell. The formation of the eyes is also, of course, an important element in the embryonic development of the face. Their placement within the developing face, laterally to the fronto-nasal primordium, is determined by the prior development of a pair of thickened regions of the outer ectoderm, so-called ectodermal placodes that will become the lenses of the eyes. The nascent eyeballs develop underneath them, growing outward from the brain. These, in turn, attract the development of the optic nerves to the developing retinas, the light-receptive surfaces of the eyes, which receive the initial stimulus of visual images. The eyes are not unique in this respect: all of the sensory organs of the head are initiated from special ectodermal placodes, becoming sites of development of key sensory and sensory-processing neural structures, and developing connections to the brain that permit their sensory information to be processed. In effect, the initial, nonsensory ectodermal placodes of the early embryo become **neural ectodermal placodes.** They are essential but not sole sources of development of the sensory organs in the head; in the full development of the eye, for example, cranial neural crest cells also contribute.

In thinking about this rather complex set of events , the crucial fact to remember is that the basic shape of the face is dictated by the detailed positioning, growth, and, finally, fusion and subsequent development of the facial prominences. It is these events that ultimately determine the placement and precise shape of the developing facial bones, which in turn determine the overall outer shape of the face, whose outer surface conceals the detailed arrangement of blood vessels, nerves, and muscles within.

When Face-Type Differences Emerge: The Second Role of the Brain in Face Development

The step of fusion of the facial prominences is particularly crucial in forming the face. It is the event in which the face first emerges as an integral entity and must do so with minimal errors. Even slight aberrancies during this stage can become amplified to become visible and significant ones. The occurrence of cleft lips and cleft palates generally reflect flaws during this process, specifically of the maxillary primordia.

It is not surprising, given the importance of facial prominence-fusion in forming the face, that this process has been found to be highly conserved among the large group of land-living vertebrates termed the *amniotes*, namely the amphibians, reptiles, birds, and mammals. Thus, during fusion of the facial prominences, there is convergence on a common shape in the embryos of these animals, as has been shown by precise measurements. Embryonic vertebrate faces are more similar at this point than earlier in development. Yet it is also from this point onward, during the course of fetal development, that the differences in facial type among the different kinds of amniote vertebrates steadily emerge. Thus, for example, bird embryos become distinguishable from mammalian embryos by an exaggerated enlargement of the face along the antero-posterior axis due to growth from the FNP of what will become the upper beak. As mentioned earlier, this develops from the premaxillary bone, a bone just in front of the main maxillary bone (the principal bone of the upper jaw), which remains small in mammals. In contrast, in mammalian embryos, in those species that show muzzle outgrowth—the great majority—the outgrowth of the upper jaw comes from the maxillary itself.[12]

As might be expected from such observations, the larger the taxonomic divergence between animals, the earlier in development the differences appear. Thus, differences between avian (chick) and mammalian (mouse) embryos, following fusion of the prominences, appear earlier than differences between two different kinds of bird embryos (for example, chick vs. duck), which in turn appear earlier than differences between two bird species within the chicken family (chickens vs. quail). Again, the most striking differences involve structures derived from the FNP.[13]

Differences of face (and head) shape even *within a species* can be detected at mid- and late-embryonic (fetal) stages. This has been shown in mice, where several mouse strains with somewhat differing facial and head morphology are known, for example the "short-faced" mouse strain, which has a muzzle that projects visibly less and a rounder head. Analysis of these strains has demonstrated measurable differences in late embryonic head shape and identified a major factor influencing the shape of the face via its effect on the developing skull. That factor is the pattern of brain growth, as manifested in its extent, or rate, or precise shape, or a combination of those variables. In the short-faced mouse strain, for example, development of a smaller-than-normal brain during embryogenesis clearly precedes the changes in face architecture. In the other strains, the changes in brain growth and facial bone pattern appear to proceed more closely in synchrony. The simplest explanation, though others are not excluded, is that the dynamics and pattern of brain growth have a direct physical influence on the detailed shaping of facial bones and thereby the face.[14]

That influence probably precedes fusion of the facial prominences or at least is believed to do so. The reason is that while the prominences grow outward and later together, the whole head behind the face is expanding, in large part due to brain growth, and that growth tends to push them apart. That should produce a wider face, while conversely a slower rate of growth should produce a narrower face and quite possibly a muzzle that projects more. Although these observations stem from the study of mouse embryos, they are almost certainly relevant to humans. It has been recognized for centuries, for instance, that some people have rounder heads and broader faces, the *brachycephalic* type, while others have longer and thinner faces, the *dolichocephalic* type.

The reason for stressing how much facial shape is determined by early developmental processes in the embryo and fetus is that many of the genetic factors that affect head and face growth and development during these stages are known. To the extent that facial characteristics are determined prenatally, evolution would be expected to "work" with those materials to yield the changes in face shape characteristic of different animal types. Of course, in humans and other anthropoid primates, there is much facial maturation, involving considerable change from infancy through the teenage years. Relatively little is known to date about the genetic foundations of those changes, and they could in principle be quite different from those in the earlier stages. It is equally possible, however, that the "template" for those genetic changes is already essentially set by the time of birth and that those maturational changes are largely a developmental "readout" of that template as it exists at birth. The issue is yet to be resolved.

To this point, we have been considering the physical influence of the brain on the development of the face as mediated by the brain's effects on the skull. The skull as a whole, however, merits a direct look in its own right.

The Skull and the Face

We have been focusing on the development of the head and face, but that focus is directly relevant to the evolutionary origins of the animals we call the *vertebrates* or, more formally, the Vertebrata. That name codifies the defining feature of the group as their backbones. Yet all of the fossil evidence indicates that the founding members of this group, tiny jawless fish, lacked a bony backbone and probably did not have a soft-tissue one either. The first "vertebrates," however, did possess a distinctive head, almost certainly including some form of skull, albeit one made of softer, collagen-rich, and possibly cartilaginous material rather than bone. Hence, it is probably more appropriate to designate them as **craniates**—animals with complex heads—a group that includes the most primitive surviving forms, two kinds of jawless fish—the hagfish and lampreys. Hence, while the words *vertebrate* and *craniate* are essentially synonyms and will be used interchangeably in this book, *craniates* is probably preferable in terms of evolutionary history.[15]

The particular form of the adult head of all modern craniate species is ultimately determined by its skull. The human skull, as seen from two angles, is diagrammed in Figure 2.9. It consists of three parts: the **neuro-cranium,** the **viscerocranium** and the **basicranium.** The neurocranium, as mentioned, makes up the arched roof or *vault* (roof) of the skull, which houses the brain; it is shown from the top and the side in Figures 2.9A and B, respectively. It consists of several major bones (the frontal, temporal, parietal, and occipital) and extends from the level of the eyes over the vault (the frontal bones) to the most posterior bones at the back of the skull (the occipital bones). The viscerocranium consists of essentially the bones of the face and a few other internal bones, including the palatal and pharyn-geal bones; the structure of the viscerocranium and the positions of its constituent bones is shown in Figure 2.9C. (In the embryo, the viscero-cranium is sometimes designated the *splanchnocranium*, where *splanchno* indicates the gut, a reference to the fact that the face forms around the mouth, the terminus of the gut.) Lastly, the basicranium consists of the bones that form the base of the skull, hence lying beneath the brain, which rests on it, but above the face. (These bones are not shown in the figure.) In effect, the basicranium may be considered the "roof" of the face (but not the "full face"), though it is internal to the face. This positioning of the adult face beneath the brain makes for a singular difference from other mammals, including the great apes, where the face lies in front of the brain. One major reason for the difference is the greatly expanded neurocranium in humans, which partially reflects the disproportionate size of the **cerebral**

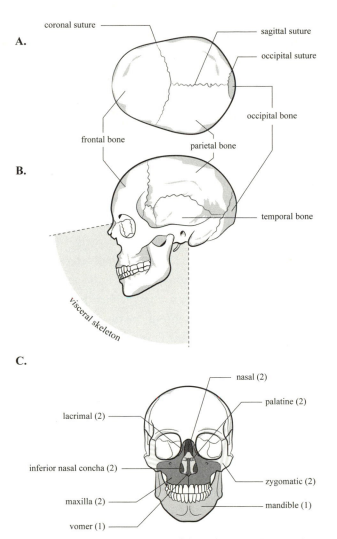

A.
coronal suture
sagittal suture
occipital suture
occipital bone
frontal bone
parietal bone

B.
temporal bone
visceral skeleton

C.
nasal (2)
palatine (2)
lacrimal (2)
inferior nasal concha (2)
zygomatic (2)
maxilla (2)
mandible (1)
vomer (1)

FIGURE 2.9 A. Dorsal view of the neurocranium and three of its major divisions: the parietal, frontal, and occipital bones. Anterior is to the left. Also indicated are the *sutures*, or dividing grooves between the major bony elements. **B.** Lateral view of the skull, providing a side view of the bones of the neurocranium and the visceral skeleton (indicated by dashed lines), the bony architecture of the face. The temporal bones of the neurocranium shown here are visible from this angle. **C.** A front view of the visceral skeleton, indicating the six pairs of bones and two unpaired bones (as indicated by the numbers) from which it is constituted: maxillary bones, nasal bones, zygomatic bones, lacrimal bones, nasal concha bones, the vomer, palatine bones, and the mandibular (lower jaw) bone.

cortex (or cerebrum), the major derivative of the telencephalon, along with geometric aspects of human skull development, to which we shall return.

Altogether, there are twenty-two bones in the human skull, most of them paired, reflecting the bilateral symmetry of the skull; many are small and not indicated in the figure. (The other hard-tissue parts of the skull are the teeth, which develop within the main maxillary and mandibular bones). The majority of the skull bones, including those of the face, are ultimately derived from osteoblasts, most of which are derived from neural crest progenitor cells. The two occipital bones located at the back of the skull, however, derive from mesodermal tissue, which is also the source of the bones of the main part of the skeleton, namely, the vertebrae, ribs, collar bone, and limb bones. All such mesodermally derived bones—which are the majority type of the skeleton—form around and within an initial **cartilage** precursor or *model*, and are termed **endochondral bones.** In contrast to endochondral bones, most of the bones of the head, including those of the face (but not those of the cranial base) form by deposition directly between membranes, involving cartilaginous material but not a rigid cartilage model, and are termed **membranous** (or **dermal**) **bones.**

While one thinks of skulls as rigid, unitary structures—as they are in mature adults—that is not the case for skulls of newborns or young children. Indeed, there must be room for growth of both the brain and the skull itself in the passage from birth to adulthood. At birth, the brain is only about 25 percent of its adult size though it contains nearly its full complement of neurons. The source of the remaining 75 percent mass is produced by the multiplication of the helper or **glial cells** and by the production of massive amounts of the protective myelin sheet, secreted by the glial cells, which surrounds many neurons; that sheathing is the so-called white matter of the brain. This material wraps around and protects neurons, providing insulation that allows the transmission of electrical signals in the neurons. By the time a child has reached age eight, her or his brain has achieved its final size and shape but not necessarily its final content of neurons.[16]

While the individual bones of the neurocranium have partially ossified at birth, their edges have not. This makes possible skull expansion with brain growth during development because the separate bones of the neurocranium can continue to grow from their edges. They are kept separate by open regions, the **cranial sutures,** which exist between the individual bones. Each bony plate has at its growing edge soft tissue that produces the cells that generate the bone, and each suture shows two of these soft-tissue regions confronting one another across a narrow space. After brain growth is complete, the process of fusion of neighboring bones takes place, obliterating the sutures, with each one closing at a characteristic time; the last neurocranial sutures fuse around age twenty. (For the sutures between facial bones, this process may not be complete until the eighties.) The whole process of suture growth and eventual fusion is ultimately under genetic control. If it happens prematurely, however, this interferes with normal brain growth and can be fatal. Such premature closures do, indeed, sometimes take place through particular genetic defects, the resulting premature suture closures termed **craniosynostoses**.

This relationship between the developing brain and the developing skull has further significance: because the dermal bones take shape within soft tissues being pushed outward by the developing and growing brain, the inside of the vault of the skull is molded on those regions of the brain that it contacts. As the membranous bones of the cranial vault form over the brain, their still soft inner surface takes on the shape of the outer surface of the brain that it directly contacts. As skull growth and ossification are completed, that initial soft mold becomes part of the mineralizing hard bony structure. One consequence is that well-preserved fossil skulls can yield clues to relative sizes of different outer regions of the brain in extinct hominins. Since the functions of these regions in modern humans have been well explored, one can make inferences from some of these skulls about certain features of the brain that lay just beneath, in these hominins, as will be discussed in Chapter 6.

Another distinctive feature of the human head, one rooted in brain and skull development, is the roundedness of the human skull and the corresponding relative forward position of the anterior part of the neurocranium, the forehead, which contributes—along with the absence of a muzzle—to the essentially vertical characteristic of the human forehead and the face as a whole.

This rounded vault of the human skull is thus, in part, a consequence of brain size. The extensive growth of the human brain in hominin evolution, however, was not only a key feature of their morphological evolution but also intimately related to expanding mental and behavioral capacities, including aspects of the functions and roles of the face in human social existence, as we will come to. Although brain size was one element in the roundedness of the human skull, and thus a contributor to shaping the corresponding smaller, more vertical face of *Homo sapiens,* it could not have been the sole determining feature of that shape. Neanderthals, who were probably our last surviving and closest hominin relatives, had even slightly larger brains than ours, about 1,450 cubic centimeters versus the average 1,350 cubic cm volume typical of adult modern humans. (Relative to body size, however, Neanderthal brains were not larger; the net difference in brain size was in proportion to their larger bodies.) Nevertheless, these large-brained hominins showed the more backward-sloping foreheads typical of earlier hominins. Several features of head and skull development can contribute to more rounded skulls, and it is still not clear which are the most important. There are, however, indications that an enlargement of the temporal lobes may have been the most critical factor, which may, in turn, be connected with some of the special mental functions associated with those lobes in our species.[17]

One final point about the head: its functioning requires an intricate system of blood vessels and neural connections—within the face, between the brain and the face, and, of course between the brain and the rest of the body. Perhaps the most dramatic of these connections is that between the spinal cord and the brain through a centrally located opening in the base of the brain, the **foramen magnum**. That placement permits a balancing of the head directly above the spinal column rather than slightly forward of that junction, as in the great apes, permitting a fully bipedal, stable form of movement. In considering the development of the soft tissues in and around the developing skull, it is important to remember that the soft, cartilaginous precursor structures of the neurocranium and viscerocranium must form *around* the key vascular and neural connections between head and body. Thus, for example, the cartilaginous precursor of the base of the skull forms around the developing spinal cord to create the foramen magnum, and that cartilage model is later replaced by bone.

Facial Development from Birth Onward

In humans, the transition from the juvenile to the mature adult body is prolonged. Different parts of the body generally grow and mature at different rates, the particular pattern and rates being characteristic of the different body regions and, of course, the particular species. Thus, as mentioned earlier, the human brain grows particularly rapidly in the first two years of life, then more slowly, completing its growth by age eight. The face and bones of the body grow more slowly and only complete their growth later, usually between ages sixteen and eighteen. The first phase of strong facial individuation takes place in the beginning years of childhood: final eye color develops rapidly, usually within the first month; the baby's round face gives way to the child's somewhat longer, more oval, and more individual face; skin pigmentation in sub-Saharan Africans and Melanesians deepens significantly, with exposure to sun playing an important part; and hair develops, framing the upper face and, in so doing, changing its appearance.

While the face of the older child between ages eight and ten has already acquired its basic individual characteristics, clearly foreshadowing the face of the adult, the final touches take place after puberty roughly between ages twelve and eighteen. This is often most strikingly apparent in the growth of the nose, based on the growth of the nasal septum, which is derived from the fronto-nasal primordium. The bony part of the nose and the structure of the facial bones generally determine the final shape of the nose: for example, a relatively large bridge of the nose in a child almost certainly foretells a larger nose in the late teenager. Ultimately, the growth of the skull determines the detailed pattern of connective tissues, muscles, and blood vessels, hence the final morphology of the face, although it takes about eighteen years for the fully mature face to develop.

Nose shape is perhaps the facial feature that most strikingly contributes to the perceived individuality of faces. Furthermore, the differences in nose shape almost certainly reflect in-built genetic specification and not, for example, nutritional or other environmental influences. We can be certain of this in light of the near identical quality of the faces of identical twins. Given the foundational genetic differences in nose shape between individuals, one can further infer that average (and readily observable) differences in nose type between the major "racial" groups also reflect genetic differences. Thus, in general, people of largely Japanese or Chinese descent have smaller noses and less rounded eyes than individuals of primarily European descent. It is not "racist" to say this because there are no superior nose or eye shapes, just differences, which are simply a fact.

We will return to matters of how genetic inheritance affects the details of face shape in Chapters 4 and 9.

Yet not all physical features of the face are tightly specified by heredity: the size and shape of the lower jaw is distinctly influenced by diet. From the time of development of the baby teeth, a diet that consists of highly fibrous material such as tough meat that requires more chewing will lead to a more highly developed lower jaw in the child, and that difference will be reflected in the teenager and the adult that child becomes. While the image of distinctively prognathous "cavemen" returning to their families with their fresh kills of hunted animals is a long-standing cultural joke: we equate strong jaws with brute strength and early humans with such strength. In actuality, that facial image may reflect, in part, the biological reality of a diet heavy in meat. This ability of jaw bones to grow in response to muscular stress reflects a general responsiveness of bone growth to tensile stress.[18]

One more set of differences emerges following puberty: the development of gender characteristics. Many primates, including our nearest great ape relatives, have rather strongly pronounced gender differences in their faces, as well as other body characteristics. Think, for example, of silverback senior male gorillas with their pronounced bony sagittal crests as compared to their smaller mates who lack them. Such strong facial **sexual dimorphism,** however, is not the case for humans. In humans, the facial differences—excluding facial hair—are comparatively slight. Nevertheless, they are significant, leading to faces that are generally either feminized or masculinized. This matter, however, presents an interesting puzzle: most people can identify the gender of a person from photographs of that person's face, even when the hair is excluded, yet it is not entirely clear just how we do this. Women tend to have slightly longer and more oval faces than men, whose faces are somewhat more square with eyes that are slightly farther apart, while chins and noses tend to be more prominent in men than women. Although no single feature is unambiguously diagnostic of gender because, for any specific characteristic, there is a substantial spread of values, tests have shown that many individual features carry gender information. In addition, however, there is a general, "gestalt" impression, such that it is the *ensemble* of features and their degree of gender specificity that creates an impression of maleness or femaleness of the face. The surprising aspect of this is the speed and general accuracy of this capacity for gender recognition, given the subtlety of the clues.[19]

Precisely what the developmental basis of facial sexualization consists of is unknown. It probably involves slight differential responses in cell proliferation in different areas of the face in response to the different hormonal milieu of the bodies of women and men. Relatively high testosterone levels promote these facial-region growth differences toward the more masculine range of features, while relatively high estrogen levels would do the reverse. This possibility accords with the fact that faces acquire a more certain gender identification after puberty when the sex hormone levels increase. Evidence does support this relationship.[20]

These differences in general facial configuration are almost certainly neither accidental nor without some functional significance. In general, more feminine faces are attractive to men, and more masculine faces to women. Sexual selection for such differences during the evolution of our species has almost certainly played a part in shaping them, as shall be discussed in Chapter 8.[21]

Summing up the Key Features of Face Development

We have covered much ground in this chapter, but the gist is readily summarized. Essentially, after the embryo has achieved its elongated shape, consisting of "two tubes within a tube" (the neural tube and the future gut within the elongated embryonic body), its face develops at the anterior end of the embryo from the facial primordia, which lie opposite the ventral surface of the folded-over telencephalon. Six of the eight facial primordia receive their start by acting as sites that have been populated by cranial neural crest cells; the cells home in on the sites in response to a molecule, sonic hedgehog, which is emitted by the ectodermal site in which they arise. The remaining two primordia start out as endodermal sites—that is, the first branchial arches—and use the same chemical signal to attract neural crest cells. All these sites grow, expand, and undergo a characteristic sequence of fusions to yield a unified structure that encompasses both the mouth and the rudimentary eyes. That structure will itself grow and expand farther to yield the fully developed face. Following the completion of the fusion events, near the end of week ten of development, there is substantial growth and further development of the head and face such that by week twenty there is not only a characteristic human face but also a distinctly larger one that has individuated to some extent.

The entire sequence of steps takes place in every normally developing human baby. Given their complexity and their vulnerability to environmental insults and genetic alteration, the wonder is not that some human embryos die in utero or develop into babies with severe defects, but that the overwhelming majority of live births yield fully normal and healthy babies.

What processes drive these events and ensure that they happen in the right sequence to produce (with high frequency) the right outcome? It cannot be the environment of the mother's womb, which simply provides an essential sheltering and nutrient-rich environment, that allows these changes to take place but in no way directs them. In the language of developmental biology, the womb is a "permissive" environment for development of the embryo and fetus, not an "instructive" one. Yet if those instructions do not come from outside the embryo or fetus, then they must come from within—and they do. Certainly, the physical dynamics of the various developmental events, as shown in particular by the later brain–face interactions, constitute one crucial component. Ultimately, however, the source of all the developmental changes is the activities of the genes within the cells of the developing embryo. The characteristic differences between the faces of different kinds of animals, which often become apparent during the fetal stage, as discussed earlier, are ultimately if not directly the result of differences in those gene activities.

Simply saying that "genes are responsible" for the morphological differences, however, is too general an answer to give any genuine insight. It is important to know which particular genes are most important to, indeed necessary for, the development of the face; in what temporal sequences they are switched on; what their gene products actually do; and where and how those products function specifically in cell and tissue interactions to generate the face. We have already encountered two of these key gene products, FGF8 and SHH, but they do their work within a large context of other genes and developmental processes. Understanding that context is vital to understanding the evolutionary roots of the human face. In the next chapter, we will look at what is currently known about these genes and the ways they contribute to form the human face.

— three —

THE GENETIC FOUNDATIONS OF THE FACE

Introduction: On Genetic "Determination" of Traits

In thinking about the relationship between genes and faces, it makes sense to begin with a fundamental and general question: how exactly are an organism's physical characteristics "specified" by its genetic material? Although vast amounts are now known about the details of development in many animals and plants, there is still no satisfying answer to that general question. To be sure, there is an explanatory framework that attempts to serve, which we will come to shortly, but it lacks depth and predictive power. Furthermore, in considering this basic issue, one soon encounters a terminological problem that only confuses matters: geneticists often describe the total genetic material of an organism, its **genome**, as "determining" that organism. Crudely speaking, this is true: ultimately, house cats are not tigers and poodles are not wolves because the respective pairs of genomes of these beasts differ in crucial respects. Yet the word *determines* implies far greater directness and sufficiency than the genome possesses or could ever deliver.

In the 1980s and 1990s, there was a corresponding vogue for smaller claims that specific genes "controlled" particular traits, in particular personality traits in humans, such as sexual orientation, or the propensity for "criminality" or that for taking risks. Invariably, such statements were based on a nonrandom association of a particular genetic variant with a particular characteristic. But the correlation between possessing the variant and showing the trait was always imperfect. Such indeterminacy is the norm; in general, gene variants in plants and animals exhibit some variability in the presence or extent of an associated trait. The sources of such variability are either other genetic differences in the organism— so-called **genetic background** effects—or environmental effects, or both. Whenever a genetic variant is associated with such variable outcomes, however, one cannot say that it "controls" a particular trait.

Geneticists have long recognized that altered genes do not always reveal their presence. In effect, the total genetic information of an organism, its **genotype**, is not always fully reflected in its appearance or **phenotype**. In the next chapter, we will look at the ways in which this masking of genetic effects takes place, but here suffice it to say that the notion of strict, invariant genetic "control" of traits by genes is simply untrue.[1]

Nevertheless, the visible and highly characteristic differences between species, traceable to their different genomes, and the fact that certain genetic variants are often associated with a particular phenotype, argues that there is *something* to the notion of specification of organismal properties by genomes. How can we reconcile these seemingly disparate facts about genetic "control"? It becomes possible to do so by putting biological development back into the picture, just as doing so deepens understanding of evolutionary change, as we have seen. The key consideration is that between particular sets of genes and the traits they affect is a long intermediary train of processes and events that can be represented as follows: A (genes) $\rightarrow$ B (proteins) $\rightarrow$ C (protein actions) $\rightarrow$ D (cell properties) $\rightarrow$ E (tissue properties) $\rightarrow$ F (tissue interactions) $\rightarrow$ G (organ or regional properties) $\rightarrow$ H (final trait).

It looks simple in its linearity and sequence of bottom-to-top causation, if not in its details. The interpretative difficulties tend to cluster around how exactly new tissue and organ properties arise; this is an example of the general phenomenon of "emergence" (which we shall discuss later.) Additional complexity in the process derives from the fact that new cell and tissue properties (steps D and E) cause new genes to be "activated" (step A), which causes further changes in those properties and makes the causal sequence loop back on itself in a form of top-to-bottom "downward causation." Here, however, we will ignore those aspects and focus simply on the apparent inexorability of the genes-to-traits progression. The crucial point is that the seeming direct specification of traits by genes is a function of the high degree of reproducibility of each step in the chain. Thus, if each step operates with high determinacy to trigger the next, say 99 percent or more of the time, then the properties of A (the genes) will translate with high predictability into specific phenotypic features, traits (H). Increasing that reliability is the tendency of similar cells in tight groups to march in lockstep as they go through development, the so-called "community effect", which enforces similar behavior within the group.

Such cellular self-policing, along with (frequently) a degree of redundancy in the genetic input information, serves to further diminish the potential variability. Such reliability, often in the face of perturbation from outside factors, is denoted *robustness,* and developmental processes tend to be highly robust. Yet the existence of many steps between genes and traits itself increases the possibility of deviation at different steps, hence variability in the outcomes. Within this nexus of causality and potential error, one can begin to understand why mutant effects so often show variability. Most mutations cause a slight reduction in the amount of active gene product and can be described as partial **loss-of-function** mutations. That slight loss of active gene product, however, translates into a lower probability of transition to the next step in sufficient cells, hence a reduction in successful completion of the chain, with consequent increased variability in outcome.

Beyond such effects due to mutations, various external inputs can influence particular steps, hence affecting whether the sequence is completed or the pathway is diverted into another channel and a slightly different outcome. Indeed, in many species, there are particular points in the life cycle at which a specific environmental variable can switch the developmental path to a different outcome, and these are often crucial to the life cycle of the organism; such nodal points are themselves products of evolution.[2]

In this chapter, as we look at the genetic foundations of the face, we should bear in mind the complexity and intricacy of the events and processes connecting genes with the final developmental outcomes. That complexity virtually guarantees the occurrence of occasional errors in craniofacial development that, as we have seen, are the focus of much pediatric medicine. Here, however, we will be examining the typical events of face formation as they occur in nonmutant, healthily developing embryos from the perspective of their genetic underpinnings. To understand this story, however, a brief glance at the basic facts of gene action and expression, and the history of their discovery, may provide a useful prelude.

From Genes and Gene Activities to the Concept of Gene Regulation

In 1950, the midpoint of the twentieth century, evolutionary biology was in a most peculiar state. What had been achieved through a decades-long collective effort by dozens of gifted people was a modern theory of evolution. Indeed, that was no small matter. Variously termed **neo-Darwinism**, the *modern synthesis,* and the *evolutionary synthesis,* it was widely acclaimed as a convincing and complete explanation. Looked at closely, however, the theory's claim to completeness could be seen to be false; at best, it was a scaffold on which a later, more complete theory could be built. At its heart were ideas of the gene and of genetic change, yet these matters were swaddled in ignorance. In particular, the chemical identity of genes was unknown (although one experiment published in 1944 had given a provisional answer). This uncertainty, in turn, made it impossible to understand the nature of genetic variation, which was central to the theory as one postulated essential condition for evolutionary change. Furthermore, the ways genes actually worked was also unknown. The only certainty to cling to in this sea of ignorance was that some genes seemed to act by specifying particular proteins, enzymes, the molecules that carry out the metabolism reactions of living cells.

Within slightly more than a dozen years, a series of discoveries illuminated these basic issues in what was probably the most dramatic period of scientific discovery in the history of biology. Initially, that work in the then-nascent field of molecular biology had little impact on the ways evolutionary biologists thought or worked, but its potential was clear. By the 1980s, however, evolutionary biology had been transformed by the conceptual revolution wrought by molecular biology. In the few pages that follow, I will give a simplified version of our understanding of the basic molecular biology of the gene; it does not do full justice to the subject itself or its history, but there are excellent accounts that do.[3]

Genes are composed of the molecule **deoxyribonucleic acid** (DNA), which consists of two long polymers of complex molecular units termed **nucleotides,** with the two chains wrapped around in each other in a configuration known as a *double helix.* A schematic of a short section of a DNA molecule is shown in Figure 3.1, with the details described in the legend. The gist of the matter is that DNA consists of a four-letter chemical alphabet—the nucleotides, which come in four different kinds, abbreviated here T, A, C, and G—and that a gene does its work by specifying a proteinaceous molecule termed a **polypeptide chain.** (Each protein in the cell consists of either a single polypeptide chain or, more commonly, a combination of two or more such chains, either identical or different.) As

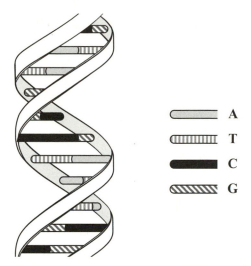

A

T

C

G

FIGURE 3.1 Schematic diagram of the double-stranded helical DNA molecule. Each of the two helical strands on the outside of the molecule consists of a backbone of deoxyribose sugar molecules alternating with phosphates in chains (–P–S–P–S–P, etc.). Linked to each sugar and projecting internally into the space between the two helixes are small compounds, nitrogenous bases that pair in specific fashions: A (adenine) with T (thymine), and G (guanine) with C (cytosine). (The nucleotide units mentioned in the text each consist of one S, one P, and one base.) Thus, starting with one helix, there are four possible base-pair "steps" in the helical internal ladder: A–T, T–A, C–G, and G–C. It is the sequence of the base pairs that carries the specific genetic information for each gene. (A further wrinkle, though one not important for our discussion, is that the chemical linkages in each backbone have a defined direction or polarity, and the two strands run with opposite polarity.)

described below, it is the sequence of the base pairs within a gene that specifies the sequence of the subunits of that polypeptide chain. This was one of the revolutionary insights provided by the Watson-Crick model of DNA structure, unveiled in 1953—a gene can be seen as a stretch of chemical information, with the sequence of the base pairs providing that information. Only one strand of a gene's DNA double helix, however, provides the direct information for the polypeptide chain that gene specifies.

Since there are twenty component units of proteins termed **amino acids**, a four letter-DNA alphabet has to encode a twenty-letter protein alphabet. It does so via groups of three nucleotides per amino acid. Since a three-letter code with four options at each position amounts to $4 \times 4 \times 4 = 64$ possibilities, a three-letter coding system is clearly more than enough. Indeed, most of the twenty amino acids are specified by more than one such nucleotide-triplet *codon* (with three codons providing "stop signals" for naturally ending the synthesis of a polypeptide chain). In effect, protein specification by DNA is overdetermined—there is more capacity in the system than seems to be strictly needed. (That extra capacity, however, conveys robustness to the system.)

Even as these basic features of how genes work were coming into view, they evoked a puzzle. As already noted, the genes are on the chromosomes, which reside in the nucleus. Yet it was clear by the early 1950s, using new methods of radioactive labeling of cell components, that the cells synthesize their proteins not in the nucleus, where the chromosomes are, but in the cytoplasm. How could genes act "at a distance," as it were? The answer was that a third kind of molecule was involved, a kind of nucleic acid related to DNA but differing in one of its three components, the sugar unit, and termed **ribonucleic acid** (RNA). RNA comes in several kinds, but the kind used in protein synthesis was a single-stranded copy of one of the strands of the DNA and was termed **messenger RNA** (mRNA). Thus a gene is "read" by a molecular process that copies one strand into a complementary mRNA molecule, the process being termed **transcription**, thus copying one molecular text into a slightly different kind. (Apart from the sugar difference, one of the DNA "letters," T, is replaced by a variant, U.) The mRNA then moves out of the nucleus into the cytoplasm (after some further alterations that need not detain us here) where, on the structures termed *ribosomes,* the nucleotide sequence into the mRNA is converted into an amino acid sequence of a growing polypeptide chain. That process is referred to as **translation**; in effect, it is the conversion of a chemical string of one set of chemical symbols into another set, where there is tight correspondence between symbols and final units. Crudely, the whole chain of events, from gene to protein, can be described as:

DNA (a gene) → messenger RNA → polypeptide chain

It is depicted schematically in Figure 3.2. (An extra complexity—noted in the figure, but we can ignore it for the moment—is that the original RNA copy is longer than the final mRNA; it has pieces cut out, and the remainder is then stitched together, the whole process being termed *splicing.*)

All of this had been worked out by 1964, including the detailed identification of which specific three-letter nucleic acid codons corresponded to which particular amino acids, a set of assignments that was termed the "genetic code."(Sometimes the whole genome is referred to this way, im-

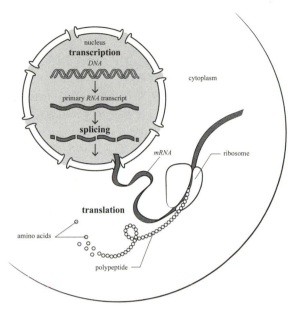

FIGURE 3.2 A diagrammatic representation of the crucial steps in gene expression: transcription of the DNA to give an RNA copy of one strand, termed a *primary transcript*; the cutting and rejoining of pieces, hence *splicing*, of that transcript to give a mature messenger RNA, which travels through the nuclear membrane via special pores to the cytoplasm; and translation of the mRNA to yield a protein chain within the cytoplasm. The nature of transcript splicing to produce mature mRNAs and another process that contributes to gene product diversity—post-translational modification of new protein chains—is discussed in Chapter 4 (Box 4.1).

plying that it "encodes" the traits, but that, as discussed earlier, is a gross oversimplification, and it is best to avoid the use of "genetic code" as a synonym for the genome.) Thus, by the mid-1960s, it looked as if all the mysteries of how the genetic material worked had finally been solved. Had they? Not really. In the human genome, as revealed by the Human Genome Project in 2003, only about 2 percent of the DNA is devoted to coding proteins. It had long been suspected that much of the DNA was not protein coding; the surprise was in just how small the fraction was. The vast majority of the DNA in the human genome must, therefore, either be functionless "junk" DNA or possess some as yet unknown function(s), or possibly both, being a mix of junk and novel functional units.

For decades, before the full extent of the noncoding portion was known, the junk DNA hypothesis held sway. It is now clear, however, that some fraction has some functionality. One portion consists of relatively short DNA sequences that are necessary to turn genes on and off (as will be described). In addition, many segments of DNA within the "junk" portion of the genome produce RNA molecules—but noncoding ones. Indeed, one whole set, *microRNAs*, are each only about twenty base pairs long, far too short to encode a polypeptide chain, whereas an extremely large set termed *long noncoding RNAs* (lncRNAs) have the requisite length to do so but do not leave the nucleus and are therefore not messenger RNAs. In addition, a whole range of small (but not micro-) noncoding RNAs also exists. It is now clear that many of these RNAs have important functions, indeed a diverse range of such. It therefore seems fair to call many of these noncoding, RNA-specifying segments "genes," but they are clearly different from the classical genes that molecular biologists had thought about and investigated in the 1950s and 1960s. Here, I will term these noncoding DNA segments collectively *nontraditional genes* in distinction to classic, protein-encoding ones. At present, nothing is known about what if any roles these nontraditional genes play in the formation of the face, but it seems highly probable that some are involved, perhaps as "regulators" of the actions of the (classical) genes we know to be involved in development of the face (see below).[4]

The general concept of regulation of genes, however, was born long before the discovery of these nontraditional genes. It arrived in the late 1950s, completing the landmark discoveries of what is seen in retrospect as the Golden Age of molecular biology (1953–1964). This concept is crucial not only to understanding how organisms function but also to how they evolve. The critical fact to note, before we enter the thicket of gene regulatory phenomena, is that not all genes are on all the time. They are turned on and expressed solely when copied into RNAs (both classic and nontraditional); only then can they exert an effect on the organism.

Gene Regulation and Genetic Regulatory Networks

The question of whether genes are turned on or off in development had long been an implicit problem in biology, but it lacked the irritating urgency of the question of just what precisely a gene was. By the early 1950s, it had been established by biochemical measurements that all cells of the body (the *somatic* cells) have essentially the same amount of DNA. The only major exceptions are the reproductive cells, sperm and eggs, which have half the amount of DNA of the somatic (nonreproductive) cells. (This twofold difference in DNA content reflects the fact that the sex cells carry only one set of chromosomes, unlike somatic cells, which have two). Yet since the somatic cells are unmistakably different from one another (e.g., Figure 2.4), the simplest inference was that cells differ not in their gene content but in the genes they express. Few scientists, however, tried to define exactly what this meant.

Three who did were French scientists working at the Institut Pasteur in Paris: Andre Lwoff (1902–1994), Francois Jacob (1920–2013), and Jacques Monod (1910–1975). The answer they produced was simplicity itself, although the work that yielded the answer was one of the most intellectually sophisticated pieces of research of modern biology. Using as their experimental organism a simple and easy-to-grow bacterium, *Escherichia coli,* Jacob and Monod focused on three genes needed to metabolize the sugar lactose, which was the sole energy source in the growth medium. (Lwoff worked on gene regulation in a virus that infected *E. coli,* but it turned out to involve the same principles.) They found that three enzymes were synthesized in bulk when lactose (or related "inducer") was added to the growth medium but that that synthesis was turned off when lactose was taken away or glucose, a more readily utilizable sugar, was added. The mechanism involved a simple molecular off switch that regulated the transcription of these genes, a site on the DNA molecule, the *operator,* near the three genes encoding those enzymes. When this site bound a certain protein, termed the *lac repressor,* the three genes were not transcribed and hence not expressed. But when lactose or other inducer was added, if no more efficient sugar or energy source was available to the cells, then the repressor binding the inducer molecule changed its shape and came off that site, allowing vigorous transcription of the three genes to take place.

Control by the *lac* repressor–DNA binding site switch was termed *negative control;* the regulatory protein's sole job was solely to turn off the expression of the genes. Within a few years of this work, however, it became clear that some bacterial gene regulation worked via *positive control,* when a regulatory protein had to bind to a special site in order to turn on transcription. In other words, selective activation of transcription of genes was as much a feature of gene regulation as selectively silencing such expression. (Indeed, in complex organisms, it is probably the more predominant mode.) Such activation involves specific proteins whose function it is to turn on the transcription of (other) specific genes. Proteins that do this are termed (reasonably enough) **transcription factors**.

Apart from establishing the reality of **gene regulation** as a phenomenon, providing a basic model as to how it occurs, and furnishing a methodology for investigating it—both huge accomplishments—this early work yielded a major general insight: genes exist in two broad classes. There are those that do the actual metabolic (or mechanical) work of the cell such as the three enzymes of lactose metabolism in *E. coli* or the hemoglobins that carry oxygen in vertebrates; these were termed **structural genes**. The second class of genes function solely to turn on or turn off (for example, the *lac* repressor) the expression of the structural genes and were designated **regulatory genes**. Both terms have gradually faded out of common use, however, although for different reasons. (One was the finding that some genes do double duty in both roles.) Nevertheless, since the broad functional distinction generally holds, some terminology is needed. In this book, I will refer to the genes that directly do the work of the cell as *worker genes* and those that control them as *manager genes.*[5]

In bacterial cells such as *E. coli*, there are far more worker genes than manager genes. This fits with the analogy of how an old-style factory operates; a relatively small number of managers directs a much larger group of workers. In the cells of plants and animals, however, the situation is the reverse: there is an overwhelming preponderance of manager genes. At first sight, this might seem a rather extraordinary situation. An analogy in human affairs would be a factory in which, for example, 2 percent of the employees are actual workers on the shop floor, while 98 percent are involved in management, managing mostly each other (!). How can one account for such a baroque state of affairs?

To understand it, one must grasp the degree of complexity of cells and what they do. It is not just that cells contain thousands of different molecules they have made themselves, each doing a specific job in one or more places within the cell, which is remarkable enough, but that cells reproduce themselves. Indeed, they may be regarded as self-replicating factories, entities that have to reproduce themselves with high accuracy and often quite rapidly. Fast-growing, rapidly dividing animal cells typically divide once every ten to twelve hours. Imagine trying to build a factory in that time! The fact that a cell is so much smaller than a factory does not in any way diminish the complexity or intrinsic difficulty of the task; if anything, that miniaturization makes it all the more remarkable.[6]

Furthermore, cells not only have to replicate themselves but also, when participating in a developmental process, modify themselves (or their daughter cells) in characteristic ways to become in effect *self-altered* factories making different products at different rates. The neural crest cells are an excellent example: they change from epithelial to migrating mesenchymal cells, and those then differentiate into any of a large number of different, specific types of cell, the particular end result depending on their final destinations. Nothing of this self-manufacturing and self-altering complexity exists in the human world of industry, even at its most impressive, whether rocket guidance systems or the modern marvels of electronic communication.

Of course, such a comparison between the achievements of evolution and human engineering is grossly unfair: cellular life on Earth has had a history of approximately 3.7 billion years. During that time, evolution has conducted countless trial-and-error experiments, discarding the failures and saving and multiplying the successes. In contrast, human civilizations are only 5,000 to 6,000 years old and industrial civilization a mere 250 years. What humans have achieved in the latter interval is extraordinary, with the rate of technological progress dwarfing that of biological evolution, but none of the products of human inventiveness matches the complexity of a single living cell and its activities. Even single-celled organisms such as bacteria and protozoa have the self-replicating, self-modifying capacities noted above. Plants and animals, however, which are composed of thousands to (literally) many trillions of cells, exhibit far greater regulatory (managerial) complexity than those organisms. The astonishing diversity of living things is the outward manifestation of the evolution of genetic regulatory systems in all their complexity.

As in the world of human manufacturing, cellular managerial systems for manufacture involve hierarchical systems and strict chains of command along with occasional nodal or *switch* points where variations and changes in the flow of instructions can be introduced. The simplest form of genetic management consists of linear strings of command, where one gene product does something to another (specific) gene, which does something to a third, and so on. Such systems can be symbolized as linear strings of genes, each giving a *command* to the next gene in the sequence. The simplest example is one in which each gene turns on the next in the sequence, as indicated by arrows:

gene $A \rightarrow$ gene $B \rightarrow$ gene $C \rightarrow$ gene $D \rightarrow$ gene E

(By convention, gene names are given in italics, as indicated here, and their protein products in roman type.[7])

Such linear sequences of gene activities are termed **genetic pathways**. Two points about them must be noted, one concerning terminology and the other biochemistry. The terminological matter is that early events (in this case, the actions of genes A and B) are termed *upstream* ones while the ones at the end, which often involve the molecules that do some direct work in the cell, "downstream" steps. The biochemical point: the commands in such a linear sequence are all molecular activities, but they can be quite different in nature from one another. Hence, a particular change could be an activation of the gene's transcription (*transcriptional control*), or it could be something that specially allows its mRNA to be translated (*translational control*) or, in other cases, alterations to the protein itself (hence *post-translational control*) that allow it to its job. For understanding the basic logic of pathways, however, those biochemical details can be ignored. We will simply take the "$\rightarrow$" symbol to mean that a gene activity is turned on. Not all steps, however, are activations. There are also inhibitory regulatory actions, one gene product blocking another, which is symbolized by a bar, "$\dashv$." We have seen one: the *lac* repressor on its own, that is one free of an inducer molecule, binds to its respective DNA binding site and blocks transcription of the lactose enzyme genes. Such an inhibitory step in a pathway will prevent any downstream activation steps unless other chemical signals (in our diagram, coming in from the top or bottom) inhibit those inhibitions, allowing the gene to be activated and the chain to continue.

In the previous chapter, we looked at two regulatory molecules, fibroblast growth factor 8 (FGF8) and sonic hedgehog (SHH), which trigger changes in cells of the developing brain or the facial primordia. Indeed, much of face development involves precisely these two molecules initiating changes in different tissues at different points. They do so through such linear chains of command of gene products. These particular genetic pathways are termed **signal transduction cascades** because each signal is altered by being *transduced* into a new one that affects the next molecule in a cascading series of such events whose end product or products changes the properties of the cell in some way. Usually, a signal transduction pathway is named for the molecule that triggers its action, and the molecule is typically named from its first identified biological role. Thus, for example, the first-discovered *fibroblast growth factor* (FGF) was named for its role in promoting growth in the cells known as **fibroblasts** on artificial media (in vitro) in the laboratory. The signal transduction pathways that FGFs initiate are collectively termed *FGF signal transduction pathways.* Like all such signalling molecules, the FGFs come in related forms, each encoded by an individual gene and associated with a specific set of effects, although the effects are often different in different tissues. (Such groups of closely related genes are termed **gene families**.) Significantly, there is no unique relationship between a particular signal transduction cascade and a specific developmental process: each signal transduction cascade is employed in many different developmental process (and most employ more than one). We have seen, for example, how FGF8 is involved in several key processes in different tissues in the development of the face. The phenomenon of signal transduction cascades is described more fully in Box 3.1.[8]

Regulation by signal transduction pathways looks very different from the kind of transcriptional regulation exemplified by the *lac* gene system in *E. coli.* Yet signal transduction and transcriptional controls are often sequentially related: signal transduction cascades typically lead to the activation or inhibition of transcription of specific genes. The set of particular genes controlled, however, is a function of the cell type, not the signal transduction cascade. Thus, the FGF8 pathway will activate different downstream genes (both manager and worker genes) in the developing telencephalon than it does in the facial prominences, although some members of those two gene sets are shared.

A further complexity is that regulation in development is usually not a simple matter of independently triggered linear pathways of gene action but of complex, reticulated webs of gene activities regulating each other. These molecular control systems involve both activating and inhibiting signals and often many different mechanisms, although transcriptional control is probably the predominant one. Such regulatory webs are termed **genetic regulatory networks** (GRNs).

Diagrams of GRNs look like complex wiring diagrams. They link the various genes with an activating (→) or inhibiting (⊣) event at each step,

---

BOX 3.1 The Phenomenon of Signal Transduction

The essence of signal transduction is that the immediate effect of a small molecule is transmitted along a chain of molecules in sequential interactions, often with amplified effect as those interactions take place. The end result is to alter the expression of genes and to produce a major effect on the cells in which the chain of events has taken place. Those biological effects are usually in one of four kinds of response: (1) the triggering of cell division (hence, cell proliferation), (2) the cessation of cell division, (3) the onset of programmed cell death (apoptosis) of those cells, and (4) differentiation of the cells. The initial chemical signal is usually a relatively small molecule such as a steroid hormone or a small protein, which are termed **ligands**. This denotes a molecule that binds to a larger molecule or structure, the latter termed a **receptor**, as the first step in the so-called signal transduction cascade. Crucially, as noted in the text, a particular ligand is not tied to a distinct biological effect. Thus, for example, many individual FGFs or Wnt's (*wingless-integrative* growth factors, whose strange compound name reflects their discovery in two different biological systems) can be involved in two, three, or all four of the general biological effects.

Like a great many discoveries, the discovery of signal transduction as a phenomenon in the early 1970s was a surprise. Up to that point, biologists were used to working with fairly direct effects, such as the substrate of an enzyme reacting with that enzyme to be changed into one or more products by that enzyme. The novel realization that molecules can initiate long chains of effects involving sequences of different molecules and resulting in major cellular effects grew in importance as this phenomenon was found to be ubiquitous. Within a few years, it had been established that these chains of effect could either be initiated at the cell membrane, with the binding of a ligand to a receptor molecule within the membrane, or internally in the cell when the ligand can travel through the cell membrane (as, for example, steroid hormones do.) Protein ligands such as FGFs or Wnts cannot go through the membrane, so their effects are always initiated at the membrane when they bind to a specific receptor within the membrane, initiating a new signal on the inside of the membrane or directly in the cytoplasm. Both kinds of signal transduction are ubiquitous in developmental systems, and much evolutionary change in development involves genetic change that alters the deployment of one or more of these systems.

---

often with a gene exhibiting multiple links to others. While wiring diagrams depict static situations, however, gene networks are inherently dynamically changing sequences of activities. Correspondingly, the more realistic diagrams of genetic networks have a built-in temporal dimension, tracing the sequence of events—activations and inhibitions—through time. Each part of the network that regulates a particular stage or event of a developmental process can be designated as a network **module**. The functioning of a module

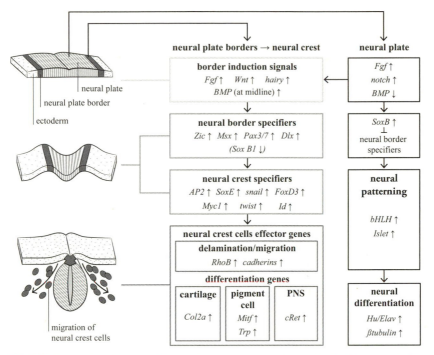

FIGURE 3.3 A simplified diagram of the genetic regulatory network governing the development of the neural crest and the cells derived from it. It is composed of four major modules that govern four sequential stages: (1) the induction of borders delimiting the future neural crest, (2) the acquisition of specific properties that confirm those borders, (3) the specification of neural crest cells within the delimited regions, and (4) the differentiation of both neural crest cells and neural cells (in the neural tube). Modules for the neighboring neural plate, which will give rise to the rest of the neural tube, are shown for comparison. Some of the key gene activities active in each module are shown although not their interactions. Upward arrows indicate that the gene is activated in the particular region; downward arrows that the gene is repressed. The inhibitory action of *Sox B* (the bar) indicates repression of the neural border specifier genes within the neural plate. (Modified from Yu 2010.)

produces a relatively small set of output molecules that change the cells' properties or those of their neighbors. Those changes, in turn, set the stage for the workings of the next module. Thus, the entire genetic network can be depicted as *a hierarchical sequence of modules,* with connecting links, running from top to bottom and unfolding in a set temporal sequence.

An example of a genetic network and its modules is that which governs the formation and subsequent activities of neural crest cells in the embryo, from their first formation at the dorsal edge of the neural tube to the diverse activity of their descendant cells. This schematic depiction is given in Figure 3.3 as a set of boxes (modules) that act in temporal sequence from top to bottom in the diagram. It is the overall structure that the diagram emphasizes; for simplicity, I have left out the detailed interactions, activations, and inhibitions between the genes within each module. The sequence runs from the module of interacting genes that delimit the neural crest region on the dorsal side of the neural tube to the next module, which imposes the identity of neural crest cells on them, to the successive modules governing their migration, and ultimately the various states of differentiation of the cells they give rise to in their new locations.

Now, with these basic concepts about GRNs in mind, we can return to the face and discuss its specific genetic underpinnings. A surprising aspect of the story is that the development of the face and the limbs have been found to be connected: they share a major, complex module at the heart of their respective GRNs. Indeed, it was the prior analysis of limb development that provided major clues to understanding the genetic foundations of the face. Thus, the connections between limbs and faces lie both in their biology and in the history of their investigation. The biological links provide clues to the ways that complex new structures arise in evolution.

From Limbs to Faces: Unexpected Links in Development and Evolution

Limbs and faces are, of course, very different entities: they look different, they perform different functions, and, not least, they differ greatly in complexity, with faces being far more complicated than limbs. (Nevertheless, the complexities of limb structure are such that robotics engineers have yet to duplicate limb movement except in the crudest way.) The relationship, a genetic one, cannot be deduced from their appearance or functions.

What they visibly have in common is a rather abstract property: both are stereotypically organized three-dimensional spatial *patterns* in which the elements have fixed and characteristic physical positions with respect to one another. Creating such patterns is an intrinsic and essential part of biological development, yet how exactly does development achieve them? The question is fundamental but was hardly perceived until the late 1960s. Following the promulgation of the Jacob-Monod model of gene regulation in the early 1960s, developmental biologists concentrated on a different problem, that of cellular differentiation and how genetic regulatory mechanisms might produce different kinds of cells and tissues during embryonic development. How, for instance, does the embryonic ectoderm give rise to both neurons and skin cells? These two cell types differ strikingly in appearance, biochemistry, and function. With roughly 400 different cell types in the human body, the general phenomenon of cellular differentiation was a major problem to be solved. Solving it, however, would not have simultaneously resolved the problem of **pattern formation**. The latter must involve the ways that cells in different parts of a developing structure "talk" to each other in order to set up their precise spatial relationships. It was only with the publication in 1969 of a seminal paper by South African–British developmental biologist Lewis Wolpert that developmental biologists began to focus on the problem of pattern formation—and many have been doing so ever since.[9]

Nevertheless, being patterned is a relatively superficial connection between limbs and faces, one shared with countless other structures. A more significant link would be if there were similarities between the genes employed, and the manner of their use, in the development of these two structures. That there might be such hereditary links had been suspected for some time. Darwin suggested it initially in 1868 in his book on heredity, *The Variation of Animals and Plants Under Domestication.* He had observed correlations of altered faces and limbs in certain breeds of domesticated animals and cautiously speculated that certain hereditary "variations" directly affected both limbs and faces. Furthermore, it is now known that such connections are not limited to farm animals but are also indicated in several rare human mutant syndromes, where the affected individuals show both facial defects and particular aberrancies of the hands, such as extra digits (polydactyly) or identically sized digits (syndactyly).

Such mutations did not provide the entry point for investigating either limb or face development, however. That research was made possible by the material that emerged in a wealth of molecular studies of development in the 1990s. The similarities between limb and face development first emerged fortuitously, as findings from the limb work suggested clues to those working on the face. I will summarize these similarities, then discuss some salient differences, and finally try to synthesize the material in terms of genetic networks in a way that makes sense of both the similarities and the differences.

Similarity one: both limb buds and developing facial primordia grow outward from the body of the developing embryo. Each possesses a jacket of ectoderm (the mandibular primordia also have an inner endodermal lining) that encloses a core mass of proliferating mesenchymal cells. Indeed, early developing facial primordia, especially the mandibular primordia, look like limb buds. In both cases, the outer ectoderm secretes one or more growth factors that keep the mesenchymal cells in a state of proliferation. In the case of limb buds, the stimulating ectoderm covers the distal (outer) tip of the growing limb bud; this is termed the **apical ectodermal ridge** (AER). In the upper facial primordia, the **frontonasal ectodermal zone** (FEZ) plays this role.[10]

Similarity two: both FGF8 and SHH are the specific molecules that promote growth in both limb buds and the facial primordia, although they do so in different ways.[11]

Similarity three: SHH, in particular, is also crucial in the patterning of both limb buds and facial primordia because it regulates differential amounts of growth within the developing structure. In the limbs, this critical growth-patterning role involves the differential growth of the digits (apart from digit one, the equivalent of the thumb in humans), with more SHH favoring longer digits, and less SHH shorter ones. Such correlations have been seen in the embryos of chickens, mice, bats, and porpoises. In the development of the face, SHH appears to be the main initial regulator of growth of the upper facial primordia, with greater overall SHH stimulation producing broader faces with wider midsections and more widely spaced eyes. Conversely, reduced SHH produces smaller, narrower faces. Both results have been shown experimentally in chick embryos by increasing or decreasing the overall amounts of SHH present in the developing facial prominences. In addition, injection of SHH into individual areas of the developing frontonasal process has shown that different regions can be stimulated to grow independently of others by high local amounts of SHH or their growth reduced by partial SHH depletion. The SHH pathway is equally crucial in face development in humans, as shown by various mutants in the SHH pathway, which display a loss of development of the midface, a condition known as **holoprosencephaly**.[12]

Similarity four: retinoic acid, a small molecule produced from vitamin A, is involved in and required for normal development of limb buds and the face. It acts upstream of SHH in both, but its precise roles may be indirect, inhibiting other processes that would otherwise interfere with the SHH pathway.[13]

Similarity five: besides sharing SHH and FGF signal transduction pathways, limb and face development both utilize at least two other signal transduction pathways. These are the *bone morphogenetic* (BMP) pathway and the *wingless-integrative* (Wnt) pathway. The role of at least one BMP in setting the size of the face in development was discussed in the first chapter in connection with the faces of Darwin's finches. In mammals, three BMPs are involved in face development, all stimulating growth and hence the size of the face, and the synthesis of all three is governed by SHH; in effect, all three are downstream of SHH. The Wnt pathway is also a growth regulator and acts specifically as such for the frontonasal process, where it affects not only its outgrowth but also its detailed patterning (for example, whisker development in the muzzles of mice). It is activated by the SHH pathway, so the Wnt pathway is like the BMP pathways and is downstream of SHH. In the limb bud, different pairs of the four signal transduction pathways act in concert for many steps. It is not known whether this take place in the face to the same extent, but there are indications of such dual control in some aspects of face development.[14]

Similarity six: both limbs and faces employ some of the same upstream transcriptional factors in their development. In this aspect, the resemblances are strongest between the limb buds and the mandibular facial primordia. In particular, expression of some members of the *Distal-less* transcription factor family is required for both the outgrowth of the limb buds and for development of the jaws, but additional similarities—in particular, a gene named *HAND2*, are also known.[15]

One might argue that these resemblances between limb and face development are simply fortuitous. For example, each of the signal transduction pathways shared between limb buds and facial primordia is also utilized in the development of other structures. Yet while coincidental similarity is possible, it seems improbable. Even with the signal transduction pathways, it seems hard to accept coincidence as the explanation for all the resemblances listed in the preceding. In particular, for the FGF pathway, it is striking that the same FGF—namely, FGF8—out of more than twenty different FGFs in mammals, is the one used in both limb and face development. Altogether, the similarities of organization, growth, and molecules between developing limb buds and facial primordia seem to be too many, too close, and (in several respects) too distinctive relative to other developing structures to be purely coincidental.

The visible differences between limbs and faces, of course, must reflect developmental divergences between them. At the level of genes, one difference that became apparent in the early investigations concerned the involvement of a group of genes termed the *homeobox* or **Hox genes**. Several are required in limb development, specifically in initiating limb bud development and later in patterning the autopod. The *Hox* genes, however, play no role in face development. Hence, this is one but almost certainly not the only difference in upstream control between the two structures.[16]

Furthermore, there must be many differences in gene expression in the late stages of limb and face development when the different tissues and final patterns come into being. In the developing limb, for example, the bones are formed endochondrally—that is, a cartilage precursor or model of the bone is first formed and then replaced slowly by bone as bone-forming cells infiltrate and lay down mineralized material within the shape of the cartilage model. In contrast, the bones in the face and neurocranium are formed directly within a flattish, developing cartilage matrix via the so-called membrane bone route, allowing molding of the developing skull by the developing brain, as discussed in Chapter 2. This difference in bone development must entail some differences in the sets of worker genes involved in their construction. Furthermore, there are likely to be differences in the genes expressed in blood, muscle, nerve, and connective tissues between the limbs and the face. In addition, it is highly likely that additional, different transcription factors regulate the genes involved in the late steps of limb and face formation.

How does one put together these disparate observations about the genetic foundations of limb and face development in a way that accommodates both the similarities and the differences? The simplest solution is to posit the existence of a shared core genetic module in their construction, which is situated between differing upstream and downstream modules. This hypothesis is diagrammed in Figure 3.4. Of course, even the shared core module need not be—and almost certainly is not—identical between limbs and faces. There could well be qualitative differences in the sets of molecular players in that shared module and, almost certainly, quantitative differences in expression of many genes within that module between the two developmental processes. Core modules can and do

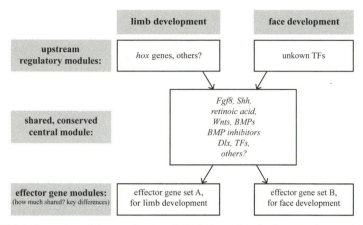

FIGURE 3.4 Comparison of the genetic networks for limb and face development, showing the central conserved module shared between them and indicating the differences in the two networks both upstream and downstream from that conserved module. See text for a fuller description.

evolve over time, coming to be different in either different sites or different organisms.[17]

The existence of a (largely) shared module between developmental processes that produce different outcomes provokes the question of how this came to be. The module is too complex to have evolved independently twice. Rather, the sharing of a complex module between the two structures almost certainly reflects the fact that once such a network element for the development of one feature has evolved, it can later be recruited for the evolution of one or more other features. In effect, the building of a new genetic network often involves the incorporation of a part of a preexisting one.[18]

In the case of limbs and faces, we can venture a reasonable guess as to where this genetic machinery evolved first. It was almost certainly during the origins of the vertebrate face, not the limbs. After all, the shared and original feature that ties together all the animals that we term *vertebrates* is their heads, not their backbones, hence their alternate designation as *craniates* as previously discussed. One can readily imagine the earliest members of this group, let us call them *proto-craniates,* possessing simple heads and faces but lacking appendages for motion such as even rudimentary fins. Such animals could have moved by undulatory eel-like movements (noncraniate motile chordates swim this way). The reverse, however, is inconceivable: headless, faceless animals, happily moving along by fins.

Hence, it seems most likely that the hypothesized core genetic module used in limbs and faces first evolved in the development of heads first and was later recruited for the evolution of appendages, initially in fins and later in the limbs of land-living, four-limbed animals: **tetrapods**.[19]

An even more specific suggestion for the relationship between face and limb development is possible: it was the first branchial arches in the proto-craniates, which would later evolve into the mandibular facial primordia, where the core genetic module was first assembled. This possibility is suggested by the morphological similarities of the mandibular arches and limb buds and the fact that the branchial arches must have been one of the earliest distinctive structures to evolve in the craniates. It is strengthened by the discovery of some specific transcription factors that are required in and shared between the mandibular facial primordia and the developing limb buds. This genetic architecture would later have been recruited for the other facial primordia and in both the facial primordia and the developing limb buds, new upstream controlling modules and downstream *effector* modules (for the elaboration of specific structures) would have been added and modified during the further evolution of faces and limbs.[20]

This discussion about faces and limbs began with how clues from the study of limb development helped elucidate face development, but, as we have seen, the evolutionary sequence was most probably in the reverse direction. The vertebrate face evolved before fins or limbs, and the latter were able to evolve because a key part of the existing genetic network—what would become the central core module shared between faces and limbs—had already been assembled and deployed in the development of the face. This is still only a hypothesis, but it synthesizes the data coherently and does not contradict any known facts. More generally, it would be an instance of a general feature of evolution: evolutionary processes often yield something new by borrowing and adapting some previously existing bit of genetic machinery—whether individual genes or GRN modules. Such evolutionary constructions do not follow the modus operandi of an engineer—using first principles to design the most efficient device—but that of the tinkerer—making do with what is to hand. Building by tinkering, it turns out, is the essence of evolutionary invention, though optimization via natural selection, in line with engineering-type thinking, is often equally important to maximize the fit of the new part with the preexisting structure.[21]

"Vertical" Faces and Brain Development

When we look at human faces in profile and compare them to the profiles of other mammalian faces, including our closest animal relatives, the great apes, or to reconstructed faces of our hominin ancestors, we are immediately struck by a distinctive difference between the human face and the others: it is flatter and more vertical. This trait also makes for a smaller face relative to mammals with more typically relatively elongated heads. This change in conformation results from two changes in the typical mammalian face: the reduction of projecting jaws, eliminating any trace of the mammalian muzzle, and the presence of a true forehead above the eyes, the latter a partial consequence of the larger human brain. The reduction in the muzzle reflects evolutionary changes in jaw outgrowth and quite possibly changes set fairly early in the development of the maxillary and mandibular prominences; we will defer discussion of this aspect, however, until Chapter 6. Here we will look at the production of a forehead.

The human forehead exists because the brain pushes the frontal bones of the neurocranium forward to create a wall above the face. That forward and upward positioning of the brain reflects in part the increased size of the human brain, both relatively and absolutely, to that of our nearest animal cousins, the chimpanzees. Since the developing neurocranium is molded on the growing brain, one result of a larger brain is the development of frontal bones that make a more vertical wall above the eyes. As discussed previously, overall brain size alone does not account solely for the special roundedness of the head and the near-vertical forehead, but it is a major factor.

The large size of the human brain reflects specifically the disproportionate growth of the telencephalon, the most anterior part, especially that part that becomes the cerebral cortex (or cerebrum). As we saw previously, the telencephalon becomes folded over during its early development at its anterior end, yielding ventral and dorsal sides, with the ventral side being the initial source of SHH to stimulate growth of the facial primordia. With respect to brain development itself, however, the ventral telencephalon is the source of structures termed the *basal ganglia,* which lie within the brain, beneath the cerebral cortex. In contrast, the dorsal side of the telencephalon of the early embryo is the precursor of the cerebrum and ultimately of the cerebral cortex, which comes to envelop much of the brain during fetal development.

The cerebral cortex begins in the early embryo as a single epithelial cell layer on the dorsal side of the telencephalon. This epithelium subsequently develops, however, into a structure consisting of six layers, each possessing characteristic cells and thickness. The relative enlargement of the brain seen in larger-brained mammals comes from a great expansion of the cerebral cortex (relative to ancestors), which comes not from an increase in number or depth of its six layers but from a great extension of its length and width. That extension of area of the cerebral cortical sheet without an accompanying increase in depth or thickness and within a confined space yields a highly folded sheet that features numerous folds (*gyres*) separated by grooves (*sulci*). In the form seen in humans, the cerebral cortex is designated as either the **neocortex** (the traditional term) or the **isocortex** (the more recent term). One can think of the cerebral cortex as an intricately folded and compact sheet of cells; were the cerebral cortex of an adult human completely unfolded and spread out, it would be about the size of a double-bed sheet, albeit of extreme thinness.

The six-layered structure of the cerebral cortex is a feature that is both unique to and general among mammals. It is a mammalian specialty that arose sometime during the long period of evolution in which the evolutionary sequence starting with ancestral *synapsids* (premammalian forms) gave rise to the true mammals (to be described later in this book). In embryonic development, the layers of the cerebral cortex develop from the inside to the outside, with the initial neuroepithelial layer facing the open space, the *ventricle* within the telencephalon. This initial layer of cells is, accordingly, termed the *ventricular layer,* and its cells are termed **primary neural progenitor cells**. These divide to yield second-generation or **secondary neural progenitor cells**, which form the next layer above, the *subventricular layer.* (This name may seem somewhat confusing since this layer of cells lies just *above* the ventricular layer; but as such, it lies deeper within the six-layer structure). The sole function of the neural progenitor cells is, as their name indicates, to produce those cells that will ultimately carry out the work of the brain, namely, the differentiated neurons. It is the secondary neural progenitor cells that repeatedly divide to generate the neuronal cells that come to form the remaining layers above. These neuronal daughter cells form the outer layers by migrating dorsally, away from

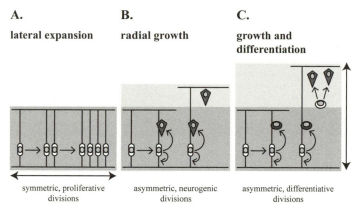

A.

lateral expansion

B.

radial growth

C.

growth and
differentiation

symmetric, proliferative
divisions

asymmetric, neurogenic
divisions

asymmetric, differentiative
divisions

FIGURE 3.5 Simplified schematic of the basic events of neurogenesis in the cerebral cortex. On the left, the first symmetrical cell divisions, which increase the number of primary neural progenitor cells in the ventricular layer; in the center, the secondary (or intermediate) neural progenitor cells, which help form the subventricular layer. Finally, the production of differentiated neurons from the secondary neural progenitor cells is indicated on the right. These latter divisions create the neurons that migrate upward and eventually form the six-layered cerebral cortex.

the ventricular layer, passing between cells in the already formed layers, with the topmost layer forming last. Each distinctive layer is composed of one or more characteristic kinds of cell and is several cell layers thick. The process of generating neural progenitor cells and then the differentiated neural cells above them is diagrammed in Figure 3.5.

A vertical, cylindrical stack of cells within the six layers, from bottom to top, is generated from a small group of neural progenitor cells and is termed a **radial unit**. Its development is diagrammed in Figure 3.6. The term *radial* reflects the geometry of the brain. Specifically, if the curved surface of the telencephalon is visualized in simplified and abstracted form as a sphere, then the cells layered on top of each other in the six-layered structure can be regarded as positioned along a virtual radius that runs from the center to the outside of of the telencephalic sphere. Whatever the size of a particular mammal's brain, the radial unit stays essentially the same length along its axis; in other words, it has the same depth throughout the whole cerebral cortex. Within all mammals, the actual physical depth of the radial unit varies within only a twofold range—a tiny difference relative to those in absolute brain size, for example, between a mouse and a blue whale.

In other words, neither additional layers nor substantial thickening of individual layers within the cerebral cortex, once they have formed, takes

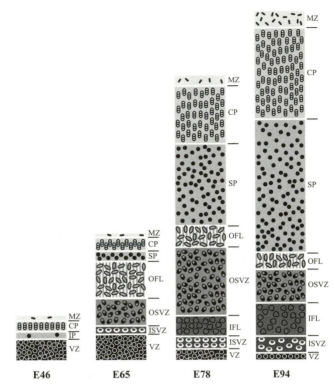

FIGURE 3.6 Sequence of the formation of the six-layered mammalian cerebral cortex during embryogenesis. The diagram shows the increase in layer number and complexity during embryonic development of a monkey, the rhesus macaque. Note that, with time, some layers increase in size and depth while others diminish slightly. The numbers beneath each column refer to the days of embryogenesis from conception. The sequence of development is broadly similar in the human embryo though the timing is different. The six main layers in the embryo from bottom to top are: (1) the ventricular zone, VZ; (2) the subventricular zone, SVZ (which consists of a lower inner subventricular zone, the ISVZ, and an outer one, the OSVZ, separated by the inner fiber layer, the IFL); (3) the outer fiber layer, OFL; (4) the subplate, SP; (5) the cortical plate, the CP; and (6) the marginal zone, MZ. (Modified from Dehay and Kennedy, 2007.)

place during brain development. These characteristics have evidently remained stable over 125 million to 150 million years, possibly 190 million years or more, of mammalian evolution. (We will come to the question of how long true mammals have existed in Chapter 5.) What changed in those mammalian lineages that developed relatively larger brains—namely, the elephants, cetaceans, and primates—was the *extent* of the surface sheet that the cerebral cortex constitutes. That dimension is a direct function of the number of radial units generated during embryonic development: the greater the number of radial units that are formed, the greater the extent of the cerebral cortical sheet, hence of the mature cerebral cortex itself.[22]

The fundamental source of that change, in turn, was an increase in the absolute and relative number of neural progenitor cells. Thus, as radial units are added through increasing the number of those cells, the result is an increase in the length and width of the sheet, not its depth. The secondary result, since this increase takes place within a relatively confined space, is an increasingly folded-over and folded-up structure, as seen in the adult human brain, with its beautiful, otherworldly, corrugated sculpted dome of gyres and sulci.

The developmental origins of larger brains—in particular, the outsize brain of human beings—thus lies in the mechanisms by which more neural progenitor cells are produced. To understand this, however, one more concept is needed. Cell biologists speak of a dividing cell as a *mother cell* that gives rise to two *daughter cells*. Those daughter cells can either have the same "fate," revealed retrospectively by the kinds of cells they produce, or different ones. When a cell gives rise to two daughter cells of similar or identical fate, the cell division is said to be *symmetrical*, but when the daughters of a cell go on to experience different fates the division is denoted as *asymmetrical*. In fact, there is almost always a difference in the size of the daughter cells produced in **asymmetric cell division**. Hence, such divisions are indeed unequal or asymmetric. The terminology, however, refers primarily to the disparity in developmental fates of the daughter cells. This does not entail any difference in the chromosome numbers or DNA content of the daughter cells, however; their mitotic divisions, like all such, maintain genetic parity between the daughter cells. The difference in fates reflects genetic regulatory differences between the daughter cells.

Since the job of the secondary neural progenitor cells is to produce the neurons, which make up the upper layers of the radial units, many of their divisions are asymmetric: each divides to give one neural progenitor daughter (which will go on to produce more progenitor cells) and one neuronal daughter (a differentiated neural cell that will not produce new cells). Neuronal progenitor cells, however, can also undergo symmetric divisions, yielding two neural progenitor daughters. When this occurs, both daughters stay in the same layer as their mother cell, whether this be the ventricular layer (the one facing the open space, the ventricle) or the subventricular layer, and thus expand this layer. The more neural progenitor cells produced, the more radial units that can be generated.

The "choice" between symmetric and asymmetric division events of the neuronal precursor cells may hinge on the location of the plane of division of that cell. Imagine a long rectangular cell that undergoes cell division in the plane of its longer axis directly in the middle of the cell: the result will be two long, thin daughter cells lying side by side and nearly equal in size. If, however, the division plane is at right angles to or even at a somewhat skewed angle to the long axis of the cell, then the result will usually be an asymmetric division, producing two unequal-sized cells. If the division plane is slightly skewed from the vertical, one daughter cell will be positioned slightly above the other. For neural progenitor cells, this kind of altered division will lead to the production of cells that contribute to one of the radial layers.

One can now begin to see how the production of larger brains may be tied to something as simple as the plane of cell divisions in the developing cerebral cortex. If the generation of larger brains is ultimately the result of an increasingly larger sheet of cells making up the cerebral cortex with its ever more folded structure, the source of those changes would be mutations that increase the number of symmetrical divisions of the neural progenitor cells. The genetic evidence concerning the genes required to give normal brain size supports this picture. This evidence comes from study of mutations in eight known human genes that sharply reduce brain size, the condition known as *microcephaly* (literally "small head"). All of these mutations are of the typical "loss-of-function" variety; they make less active gene product, ranging from slightly less to essentially none.[23]

In the brain cells of the microcephalic mutants, fewer cell divisions during the genesis of the cerebral cortex take place, with the result that these brains contain far fewer neurons. Significantly, the evidence suggests that the smaller brain sizes of microcephalics results from fewer symmetrical divisions of the neural progenitor cells during embryogenesis, with the inevitable consequence that fewer radial units are generated than in normally developing brains. For the best characterized of these genes, designated the *ASPM* gene, it has been shown that the commonest or **wild-type** form of the protein associates with both poles of the mitotic spindle—toward which the two sets of chromatids move—and the spindle material at the equator, the *midbody*, that remains after the chromatids have begun to separate. That wild-type protein helps stabilize the position or functioning of the spindle within the cell, ensuring correct symmetrical divisions. In the absence of the fully functional version of the protein, the neural progenitors tend to slip into asymmetrical divisions. The consequence is fewer neural progenitor cells generated, leading to fewer radial units and, ultimately, smaller brains. When mutant, the normal or wild-type versions of several of the genes give the microcephalic condition function in a similar way to the wild-type *ASPM* gene product. Thus, something as basic as failure to maintain correct placement of the mitotic spindle within a dividing cell can lead to a major change—in this case, the smaller brains of microcephalic individuals.

These findings also provide a clue to the nature of brain expansion in human evolution: the existence of mechanisms to ensure a longer period of generation of neural progenitor cells in either the ventricular or subventricular layer (or both) of the neocortex. That period of extended generation of neural progenitor cells, however, must involve gene activities other than the so-called microcephalic genes—specifically, transcription factor genes that regulate these genes and keep them expressed longer to produce more neural progenitor cells. In effect, the evolution of brain size, and ultimately the rounder head and existence of the forehead, probably derive from altered regulation of the period in which the wild-type forms of the *microcephalic genes* are expressed, ensuring more rounds of symmetric division of the neural progenitor cells to generate more radial units. We will return to this evolutionary dimension of the subject in Chapter 6 in discussing the special features of human evolution from earlier hominins. Specifically, we will look at some of the genes that are probably the key ones involved in lengthening the period of symmetrical division of the neural progenitor cells in human brain development.

Conclusions: Gene Activities as Prime Agents in the Unfolding Development of the Face

In this chapter, we have concentrated on the development of the face with respect to some of the key gene activities that underlie its development. Thus, we examined not only the special role of SHH and Fgf8 and BMPs and Wnts in the development of the facial primordia but also the multiple roles of other genes, particularly those that act downstream of those molecules, in governing the growth and development of the facial primordia during the first four to ten weeks in human development, when the face literally takes shape. These gene activities change dynamically, and the outcome of those dynamics is the rapid sequences of growth and shape changes that take place in the facial primordia. Each change provides a platform for the next set, the developmental process moving inexorably forward. This is as true for brain development, whose role in the development of the forehead we have examined, as it is for the face.

In general, gene activities not only drive a developmental process but also provide stereotypic character and stability. One mechanism is that of *negative feedback* control in which the end product of a certain process has reached a certain level and then inhibits its own further production. Subtle mutations that influence the rates of these processes can, in principle, tightly regulate how much is produced at specific times and places, which will contribute to the detailed shaping of the size and morphology of the final structure. Whether and how such negative feedback processes operate in the growth and development of the face is unknown, but they almost certainly exist. A well-characterized example in limb development, which may have relevance to face development, involves production of SHH in the limb bud: SHH induces one or more BMPs, which build up in concentration within the cells until they inhibit further SHH production. The latter then declines in concentration locally until less BMP is produced; with the decline of BMP concentrations, SHH, induced by FGFs, can start to be produced again.[24]

The particular genetic actors we have discussed in this chapter are but a tiny fraction of the total ensemble of gene activities required for the development of the face. These include all the genes required for neural crest cell formation and migration in the appropriate numbers to the correct destinations in head and trunk; the genes that regulate the early development of the brain, first into its three primary divisions and then the genes required specifically for the development of the forebrain; and, ultimately, many thousands of worker genes in the facial prominences.

In considering the genetic foundations of face—and brain—development, three general points should be remembered. First, no gene acts independently but only as part of a genetic network whose overall structure and operation determines exactly what gene activity takes place and when. Second, every change in cells prompted by a change in gene activity, whether of growth or basic cellular-biochemical properties, is dependent on that cell's prior history, constitution, and location, and not just on the new signals those cells encounter. Thus, no new gene activity acts on a *tabula rasa*. Rather, each creates its effects on a particular cellular landscape whose contours have been set by its prior history. Third, developmental changes, though catalyzed and in a sense "directed" by the biochemical or molecular changes wrought by specific gene activities, involves physical processes that produce changes in extent and shape of the cell groups and masses that take part. These interactions and changes are influenced by the respective masses and aggregate shapes of the interacting cells, the chemical-physical properties of their surfaces, and physical constraints in their local neighborhood. For example, the bending of the anteriormost part of the telencephalon that occurs in early development, creating its ventral and dorsal surfaces (with their distinct developmental fates), takes place because of physical constraints imposed on the growing brain by the surrounding amniotic sac. And the accuracy—or failure thereof—of fusion of the facial prominences is crucial to ensuring that neither cleft lip nor cleft palate develops. All such physical interactions are part of the indirect but highly *canalized* set of consequences that flow from particular gene actions at particular times in particular cells.[25]

Most discussions of development emphasize these three aspects, but there is a fourth general property that generally receives less attention but is highly significant for permitting evolutionary change in development. This is the degree of flexibility or "play" in the developmental process of interest. Without that built-in tolerance to a degree of variability, in particular in the numbers of cells participating at any particular point, the whole mechanism would be too rigid to work. Imagine that development of each facial primordium could only take place when a specific number of cells were present; clearly, that kind of precision would not be achievable when hundreds to thousands of cells of a particular type are involved. Hence, while a developmental sequence involves a distinct sequence of both trigger ("on") signals and blocking ("off") signals, these operate over a range of tissue primordium sizes and shapes. (A few developmental situations in the early embryos of certain invertebrates exist in which specific cells in fixed numbers do precise things at particular times, but these all involve situations in which many fewer cells are participating.)

This existence of degrees of play and flexibility in the developmental construction of living things is not only essential for development itself but also crucial in allowing evolution to take place. If evolution involves genetic changes that alter one component or process at a time, then the rest of the system must be able to tolerate and accommodate such changes, at least to a degree. It is the inherent flexibility of development involving the interactions of cells and tissues that permits evolutionary change to take place, beginning with mutations that alter the time or extent of the interactions. This property of developmental flexibility, permitting change into forms not previously seen even if still relatively modest alterations, has been termed *developmental accommodation*.[26]

Recognizing that there is a degree of play in developmental processes is helpful for understanding why a key argument against evolution is false. This argument purportedly shows that the evolution of complex systems is inherently impossible and provides the central claim against evolution by the so-called intelligent design movement, an argument that goes by the phrase "irreducible complexity." Though often presented as a profound insight, it is actually very simple: because the complexity of most structures in organisms is so large and involves so many components, then if one element is changed in an important way, the functioning of the whole system must (presumably) be undermined. Accordingly, since evolution entails changes in complex systems but such changes (by this hypothesis) almost always lead to dysfunction, then ipso facto, evolution is impossible.

This idea is not new. It is found in the writings of the man who founded the sciences of comparative functional anatomy and paleontology, Baron Georges Cuvier (1769–1832). Cuvier was a man of high intelligence, great accomplishments, and consequently much influence. He argued that the exquisite functioning of living organisms was intrinsically dependent on their (near) perfect "correlation of parts": change one structure substantially and everything else would certainly begin to malfunction. He recognized that each species shows a degree of inherent variability in its traits, but he regarded that variability as slight and unimportant, existing within especially tight limits. Given those constraints, species "transformation" (that is, evolution into new species) was simply impossible. Cuvier was, in effect, an advocate of intelligent design *avant la letter*. He did not openly bring his particular religious beliefs into his scientific arguments, and they were far from traditional biblical views, but there can be little doubt that they informed his judgments in this area.[27]

In contemporary form, "irreducible complexity" takes as its reference point manufactured products of high complexity. If, for example, any one of several hundred components in a modern automobile is removed, the car's capacity to move will be either abolished or severely reduced. This analogy, however, is seriously flawed: organisms do not develop from prefabricated parts but come into existence through long sequences of self-assembly, involving both the component parts and substructures composed of those components, with each step having a degree of tolerance and accommodation for variable inputs. In effect, development involves a series of self-checking events and built-in flexibility, which allows a degree of change. This is not just a theoretical proposition; we know it to be true. We can, for example, see such changes in the dramatic changes that were wrought by intensive selective breeding in various domesticated animals, changes that are fully compatible with viability.[28]

The point is also illustrated by the development of the human face: the enormous variety of faces that we see in our species is testimony to a large latent genetic capacity for generating visibly different but fully functional human faces. Perhaps today's intelligent design advocates would dismiss such variability as trivial, but in illustrating the plasticity of developmental processes it is anything but. Furthermore, it is functionally important: it plays a role in human social interactions, providing an immediate and accurate means of individual recognition. Later in this book, we will return to a possible source of its evolutionary origins.

In the next chapter, however, we will examine not the possible significance of human facial diversity but its genetic basis. Our focus will be on the question of the numbers of different genes that would need to be involved in generating this enormous diversity of faces.

— four —

THE GENETIC BASIS OF FACIAL DIVERSITY

Introduction: The Enormous Diversity of Human Faces

As noted at the beginning, the human face is actually one of the most peculiar of all mammalian faces, even though it is the inevitable standard by which we judge the seeming oddness or even comicality of other animals' faces. Perhaps the most surprising aspect of the human face, however, is not its unusual features but its diversity: no other animal species seems as visibly individuated in its members' faces.[1]

Of course, there are often close similarities between human faces. Such resemblances are particularly frequent within families, the most striking being the faces of identical twins and triplets. The latter represent the extreme within a range of noticeable similarities among family members. Furthermore, beyond similarities within families, there are often fortuitous close resemblances between unrelated individuals, though in such cases measurement or even close inspection reveals differences.[2]

Beyond familial or fortuitous resemblances, certain general facial features tend to be shared within ethnic groups. Even within a particular "race," however, such traits always show distinct differences, making for individuated faces within each ethnic group. Significantly, members of each supposed racial group see those differences clearly, while generally seeing members of other groups as less facially differentiated. It thus seems that perceiving facial differences is, at least in part, a matter of familiarity and experience, a conclusion confirmed by scientific testing. In general, the overwhelming impression in surveying human faces is that of individual difference, hence diversity.

Is the variety of human faces, however, a genuinely distinctive feature of our species or simply an ascertainment artifact reflecting the attention we give ourselves? If wild mice could speak, for example, would they tell us that they, too, have a huge diversity of faces, allowing instant visual identification of individuals? It seems doubtful. Most wild mammalian species have far less visible—indeed, measurable—facial diversity than our species, and for those that rely more heavily on other senses than vision such as olfaction (as in mice) or hearing (as in bats) and that typically have fewer and simpler social interactions, recognition of facial difference would be a much less useful trait.

Nevertheless, a handful of other species exhibit obvious facial diversity. In particular, there is the dog. Not only do dog breeds differ dramatically in face morphology—from bulldogs to greyhounds—but there is evident and substantial diversity of faces within breeds. Perhaps, however, it is not wholly coincidental that our "best friend" reflects our degree of facial diversity. Dogs are truly animals of our creation; by selective breeding over many thousands of years—initially for their ability to associate with humans in a friendly fashion and later as guard dogs against strangers, then intensively over the past two centuries for many different traits (both morphological and behavioral)—we have shaped them extensively. By selecting for various traits in dogs, however, we may also have inadvertently shaped their genomes in ways that contribute to their facial diversity. The idea that deliberate selection for some traits can accidentally bring others in their train originated with Charles Darwin, who termed the process of accruing such by-products **unconscious selection**. Whatever the genetic foundations of canine facial diversity, however, it seems of more interest to us than to the dogs themselves, whose introductions to each other primarily involve mutual olfactory inspection—and not of their faces.

More significant is the visibly high degree of facial diversity in the great apes, our closest animal relatives—namely, chimpanzees, bonobos, gorillas, and orangutans. Facial diversity may thus be a general characteristic of the hominoid primates: the great apes plus the hominins (the latter now consisting only of our species). Since the great apes, like ourselves, have excellent vision, especially for closeup inspection, it seems probable that the facial diversity of great ape species plays a part in their identification of individuals in social interactions, as it most certainly does in ours.[3]

The extent of facial diversity in humans and other hominoid primates raises the basic question of why it exists. Does it have sufficient value to have been actively selected in human evolution, perhaps to aid rapid visual identification? Or is it simply an accidental trait but one that happens to have a convenient side effect—in this case, rapid individual identification? This is an important and interesting evolutionary question, but we will defer discussion of it until Chapter 9. It is, however, intimately tied to an equally puzzling neurological question: how are these differ-ences perceived and how well are they remembered? After all, facial di-versity without the corresponding capacity to recognize the differences would hardly be a useful and hence potentially selectable trait. This issue, too, will be taken up later; we will return to it in Chapter 7, where we will explore the ways in which the brain and face have influenced each other's evolution.

In this chapter, however, we will focus on a more prosaic but still impor-tant question: what are the genetic foundations of human facial diversity? As shown by the faces of identical twins, whose features reflect essentially identical genomes, the generation of different faces is primarily a matter of genetics, not one of diet or other environmental influences. Although it is not yet known how many and which genetic differences are involved in generating human facial diversity—though a start has been made—the question is worth pondering. Inevitably, it is connected to the matter of whether most or all of the genes involved in human facial diversity come primarily from the same set of critical regulatory genes for building the vertebrate face that were the subject of the previous chapter or whether they are part of a much larger set of genes downstream from those.

To explore these issues sensibly, however, we need to understand the basics of heredity and the terminology used to explain it; that overview comes next.

Tracking Hereditary Differences: Chromosomes, Genes, and Alleles

As we have seen, the genome of a complex organism is split into pieces termed *chromosomes* that reside in every cell nucleus in the body. In human beings, the chromosome complement consists of forty-six chromosomes that make up two equivalent sets of twenty-three chromosomes, one inher-ited from the mother and the other from the father. This double chromo-some set condition of the somatic cells is termed the **diploid** condition.

Normally, the chromosomes exist as immensely long, thin threads within the nucleus; as a cell prepares to divide in the process known as **mitosis**, the chromosomes begin to condense and each one reaches a highly condensed state just before actual division (Figure 2.4). Each chro-mosome at this stage consists of two identical copies—**chromatids**—that are tied together at a given point, either toward the end or nearer the middle, giving respectively a V-like or an X-like appearance. Just before the cell divides, all forty-six chromosomes line up at the center of the cell; at this point they can be seen under the light microscope, photographed, and then visually matched up in the photographs and counted. For the human genome, all chromosome pairs can be arranged according to their size. The one exception to the rule of identical size among the twenty-three chromosome pairs concerns the pair of **sex chromosomes,** which in the male consist of one so-called X chromosome (a large one) and a much smaller Y chromosome. (In the cells of females, there are two X chromosomes.)

Only in the reproductive cells—namely, the sperm cells of males and the eggs of females—are the numbers of chromosomes different; in the production of these sex cells, a special division named **meiosis** takes place and reduces the chromosome number by twofold to twenty-three; this single chromosome set is termed the **haploid** condition. When a sperm unites with an egg during fertilization, their nuclei fuse and regenerate the diploid condition. Since all the somatic cells of the developing individual derive by mitosis from that fertilized egg, each body cell of the new individual with also have forty-six chromosomes. Hence, with two sets of chromosomes per body cell, each cell has two copies of each gene (apart from those on the Y chromosome), and each sex cell, a sperm or an egg, has only one. These basic facts are diagrammed in Figure 4.1.

To think about genetic inheritance, however, a few more concepts and terms are needed. A particularly important distinction is that between **gene** and **allele**. As discussed in Chapter 3, a classical *gene* is a stretch of DNA that specifies a protein product, a polypeptide chain, that has a specific function within the cell, either metabolic (as for enzymes) or contributing to a structural property for the cell (as in the proteins of the cytoskeleton). All of the slightly different hereditary variants of that gene, however, are termed *alleles*. If one particular allele is overwhelmingly common for a particular gene—say, on the order of 99 percent of all copies of that gene in a population—that version is termed the **wild-type** allele, and the gene is said to be **monomorphic** (literally, "of one form"). Yet not all genes show one overwhelmingly predominant wild-type allele. As with the genes responsible for facial diversity, many have a range of different alleles within a population, a condition termed a **polymorphism,** with the gene itself said to be *polymorphic*.[4]

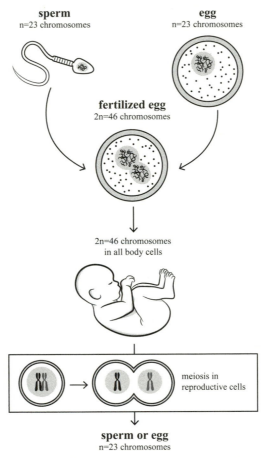

FIGURE 4.1 The basic genetics of human reproduction: sperm and egg (not shown to scale), each carrying twenty-three chromosomes fuse to give a fertilized egg, which develops into an individual whose body (somatic) cells all have forty-six chromosomes. After puberty, when sex cells are produced in the ovaries (females) and testicles (males), meiosis takes place in the sex cell precursors, halving the chromosome numbers to give gametes with, again, twenty-three chromosomes.

The preceding may sound rather abstract, but an example from classical genetics should help clarify the critical distinction between *gene* and *allele*. The first gene to be studied by modern methods was one in the garden pea, and the work was that of Gregor Mendel (1822–1882), the Moravian friar whose work inaugurated the science of genetics. This particular gene affects the shape of the pea. It was not, however, initially revealed by its wild-type form but by a variant allele that causes the peas to wrinkle up. The more typical form found in most strains of the garden pea

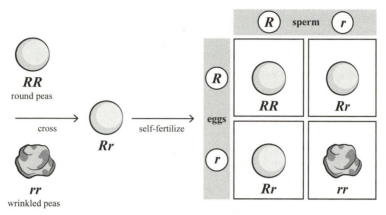

FIGURE 4.2 Mendelian genetics for a single gene difference: the figure shows the results of crossing pure-breeding round-seed peas (*RR*) and wrinkled-seed peas (*rr*) to give hybrid plants of *Rr* genetic constitution. When self-fertilized (or crossed with each other), these produce one-quarter wrinkled-seed plants and three-quarters round-seed plants. Of the latter, however, two-thirds are hybrid (*Rr*) and one-third pure-round (*RR*) plants.

produces round peas, and Mendel later symbolized this most common predominant (wild-type) form of the gene by a capital *R;* the other allele causing the wrinkled condition he denoted by a lowercase *r.* The pure-breeding round pea-bearing strains have two *R* genes in each of their cells; this diploid condition is denoted as *RR.* The wrinkled forms have two *r* alleles, and are labeled *rr.* In contrast, each of the haploid reproductive plant cells—the pollen granules and the ovules—has only one gene copy, either a single *R* or an *r,* as a consequence of meiosis. The pure-breeding round and wrinkled forms, with their identical gene compositions for this trait, *RR* and *rr,* are said to be **homozygous.**

Mendel deduced these facts from his study; nothing of this was understood prior to his work. The first step in the procedure, from which these insights flowed, involved crossing pure-breeding round and wrinkled-pea strains. To his initial surprise, all of the so-called first hybrid generation or F$_1$ plants produced round seeds that were indistinguishable from those of the homozygous *RR* parent strain. Yet the *r* allele was not missing from the F$_1$ plants because when they were allowed to self-fertilize, they produced new *rr* wrinkled-seed plants, about a quarter of the total. The experiment is diagrammed in Figure 4.2. Mendel deduced that the F$_1$ plants were truly hybrid, or **heterozygous** for the gene, and that their genotype was *Rr,* with one *R* copy enough to mask the expression of the *r* allele.

Accordingly, he termed R the **dominant** form and r the **recessive** form. Thus, a round-seeded pea strain might have either of two genotypes for that gene: RR or Rr. Evidently, its appearance or **phenotype** does not give a direct picture of its **genotype**, its genetic constitution, which can only be deduced by further breeding. In contrast, the wrinkled-pea phenotype only appears in the homozygous rr plants.[5]

All of the preceding shows the importance of the distinction between *gene* and *allele*. Yet it is often forgotten, leading to confusion. Some of these common conceptual errors are described in Box 4.1. We can now, however, turn to the genetics of the face.

On the Power of Genetic Combinatorics in Generating Different Faces

By this point, it should be apparent that the question of how many genes contribute to the differences in human faces must be a function of the number of face-contributing genes *that have allelic differences*. If everyone were monomorphic for all the potential facial-diversity genes, whatever set of genes is involved, we would all look alike (apart from differences induced by hormonal effects associated with gender difference). While all humans have essentially all the genes that contribute to the face, and in two copies (apart from differences in genes on the sex chromosomes in males), only those genes that differ in their alleles to produce a visible effect on the face will contribute to facial difference. To illustrate this idea, pick a number randomly: if, say, the number of gene activities involved in creating the face during embryonic development were 15,000 but only 2,000 had alleles that affected different parts of the face differentially, then one could ignore the 13,000 genes that were essentially monomorphic and focus on the 2,000 that contributed to making a difference.[6]

Intuitively, also, if a gene has more than one variant allele that can differentially affect one or more features, the number of such alleles must also be important. Indeed, as will be explained shortly, the numerical basis of the genetics of facial difference involves a *product* of the two numbers: the number of genes with alleles that differentially affect the face multiplied by the number of alleles for each such gene. In effect, the generation of different faces is a matter of multiplicative possibilities or **combinatorics**. When relatively small sets of entities are combined in multiple different

BOX 4.1 On the Muddling Up of Genes and Alleles

A common error when a particular human condition is being described, where some genetic basis is expected, involves hypothesizing a "gene for it," such as a gene for risk taking or same-sex preference. What such statements really mean is that there is an allele in a gene shared by all humans that (may) predispose to the condition. For all practical purposes, all genetically normal humans have essentially the same set of functional genes and thus are identical in their *gene content* (apart from the presence in males of the Y chromosome which has some unique genes) (but see note 6). Thus, the visible differences seen among people are thus caused not by differences in composition of the total gene set between different individuals but by the differences in nucleotide sequence within individual genes, namely, their *allelic differences*.

Another example is the term the *selfish gene*, which has entered the lexicon of everyday speech; it should actually be the *selfish allele*, an allele that will tend to spread in the population, usually through a net selective advantage. That spread, however, is actually conditional on other factors, hence those favored alleles do not have the autonomy and drive suggested by the original term. There are, however, truly selfish pieces of DNA, parasitic ones, that can move and spread in populations, but these make up a special class of DNA sequence termed **transposable elements**, and most or all are ultimately derived from viral elements that had earlier in evolutionary history invaded animal and plant genomes. Most normal evolutionary change involves changes in the frequency of particular alleles, relative to others, of commonly shared genes.

A third instance of confusion stems from the frequently quoted statement that "humans and chimpanzees only differ by 1 percent of their genetic material." This is frequently popularly misinterpreted to mean they differ by 1 percent in their gene composition—namely, that 1 percent of the total set of genes are found only in either humans or chimpanzees. There are a handful of genes found in one of these two species or the other, but the 1-percent difference referred to *overall average sequence* difference between the two genomes, not gene content. (That difference today is now appreciated to be larger, perhaps 3 percent to 4 percent. This is more substantial; nevertheless most of the difference involves DNA sequence that probably does not contribute directly to any of the biological or phenotypical differences between chimps and humans.[1])

---

1 The original figure of 1-percent difference between the genomes of humans and chimpanzees was based on an earlier molecular method termed *DNA hybridization*, which involved making hybrid DNA molecules, each consisting of one strand of chimpanzee DNA and one of human DNA, and then studying the physical properties to deduce the extent of difference. (For the details of how one makes such hybrid molecules, see Britten and Kohne 1968). When direct DNA sequence comparisons between the chimpanzee and human genomes were made many years later with new techniques, a whole class of DNA sequence alterations that had not been detected earlier was revealed. Most of these, however, lie in sequences that are probably not functionally important.

ways, the number of possible combinations grows exponentially. We have already seen one example in the way that genes encode protein chains: the fact that the number of possible DNA sequences grows with each additional base pair at the rate of 4^n, where n is the number of base pairs per duplex (or the number of bases for a given single DNA strand). Thus, when $n = 2$, there are only sixteen possible base-pair sequences. But, when, for example, n is increased to 6, still a low value, the number balloons to 4,096.

Let us illustrate this general point about combinatorics with a simple Mendelian example and then see how it can be adapted and extrapolated to thinking about the genetics of human facial difference. Imagine two different Mendelian traits in the garden pea, each governed by a separate gene; let us call these traits A and B after their dominant alleles. A strain with a double dominant genotype, *AABB* (for example, round seeds and purple flowers) is crossed with a double recessive, *aabb* (wrinkled seeds and white flowers) to give an F_1 generation, *AaBb,* which shows the double dominant conditions (round seeds and purple flowers). What will the progeny—the F_2 generation—be like? The frequencies will depend on whether the genes *assort independently* or show some coinheritance—as genes close together on a single chromosome pair do. Mendel, however, happened to study only the former among the seven pea genes he studied. As he found, there are four resulting different phenotypes (round and purple, round and white, wrinkled and purple, wrinkled and white) but (as shown by subsequent crossings) nine different genotypes (*AABB, AaBb, AaBB, AABb, AAbb, Aabb, aaBB, aaBb, aabb*). What about a cross involving three independently assorting genes? In the F_2 it will yield—we can omit the detailed accounting—eight phenotypes and twenty-seven genotypes.[7]

Thus, with each additional trait-affecting gene in crosses, the number of both the genotypes and phenotypes will increase exponentially but that of genotypes more rapidly. In fact, there is a simple numerical generalization. For n genes, each with a dominant and a recessive, the number of diploid genotypes goes up to as many as 3^n while the number of phenotypes goes up as 2^n. Given these simple relationships, we can ask how many genes, each gene shaping a distinctive aspect of face shape and having two alternative alleles, would be needed to generate 7 billion different human faces. The answer is that a mere thirty-three genes would suffice. This is a tiny number when compared either with the total estimated number of human genes—21,000 protein-coding genes—or the numbers of genes actually found to be expressed in the developing facial primordia, where much of the foundations for adult facial diversity are probably laid down (as previously discussed).

Of course, this Mendelian two-allele situation is a highly artificial one since most genes exhibit more than two alleles. In particular, it is most likely that the facial-diversity genes would be polymorphic, having multiple alleles in measurable frequencies. To make a genetic model of facial diversity that is more realistic, therefore, one should expand the number of alleles allowed per gene; this can dramatically increase the number of possible diploid genotypes per gene and per genome. Start by imagining there are three alleles for a gene, let us call it A, and its alleles designated by superscripts—thus A^1, A^2, and A^3. Since each gamete, whether a sperm or egg, has only one allele, there will be three haploid (sex cell) genotypes, one for each allele. For the diploid condition, however, there would be six diploid genotypes per gene: A^1A^1, A^2A^2, A^3A^3, A^1A^2, A^1A^3, and A^2A^3. How would the numbers play out with four alleles per gene? Now, there are ten possible diploid genotypes. For five alleles, there would be fifteen. Clearly, the number of diploid combinations goes up more rapidly than the number of alleles itself, though not exponentially. The following rule provides the numbers: for n alleles for a gene, the total number of possible diploid genotypes for that gene is $[n + (n - 1) + (n - 2) \ldots (n - n + 1)]$.

Now, to see how even the relatively modest increase in allele number from two to, for example, five affects matters when gene numbers increase, imagine there are three genes, each with five alleles that affect some visible aspect of the face. How many different face-affecting genotypes would there be? It would be $15 \times 15 \times 15$ or $3,375$—a large number of different facial types for only three genes and a relatively small number of alleles per gene.

Using the same logic, we find that for only eight to nine genes with five alleles each, one could create 2.56 billion and 38 billion different genotypes, respectively. Other numerical examples that illustrate the same general point for genes with other numbers of alleles are given in Table 4.1. Of course, these are all somewhat unrealistic examples; a far more plausible situation would involve genes with different numbers of alleles, The examples illustrate, however, the possibility that human facial diversity, in the billions, could be generated from a relatively small number of polymorphic genes. Of course, such large numbers of genotypes would not translate into equivalently many phenotypes, since we have seen that phenotypes increase more gradually than genotypes, even in the simple Mendelian situation. This is due not only to the complications of dominance

Table 4.1. The Multiplicative Power of Gene and Allele Numbers: A Few Examples

Number of Genes	Number of Alleles per Gene	Number of Diploid Genotypes per Gene	Total Diploid Genotypes
2	2	3	$3^2 = 9$
2	4	9	$9^2 = 81$
4	4	9	$9^4 = 6.561 \times 10^3$
4	7	28	$28^4 = 6.146 \times 10^5$
5	6	21	$21^5 = 4.084 \times 10^6$
7	7	28	$28^7 = 3.779 \times 10^{10}$
8	5	15	$15^8 = 2.56 \times 10^9$
9	5	15	$15^9 = 3.84 \times 10^{10}$

but also to the phenomenon of **epistasis** in which a mutant allele of one gene can mask the effect of an allele of a different gene. It is described more fully in Box 4.2. Still other factors, however, can *increase* the number of phenotypes relative to genotypes, a possibility that only became apparent through discoveries in molecular biology during the 1970s. These are the processes of *alternative gene splicing* and *post-translational modification*, which are described in Box 4.3.

The critical point is that when there is *independent assortment* of genes, the shuffling of genes in meiosis—the generation of new combinations of chromosomes from the original maternal (M) and paternal (P) sets—can generate huge numbers of different genotypes from a relatively small number of polymorphic genes. Thus, while the actual number of facial-diversity genes may turn out to be quite large—many thousands—it need not be; it could be relatively small. That is the counterintuitive conclusion when faced with the question of how many genes in the human genome are responsible for generating billions of different human faces.[8]

To resolve the question, however, of how many genes are involved in generating facial diversity, a priori considerations take one nowhere; actual genetic data are needed. Those research efforts have only just begun, and since the bulk of this work is probably yet to come, I will treat it as part of the future of the face, to be discussed in Chapter 9. Nevertheless, there is always some value in assessing different possible explanations in light of current ideas and information. That exercise, in turn, can establish a conceptual framework that helps make sense of new information as it materializes, even if the original hypotheses fall by the wayside.

BOX 4.2   The Phenomenon of Epistasis

The term *epistasis* refers to the modified expression of a mutant phenotype by a second mutation that is within a different gene to create a new phenotype, often a masking of the first mutant's phenotype. Thus, for example, if a mutation in gene A gives a new mutant phenotype when homozygous (a^- / a^-) and a mutation in B gives a different phenotype when homozygous (b^- / b^-), then if the double homozygote (a / a, b^- / b^-) shows the phenotype of the B mutant, one says that gene B is *epistatic* (covers up) gene A. In turn, the gene whose activity is hidden is said to be *hypostatic* to the other; in this case, A is hypostatic to B.

There are many different routes to epistasis, depending on the nature of the activities of the two genes. One simple case can make the concept clearer: a genetic pathway for the production of a flower pigment in which each enzymatic step, specified by a particular gene, creates a product or substrate for the next gene in the sequence. Imagine it as symbolized in the following way, with the action of each gene modifying the exact color of the pigment in a distinct way:

$$C \rightarrow B \rightarrow D \rightarrow E \rightarrow A \rightarrow \text{final pigment}$$

Thus, the homozygous B mutant (b^- / b^-), if it abolishes the activity of B, will leave the flowers colored as if they only have C activity (the step before B), while flowers homozygous for a mutation deficient in A activity (a^- / a^-) will have the pigmentation associated with the action of the E gene. If one breeds the two strains together to produce the double homozygote, what will be the result? Because the enzyme encoded by B works earlier in the sequence than A, the sequence is blocked at the B step and the flowers will have the B mutant phenotype (producing only the color associated with the action of C at the beginning of the sequence). In this case, as in our abstract example, B is epistatic to A.

In the genomes of complex animals and plants, which have many interacting genes in gene pathways and networks and a normal sprinkling of nonstandard alleles—many of which will be only mildly defective—there is ample scope in principle for the existence of epistatic interactions. Indeed, much is found.

What Are the Possible Candidate Genes Affecting Facial Diversity and How Many Are There?

When geneticists encounter an interesting biological phenomenon whose genetic basis is unknown, they often try to imagine what genes might be involved in its foundations. In effect, there is an attempt to find likely *candidate* genes for those that underlie the phenomenon. Those choices depend on a little prior knowledge of genes implicated in the phenomenon—and

BOX 4.3 Generating Multiple Different Gene Products from Single Genes: Alternative Gene Splicing and Post-Translational Modification

To understand *alternative gene splicing*, one must first be aware of the basic fact of gene splicing itself. In the classic view of genes and gene actions, genes were thought to be DNA sequences in which each nucleotide helped encode an amino acid in the protein encoded by that gene. This was expressed in the term *gene-protein co-linearity*. In the late 1970s, however, it was discovered first in certain animal viruses and then later to be common—indeed, the overwhelming rule for animal and plant genes—that the RNA transcript for most genes is longer than necessary for the encoded protein chain. The reason is that the coding portion of the transcript comes in pieces termed *exons*, which are interspersed with noncoding pieces called *introns*. After the transcript has been made but before it has been exported to the cytoplasm for translation into a polypeptide chain, a special and complex molecular machine cuts out the introns specifically, tying the exons together in just the right sequence. This process is termed *gene splicing*. Alternative gene splicing refers to the fact that, for many genes, some of the exons are also removed to yield a shorter polypeptide chain. This is not an error on the part of the biochemical machinery but a normal property of the molecular machinery that governs the splicing of the particular gene. Often, the different, alternative transcripts make polypeptide chains that have somewhat different properties and functions, sometimes dramatically so. This mechanism amplifies the number of possible gene products relative to the number of genes.

Post-translational modification also can expand the number of gene products relative to the number of genes, but it does so in a completely different way and at a later step—after the polypeptide chain has been made. It involves the enzymatic addition of a small molecule such as a phosphate, sugar moiety, or small fatty chain to a specific amino acid within the polypeptide chain. Those chemically modified proteins often have different activities than the unmodified copies. An example of a post-translational modification of a gene product that creates a new activity of a gene product is provided by sonic hedgehog (SHH) itself. When a certain small chemical moiety termed a *proteoglycan* is attached to SHH, it promotes cell proliferation specifically (Chan et al. 2009). Without the proteoglycan addition, the SHH molecule promotes other biological responses, in particular so-called pattern formation, in which a particular area within the developing embryo is given a prospective fate in development; an example is the way the different areas of the cerebral cortex are divided up early in development, which we will come to later.

some educated guesses. In the following, I will sketch three sets of genes defined by different criteria, each of which could be a major source of facial-diversity genes. These sets, of course, overlap and the hypotheses obviously do not exhaust the possibilities. Sketching them, however, helps to illuminate the issues.

Hypothesis One: Facial Diversity Is Governed by Variant Alleles of the Genes Known to Be Involved in Basic Facial Construction and Evolution

In principle, as we have just seen, all human facial diversity could be accounted for by variant alleles in a relatively small number of genes. What kind of genes might have this potential? The obvious candidates are the genes already known to be employed within vertebrates to build the face and whose genetic alterations in different ways have undoubtedly played some part, perhaps a major part, in the different evolutionary paths of faces in the vertebrates; these genes were introduced in the previous chapter. If diverse evolutionary trajectories for vertebrate faces have involved differential regulation of these genes, it is possible that smaller scale tweakings of these same major face-patterning genes, either their characters or their amounts—by variant alleles—could have a major role in influencing the degree of facial diversity in our species.[9]

One set of candidate genes in particular are the genes of the SHH pathway, starting with *Shh* itself. The pathway is complicated, and most of its details need not detain us here, but it is worth noting that its action begins with the binding of SHH protein to a special receptor, a membrane protein encoded by a gene named *Patch;* SHH binding to *Patch* activates the pathway. Like most genetic pathways, the SHH pathway involves both inhibitory and activating steps, the sequence serving in the end to deliver an ultimate signal that turns on the transcription of target genes. In the development of the face, those target genes influence the amount of growth in the facial primordia. (In other tissues, other sets of target genes will be activated.)

With respect to the possibility that variant alleles of genes in this pathway might contribute to facial diversity, the crucial point is that genetically caused alterations at any step should either increase or decrease the ultimate output of the pathway, the amount of growth in the different facial prominences. Thus, for any gene whose product normally inhibits the next step of the pathway, a loss-of-function mutation will reduce its inhibitory effect and therefore increase the pathway outputs, while rare mutations that increase the activity of that inhibitory step might further reduce those outputs. Conversely, loss-of-function mutations in activator steps will diminish that activity and hence reduce the pathway outputs, while rarer gain-of-function mutations might increase such. Thus, for the SHH pathway acting in the facial primordia, an increase in pathway activity should lead to increased cell proliferation, hence in increased growth and size of the region in which those cell divisions are stimulated, while a decrease in pathway activity should yield a reduced number of cell divisions and thus diminished growth and size. (Both loss- and gain-of-function mutants in the SHH pathway are known and have been found to act in these ways.)

Such is the theory, but is there any evidence that mutations in genes of the SHH pathway ever contribute to the normal range of facial diversity? The strongest piece of evidence so far comes from some intriguing work not in humans but in fish. These results involve a large family of fishes known as the *cichlids*, which are characterized by a second set of jaws located in the pharynx for processing food. While that feature unites and defines them, there is a huge diversity of different cichlid species. Indeed, it is the most species-rich fish family known. Their variety is particularly apparent in three large lakes in sub-Saharan Africa (Lakes Malawi, Victoria, and Tanganyika), each of which has several hundred to more than a thousand different species differing widely in body size, color pattern, shape of head, and kind of food eaten. Those last two properties—head shape and diet—are intimately connected, however, because particular jaw shapes are suitable for certain kinds of food and not others and jaw shape is a major contributor to head shape.

The work that implicates a SHH pathway gene in generating different facial morphologies concerns three distinct but related species groups or **genera** of cichlids found in Lake Malawi. These three groups live near the shore and around rocks but earn their nutritional sustenance in different ways. Members of the genus *Labeotropheus* have larger, stronger lower jaws that are adapted for biting—especially, for scraping algae off rocks with their teeth. *Metriaclima,* in contrast, has more extended (*gracile*) lower jaws, which are suited for suction and filtration feeding. The difference in appearance of the two fish is shown in **Plate 5.**

That overall difference in jaw morphology traces to a difference in a particular bony element at the back of the jaws, part of the hinge between the jaws, the retroarticular joint, whose length determines the force of biting. The long form, associated with *Labeotropheus,* gives a stronger bite and is associated with a greater jaw strength, one suitable for scraping algae off rocks. The shorter form of this jaw element gives a less forceful but more rapid bite and is found in *Metriaclima.* Genetic analysis of the basis of this anatomical difference localized it to an allelic difference in a gene for the receptor, *Patch-1,* one of two *Patch* genes in the cichlid genome. *Labeotropheus* possesses the so-called long form, and *Metriaclima* the short allelic form.

The finding that makes the connection between this allelic difference in *Patch-1* and the respective jaw shapes significant is that both alleles are found in a third genus, *Tropheops,* with different species adapted to different feeding strategies but employing the same jaw shapes. There is a perfect correlation between allele type of *Ptc-1,* diet, and jaw shape in the different *ecomorphs* of *Tropheops.* The total evidence shows that a mutation in a SHH pathway gene can contribute to visible—indeed striking—facial diversity.[10]

A question provoked by these findings concerns the viability of mutations in widely used essential genes such as *Patch:* how can a mutation in a gene of this kind yield such a fully functional and specific effect in development? Would it not interfere with development in all the other places in the developing body where the SHH pathway operates and therefore be dysfunctional? Such mutations with widespread effects are a well-known phenomenon in development and are termed *pleiotropic mutations.* There is, however, one way in which a widely used (indeed, *pleiotropic* gene) can be mutated in such a way to create localized effects. It involves a structural and organizational aspect of DNA molecules. For most genes, there are short DNA sequences on the same DNA molecule as the gene itself that can independently switch a gene's activity on or off. (The binding site for the *lac* repressor, the first such discovered, is an example of a DNA binding site off switch.)

Collectively, these short DNA sequence regions are termed **cis-regulatory sites**, the Latin prefix indicating that they exert their effect only on the gene copy on the same chromosome (hence not on the same gene on the homologous chromosome). The best characterized of these regions govern the transcription of genes and are termed **enhancers** and range from 200 to 500 base pairs long, with each containing multiple binding sites for transcription factors. While there are enhancers that bind factors that turn off transcription, the great majority activate transcription when the requisite transcription factor is bound. Intriguingly, some enhancers are at considerable distance from the genes they regulate—up to millions of base pairs—or clustered near the gene in front of its transcription

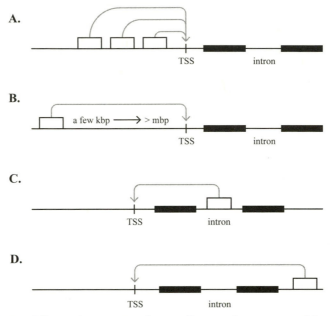

FIGURE 4.3 Four different enhancer patterns for controlling a gene's expression. In all four examples, the black boxes represent the protein-coding gene, which, in this simplified example, consists of two parts (exons) separated by an intron; TSS stands for *transcriptional start site*, the place where transcription begins when activated by an enhancer; and the enhancers are indicated by white boxes. **A.** Several different enhancers are immediately upstream (5′) of the gene. **B.** An enhancer that is distant from the gene—the distances can be from tens of thousands of base pairs (kbp) to more than a million base pairs (mbp) upstream. **C.** An enhancer within an intron. **D.** An enhancer downstream (3′) of the gene.

start site or, in some cases, located even within introns. The reason that distant enhancers can regulate specific genes is that DNA has an enormously variable capacity for looping back on itself, allowing a primed enhancer, even if at some distance, to find its target gene. Several possible arrangements of enhancers are illustrated in Figure 4.3, and many genes probably have more than one kind.[11]

Since enhancers are often specific in their use to particular tissues, a mutation that inactivates or otherwise alters the activity of one of them can give highly specific cell or tissue effects without affecting expression in other tissues. This is the probable genetic basis of the cichlid *Patch-1* effects: in the development of the more robust jaw type *(Labeotropheus)*, there is more *Patch-1* transcription in the crucial region of the lower jaw (the retroarcticular joint) than in embryos possessing the short allele. Stronger *Patch-1* expression in turn leads to more extensive growth and bone development in the lower jaw. The critical allelic difference is thus in an enhancer sequence governing the amount of transcription of *Patch-1*.[12]

These findings show only that mutations in a gene of the SHH pathway *can* contribute to normal facial diversity. It is unknown how widespread the involvement of mutations in the SHH pathway or the other major face-building pathways in creating human facial diversity actually are, though recent studies in humans are beginning to implicate them, as we shall see.

Hypothesis Two: Facial Diversity Is Governed Primarily by a Set of Genes Already Revealed by Their Clinical Effects on Craniofacial Development

If one approach to finding the genes that underlie facial diversity involves educated guesses based on the known genes involved in facial development, a second avenue involves looking at mutants already known to affect the development of the head and face. There are more than one might expect: several hundred genes in humans have been identified by mutant alleles that create some visible and frequently marked effects on facial development. They have been collected in a database available on the Web, the **Online Mendelian Inheritance in Man** (OMIM) database. If one goes to the OMIM Web site and types in "craniofacial abnormalities," a list of more than 320 genes immediately appears. Each was identified initially by mutations that generate some visible abnormality in the face or the head (the latter often associated with facial effects). Each entry describes the phenotypes, including the relevant clinical information, and many list additional genetic and molecular details.

Some of the listings involve genes and conditions that have already been described—holoprosencephaly and the various craniosyntoses. Hundreds of the entries, however, involve genes that seem to have no specific relationship to the head or the face. Many are genes for general cellular functions—so-called **housekeeping genes**—that are widely expressed in many cell and tissue types. As we might expect, a high proportion of the mutants of these essential functions have severe clinical effects, either early death or severe disability. Others, however, are associated with milder conditions. The difference between life and death (or severe disability) often depends on the extent of residual activity of the genes—if the mutations cause only partial loss of gene activity—and the extent to which their deficiencies can be compensated by other similar gene activities. Such compensation and mutual backup of gene functions by other genes is probably quite common.[13]

There are numerous puzzles about this list of genes over the ways in which the mutant alleles for many of the genes result in the particular visible physical abnormalities associated with them. One general feature, however, is that few of the mutations are associated with a single or limited effect on the head or face. The great majority not only display multiple craniofacial effects but also abnormalities in other organs or tissues. A few are connected with specific defects in both the face and limbs, but many of these links can now be understood in terms of the shared genetic machinery between face and limb development described previously.

As already remarked, multiple usage or **pleiotropy** is the rule for genes, not the exception. Correspondingly, most of the 320+ genes listed in the OMIM database under *craniofacial abnormalities* are pleiotropic genes. Hence, while we may speak of genes "for" craniofacial diversity, remember that nearly all such genes have other roles in other tissues and other places in the development or functioning of the body. All of the genes that are jointly employed to help build both faces and limbs—the presumptive core module in these genetic networks (see Fig. 3.4)—are examples of well-understood pleiotropic genes.

This matter of multiple use bears directly on the question of whether the OMIM-listed genes for craniofacial abnormalities reliably reflect the genes involved in generating *normal* facial diversity. If a particular mutation of a pleiotropic gene could contribute to that diversity but has other serious detrimental effects—even when present in only one copy, as many do—it would be strongly selected against and not likely to be frequent in the population. Hence, most mutations that contribute to the normal diversity of faces are probably selectively neutral or nearly neutral (see Box 1.2). That consideration directly rules out most of the mutations listed in the OMIM database. Indeed, most are. associated with clinical defects., and hence are unlikely to survive long in populations. It is quite possible, however, that other, less severely defective., alleles of those same genes might be contributors to facial diversity. Such variants, however, would also have to be selectively neutral or nearly neutral for the other traits to which they contribute.

Beyond the absence of evidence that many of the OMIM-listed genes for craniofacial defects are the primary gene set for normal facial diversity, the basic premise that they constitute candidate genes for normal facial diversity is questionable. For example, take mutant genes that generate a craniosyntosis. Such premature fusions of cranial sutures lead to alterations of the shape of the skull as a whole, with corresponding subsequent distortions of the face as secondary consequences. Such effects are indirect, however, while normal facial variation is not associated with any degree of craniosyntosis. Hence, there is no reason to suspect that subtle genetic variation of the *craniosyntosis genes*—genes whose normal products help ensure that premature suture closure does not take place—would play a part in influencing normal facial diversity.

Even though the OMIM list of genes whose severe loss-of-function creates craniofacial abnormality is not a reliable guide to the set of genes that generate normal facial diversity, it may include such genes as a fraction of the total. In particular, these groups might contain slight loss-of-function mutations of the same genes whose strong loss of activity leads to major defects.. Recent evidence suggests that some genes in the OMIM list may possess mild variants that contribute to normal facial diversity, as will be discussed in Chapter 9.

Hypothesis Three: Facial Shape Is Influenced by Many Thousands of Genes

Might the differences between human faces reflect allelic differences among, not dozens or hundreds of genes, but many thousands of different genes? If so, each gene would contribute a small bit of difference to one or more parts or regions of the face during its formation; correspondingly, the alleles of such genes would make a relatively smaller contribution to the final phenotype. In this view, the identity of the specific genes is far less important than the sheer numbers of genes involved and the simple additive effects of their alleles. In theory, this is possible and fits the traditional Darwinian belief that complex traits are built on the additive effects of many different gene variants, each of relatively small effect. Were Darwin alive today, he would undoubtedly favor this explanation.

It also seems consistent with experiments that show that many thousands of genes—on the order of 15,000 to 20,000—are expressed in the development of the head and face in embryonic and fetal development. There are two main sites of gene expression that have been investigated in this connection, namely, the brain and the facial primordia. The developing brain is relevant because, as will be recalled, its growth and development can affect the shaping of the face. Furthermore, both the developing brain and the facial prominences show abundant gene expression. Indeed, in mouse embryos, most of the genes of the mouse genome are expressed in one or the other or both places during early development. Given the large degree of conservation of common developmental plans in vertebrates, it is probable that similar numbers apply to the developing human brain. Given the brain's roles in face development, this raises the theoretical possibility that around 20,000 genes might have some part in shaping the face indirectly through their effects on the brain. There are two reasons to doubt this, however.

First, genes expressed in the brain that influence face development should be primarily those required specifically for brain growth by affecting the size and shape of the brain. Those genes, however, are probably a minority of the total of the genes expressed in the brain, with the great majority involved either directly in aspects of brain physiology and functioning or in creating the foundations for those capacities. A second reason for discounting a substantial role for most of these brain genes in face development, perhaps a surprising one, is that a large proportion of the genes expressed there may play little role. Despite the common assumption that biological processes are inherently economical and do not use more resources than are needed, much evidence suggests that this is often not the case. While it seems likely that most genes expressed in the brain play roles in its development, the actual proportion of the genes expressed there that are essential for normal brain development is unknown.[14]

While the brain is important in shaping the face, there is no reason to believe that most of its expressed genes contribute to normal facial diversity. (That *some*, perhaps a tiny proportion, do so is indicated by the mouse head-shape mutants described earlier.) In contrast, those genes expressed in the facial primordia may include many of the facial-diversity genes. The crucial patterning genes discussed in the previous chapter would be one small subset of these genes.

Two recent studies allow some estimate of the numbers of genes expressed in the facial prominences of the mouse, and those numbers are large indeed. One analysis used gene-distinctive molecular probes to identify the genes expressed in the different facial prominences and identified on the order of 20,000 expressed in the midface primordia, with the great majority (about 15,000) expressed in all. A substantial number (about 5,000), however, were uniquely expressed in one primordium type or another, with correlations to distinctive features in the structures developing from those specific primordia (such as taste, olfactory, or muscle functions.) The other study focused on identifying particular enhancers that affect genes expressed in the facial primordia and concentrated on those that were fairly distant from the genes they controlled. It identified more than 4,000 such sites for about that many genes. Together, the two studies support the notion of the involvement of many thousands of genes that are expressed in the developing facial primordia of the mouse and, by extension, those of humans.[15]

Yet just as the expression of a gene is not an automatic signifier of the functional importance of that expression, these numbers provide only an upper limit to the numbers of those genes that might participate in generating facial diversity; the vast majority may be monomorphic (in terms of their phenotypic consequences) and thus would not contribute to that diversity. Only those genes that have variant forms in the population and where those variants have distinctive phenotypic signatures are relevant, and they may be a small proportion of the total expressed genes.

There is, indeed, one general observation that tends to argue against the possibility that thousands of genes are involved in creating human facial diversity. As appreciated for millennia, and probably far deeper in our history as a species, children often bear striking facial resemblances to one or the other of their parents for one or more features. If thousands of genes were involved in generating facial diversity, then mutant alleles of any one should individually contribute only a small net effect to development of a facial feature and, correspondingly, there would be small effects for each variant allele of those genes. The consequence would be that for a mother and father who were not closely related and who differed strongly in one or more of their facial features, their offspring should generally be intermediate in many or all of those features. Any quantitative trait that shows a bell-shaped (Gaussian) distribution in a human population is a possible instance of this kind of inheritance. The distribution of adult heights in human populations, though strongly affected by nutritional and other environmental effects, is probably an instance of such a trait.[16]

For physical shape features in the face, however, such as nose length, eye shape, or distance between the midpoints of the eyes, features that can be measured with high precision, such averaging out is usually *not* the case. In families where the mother and father differ strongly for certain traits, the offspring often resemble one or other of the parents for a particular feature. "He certainly has his mother's eyes" and "This little girl has her father's dimple" are the kinds of comment one often hears. Such evidence is anecdotal but common and strongly implies the importance of alleles of large effect from a relatively small number of genes governing those features. A famous example of a facial trait that seemed to be inherited as a simple dominant allele was the so-called Habsburg lip of the ruling family of the Austro-Hungarian Empire. Such marked traits of seemingly simple inheritance are not consistent with the classic Darwinian model of quantitative, additive small effects. The first results of modern quantitative analysis of facial traits, to be described in Chapter 9, are also inconsistent with that model. Hence, while the idea that thousands of genes underlie the facial diversity of 7 billion human beings has not been ruled out, the current verdict on it is best summed up as "possible but not likely."

On Linking Genes with Development in Thinking about Facial Diversity

In this chapter, we have asked, how many genes might be involved in generating the observed diversity of human faces? Phrased more precisely, this question becomes, how many polymorphic genes are there for the morphological facial characteristics that make up human facial diversity?

There is still no answer to this question. The gene expression data allow the possibility that it could be as many as 15,000 to 20,000, though we have reasons to doubt that. At the other extreme, at least in principle, the full range of differences could be generated by a relatively small number of genes, each with a modest number of alleles. The strongest reason for thinking that at least some facial traits are so governed is the common observation that visible family resemblances between children and parents are ubiquitous and that such resemblances are most readily explained by single gene differences instead of additive multiple gene effects. Furthermore, the cichlid jaw *Patch-1* results described in this chapter and the Habsburg lip indicate that some facial traits do, indeed, behave as simple Mendelian gene hereditary effects.

In this chapter, we have concentrated on the face as defined by ana-tomists—namely, everything beneath the eyebrows to the line of the chin. Yet we perceive the full face as including the forehead, and there will also be additional genes that affect forehead shape via their effects on brain size and quite possibly some of the bones of the neurocranium. In addi-tion, there are genes that affect facial skin color and eye color, which obvi-ously contribute to facial appearance—and differences between faces. This probably involves, however, no more than a dozen genes (as discussed in Chapter 8). Finally, there must be other genes that directly affect shape of the eyes, which do not develop from the facial primordia although, as we have seen, eye shape is affected by the pattern of growth and development of the facial prominences.

If future genetic evidence eventually establishes that a relatively small number of genes is involved in setting the diversity of facial shapes, it does not automatically follow that the set of genes involved in creating facial di-versity is the same in all human populations. In different "racial" groups, it need not be and probably is not the exact same set of genes. Different ethnic groups are undoubtedly close to monomorphic for one or more genes—in terms of phenotypic consequences, if not strictly in terms of DNA sequence—that contribute to the visible characteristic facial features of those groups, a possibility that still allows for many other genes that modify those traits to generate the (equally) visible diversity within the group. We will return to this matter in Chapter 8.

In thinking about genes, development and the formation of the human face, one last topic, touched on earlier, should be mentioned. Full devel-opment of the human face, it will be recalled, requires a long period of maturation—starting in infancy, then continuing in childhood and throughout most of the teenage years, hence long after embryonic and fetal development. All of the genes involved in face development that we have discussed are active in the embryonic and fetal periods while little is known about which gene activities contribute to shaping the face in later stages. Yet, in some way, those early developmental events set the pattern or tem-plate for what happens later from infancy through the teenage years. There must be some form of precise regulation of the pattern of differential growth in the different regions of the face during this long period of facial maturation, but exactly how this occurs is unknown. Correspondingly, how the developmental trajectory of the face from infancy through the teenage years is "set" by those earlier events is equally obscure.

In effect, the material discussed in this and Chapters 2 and 3 is. prologue to our proper subject, the evolutionary origins of the human face. That evolutionary history is the subject of the next two chapters and it begins with a genuine mystery—namely, how the first vertebrate faces came into existence more than half a billion years ago. Lacking fossil evidence on those events, one must necessarily be guided by creative scientific imagination. Slightly more than thirty years ago, a bold explanation of the origins of the vertebrate head—and face—was put forward. It has proven a durable framework for thinking about the origins not just of the vertebrate face but also of the vertebrates themselves. It involves—perhaps unsurprisingly—the origins of the neural crest cells. We will look at this matter at the start of the next chapter.

five

HISTORY OF THE FACE I: FROM EARLIEST VERTEBRATES TO THE FIRST PRIMATES

Introduction: On the Face as an Evolutionary Novelty

If you conducted a spot quiz on the street and asked passersby whether animals have faces, the answers would overwhelmingly be "Yes" or "Of course," perhaps accompanied by a slight smile at such a simple-minded question. For most people, after all, "animal" means "mammal," and mammals clearly have faces. Who, for instance, doubts that her or his beloved dog or cat has a face? Yet even the most exotic mammals, such as porpoises and sloths, also have faces. For those respondents who realize that the term *animals* includes far more than just mammals, the question might also evoke images of fish, crocodiles, pigeons, or frogs—but these creatures have faces, too, and the answer would still be "Yes."

All of the mentioned animals, however, are vertebrates, and vertebrates make up only a small portion of the entire Animal Kingdom, probably no more than 1 percent of all animal species. Furthermore, while taxonomists divide the Animal Kingdom into about thirty major groups termed **phyla**, the species of the great majority of phyla lack faces. The most primitive animals, those of the **Placozoa** and the **Porifera**, do not even possess mouths; they acquire their food by absorbing food particles from the water directly into their cells. The great majority of animal types, of course, possess mouths—but they lack pairs of eyes, which, along with the mouth, make up the signature of a face.[1]

Indeed, only two of all the recognized major groups consist of species with faces. These are the two most biologically complex kinds, which also possess the most complex behaviors: the Arthopoda—which includes the crustaceans and insects, with their distinctive, robotlike visages—and our own group, the Vertebrata, a subdivision of the Chordata phylum. One might also argue that the Cephalopoda, a subdivision of the phylum Mollusca, which contains its most complex members, the squids and octopuses, have faces; these animals have both paired eyes and mouths, but with their mouths hidden, they present the most peculiar kinds of faces. In addition, a few members of the Annelida, by the same criterion, would also—if only barely—qualify. The evolutionary implication of this limited taxonomic distribution of animals possessing faces, and its coincidence with most complex animal types, is obvious: the first kinds of animals to evolve almost certainly lacked faces, and only later and in a few groups did faces evolve. In effect, the face was an evolutionary novelty in the Animal Kingdom, one that arose more than once.

In this chapter and the next, we will focus on the vertebrates and their faces specifically, tracing the immensely complex evolutionary journey that led from the faces of the first ones—tiny jawless fishes that lived more than 500 million years ago—to the human face. The account necessarily omits much about the evolution of different faces in order to concentrate on the long line of descent that eventually led to humans, the long hominocentric lineage path (LHLP); nevertheless, that lineage on its own has a rich and interesting history. This first and longest part of the story, as I have divided it, will take it from those first vertebrates to the first mammals and the earliest primates, a segment of evolutionary history stretching nearly 450 million years. The next chapter will continue the narrative, tracing the evolutionary path from those first primates, with their rather generic mammalian faces, to modern humans and the distinctive faces we possess.

Something more than a recounting of changing animal faces and bodies will be attempted here, however: we will look at some of the genetic and developmental changes that underlay those morphological alterations. In this chapter, in particular, we will examine specifically four general mammalian characteristics that have importance for the human face: mammalian jaws and teeth, fur, the provisioning of milk to the young, and the mimetic muscles that endow mammals with facial expressivity. In the next chapter, we will try to deduce something about the sequence of underlying genetic changes that led from the small and rather typically mammalian faces of the first primates to the decidedly different human head and face. To navigate this complex territory, however, it might be useful to discuss two fundamental matters at the start: the concept of the *evolutionary lineage* and the usefulness of having a taxonomic system as a rough framework for structuring the evolutionary history. We begin with those two matters.

On Evolutionary Lineages and the Linnaean Taxonomic System:
Two Disparate Approaches to Biological Diversity

The notion of the *evolutionary lineage* was introduced in the first chapter but deserves more explanation because it is central to much of what follows. This term denotes the sequence of species and forms that evolved along a specific track, eventually leading to a particular group or species. Deducing the sequence is, of course, an analytical task; lineages do not appear as such in the fossil record. The mild paradox associated with the quest is that, while the goal is to see how a lineage unfolded as it moved forward in time, lineages can only be reconstructed by moving backwards into the past. In effect, we start with a particular living species or group of animals of interest (for example, chimpanzees) and then identify the probable immediate ancestors of that group as best as possible, then their ancestors, and so forth, progressing ever deeper into the past. The point at which we terminate the search for ever-more distant ancestors is either a matter of choice or a decision forced by lack of evidence. With the beginning and the end of the lineage provisionally ascertained, we can then begin to interrogate the data to see how the changes going forward probably occurred. The procedure is analogous to reconstructing the family tree of an individual: for this, too, one works backward into the past, tracing a sequence of progenitors, beginning with that individual and his or her parents, then their parents, and reiterating the process as one moves back into the ever-more distant history of the family. With the genealogy complete, one tries to make sense of the personalities and who did what and when, including of course who begat whom, but now moving forward in time.

In this chapter and the next, we will trace the evolution of the face from the first vertebrates to modern humans along the LHLP. The focus will be on changes in the face, but since the face did not evolve in isolation from the rest of the body, the larger context will be discussed at each point. Although there are many gaps and uncertainties in the reconstructed lineage, we know the identities and sequence of the main ancestral groups involved, which provides the basic framework. The main stages in this lineage, with the approximate times of origin of the different groups, are given in Figure 5.1.

Many of these names will mean little to the reader at this point, but the various groups indicated in the figure and their salient characteristics will

The Long Hominocentric Lineage Path (LHLP)

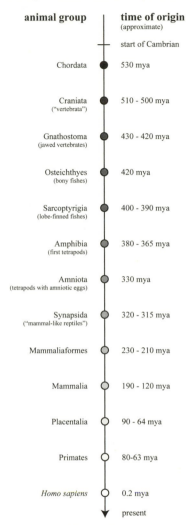

FIGURE 5.1 The LHLP, the sequence of lineages, from first vertebrates to modern humans. All dates refer to dates of origin and are approximate. Some key events in the sequence leading to the primates are as follow: chordates → craniates (first heads and faces); first craniates → gnathostomes (acquisition of jaws); gnathostomes → Osteichthyes (acquisition of bony skeleton and skull); Osteichthyes → sarcoptyrigian fishes (beginnings of limbs); sarcoptyrigians → amphibians (acquisition of true limbs for land, air breathing); amphibians → amniotes (acquisition of ability to lay eggs on land); amniotes → synapsids (start of simplifying jaws, diversified teeth); synapsids → mammaliaformes (homeothermy, mammalian gait, fur? milk production?); mammaliaformes → mammals (completion of those trends, full mammalian jaws, placentation? mammals → primates (opposable first digit, various other skeletal modifications). All of these events will be described in more detail in this chapter, while the events that led from primates to humans is discussed in Chapter 6. (Mya = millions of years.)

be described as we reach them. Nevertheless, an understanding of the general taxonomic system from which these names derive would be useful. Although that system was initially established to conceptually organize the world of present-day living things on the basis of their degrees of resemblance, it can be applied, albeit with greater uncertainty, to those extinct animals that left sufficient fossil evidence behind. The general scheme was devised by Swedish scientist Karl Gustav von Linne (1707–1778), or Linnaeus, the father of modern taxonomy. Although Linnaeus did not believe in evolution (but in the fixity of species' types), his system provides excellent evidence for evolution, as Darwin realized and explained.

The Linnaean system involves a hierarchical classification of all living things, using a sequence of ever-more inclusive categories, or **taxons**, as one goes from the species level up the hierarchical ladder. Within each level, the members resemble each other but less so than members of the previous lower category. Thus, visibly similar species are grouped into a *genus*, and the members of *species* of each genus are less similar to the members of other genera than they are to each other. The name of each species is a two-part construction, starting with that of its genus and ending with a specific name—for instance, *Mus musculus* for the common mouse. From the genus level, one proceeds to the *family* level, through to the ever-larger and successively more inclusive categories of *order*, *class*, *phylum*, and *superphylum* to the largest of all, the *Animal Kingdom*. The general system—portrayed going from the highest level at the top of the figure to the lowest (the species) at the bottom—is diagrammed in Figure 5.2 and illustrated with our own species, *Homo sapiens*.

Taxonomic thinking inherently divides things into discrete categories. In contrast, evolutionary thinking, which deals with time-dependent transitions and intermediary forms, is concerned with processes of change and therefore tends to blur the boundaries between those categories and between member groups within a category. This conceptual mismatch between the two approaches to organismal diversity often raises problems, but for the moment it suffices to know the logic of the taxonomic system, which will help us deal with the evolutionary history. The key general point to remember is that the Linnaean taxonomic hierarchy, though devised to provide a sense of order for God's "Creation," gives a general rule of thumb for evolution: the closer two species are within the hierarchy in

The Linnaean System

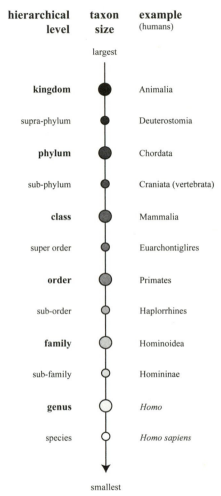

hierarchical level	taxon size	example (humans)
	largest	
kingdom		Animalia
supra-phylum		Deuterostomia
phylum		Chordata
sub-phylum		Craniata (vertebrata)
class		Mammalia
super order		Euarchontiglires
order		Primates
sub-order		Haplorrhines
family		Hominoidea
sub-family		Homininae
genus		*Homo*
species		*Homo sapiens*
	smallest	

FIGURE 5.2 The Linnaean system of classification in which organisms are grouped in ever-larger taxons from bottom to top, with decreasing degrees of resemblance among the members at each taxon level. Seven main categories of taxon were established by Linnaeus, as indicated in bold. Going from the smallest to the greatest these are species, genus, family, order, class, phylum, and kingdom. Other subdivisions have been added subsequently: subspecies (not shown), superfamily, superorder, subphylum, and supraphylum (in roman type). The system is illustrated on the right with the taxonomic categorization of *Homo sapiens* according to these rules.

general, the smaller the temporal span since their evolutionary divergence from a common ancestor; conversely, the greater the taxonomic separation, the greater the amount of time elapsed since that separation. Hence, two species of mouse in the genus *Mus* would have had a more recent common ancestor, perhaps one that existed only a few million or so years ago, while a mouse and the African lion (in different mammalian orders) would have shared a common mammalian ancestor that existed far earlier, in this case perhaps 65–80 million years ago. In this chapter, we will start with the beginnings of animal life (ignoring the long preceding history of much simpler life forms), a time that was probably on the order of 600 million years ago. In what follows, "billions of years ago" will be abbreviated "bya" and "millions of years ago" by "mya."

The Cambrian Explosion: From the First Animals to the First Vertebrates

The essential temporal framework for any long-term evolutionary process is the **geological time scale**, the dating system of Earth's history. In this construction, the total time of Earth's existence is partitioned into successively smaller intervals of time: eons are the largest, then come successively smaller time units: eras, geological periods, and epochs. Eons stretch from hundreds of millions to nearly two billion years while, at the other end of the scale, epochs may be "only" a few hundred thousand or a million or so years. The large sweep of Earth's history, as given by the geological time scale, is shown on the left side of Figure 5.3.

In normal human terms, the presence of animal life on our planet is ancient, extending back over the past 600 million years. In terms of the history of the planet, however, the existence of animals is a comparatively recent event. The first three eons—the **Hadean,** the **Archaen**, and the **Proterozoic**—make up the greatest part of the history of the planet, approximately the first 40 billion years of Earth's 4.6 billion years of existence. They were characterized, respectively, by no living forms (the Hadean, 4.6 to 3.8 bya), single-cell bacteria-like life (the Archaen, 3.8 to 2.5 bya), more complex single-cell and colonial forms, and toward the end some animal forms (the Proterozoic, 2.5 bya to 544 mya). It is only the fourth and last eon, however, the **Phanerozoic** ("recent animal life"), in which modern life forms of complex animals and plants first appeared and

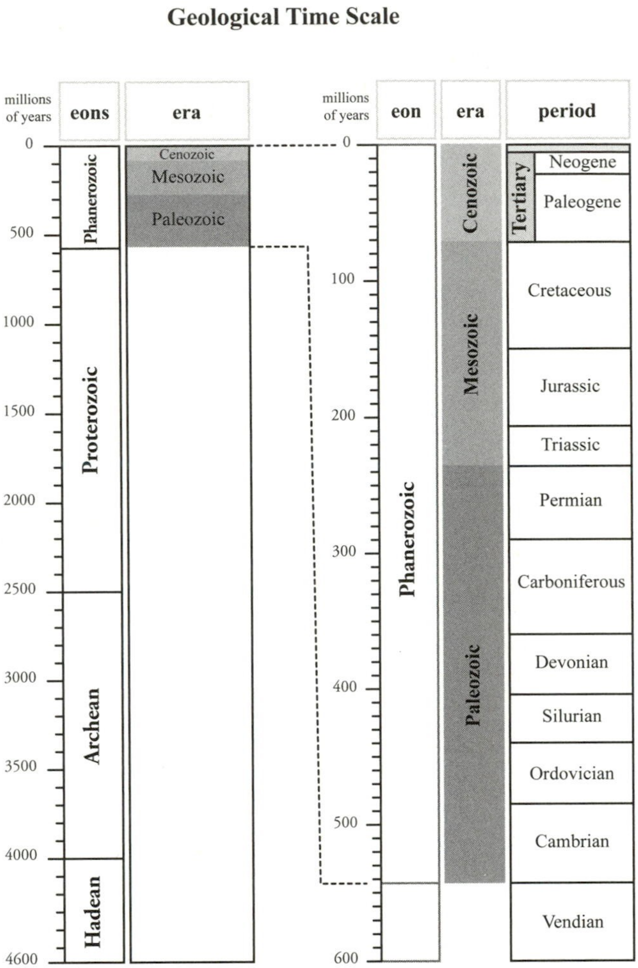

Geological Time Scale

millions of years	eons	era		millions of years	eon	era	period
0	Phanerozoic	Cenozoic		0	Cenozoic	Tertiary	Neogene
		Mesozoic					Paleogene
500		Paleozoic		100	Mesozoic		Cretaceous
1000	Proterozoic			200			Jurassic
1500							Triassic
2000				300	Phanerozoic		Permian
2500							Carboniferous
3000	Archean			400		Paleozoic	Devonian
3500							Silurian
4000				500			Ordovician
	Hadean						Cambrian
4600				600			Vendian

FIGURE 5.3 Earth's entire geological time scale (left) and the Phanerozoic ("recent animal life") eon (right). In this system, time is successively divided into *eons, eras, geological periods,* and *epochs* (the latter not shown). While the earliest cellular life arose early in the Archaeon, animal life probably first began late in the Proterozoic eon, in a period termed the *Neoproterozoic* (1,000–542 mya). Hence, while *Proterozoic* literally means "first animal life," such living forms arrived relatively late in this eon. The great majority of contemporary phyla and their subphyla, including the vertebrates, originated in the Cambrian, the first geological period of the Phanerozoic ("recent animal life").

came to dominate Earth's seas and then later its land surfaces. A detailed diagram of the Phanerozoic is shown on the right in Figure 5.3.

For the Phanerozoic, the rock layers or *strata* designated as belonging to a particular era or geological period are first identified by the fossil forms that characterize that interval; they are then dated more precisely by radiometric dating. In some cases, the dividing lines between strata that separate designated eras or geological periods mark a mass extinction of the life forms characterizing the earlier era or period, the transition sometimes signified by a striking change in chemical composition at the boundary line. For most transitions, there was a rapid shifting of many of the life forms due to climate change or other factors, without an abrupt mass extinction. In either case, however, it is often new forms of life appearing above a particular boundary line that give that boundary significance in defining the start of a new period. Altogether, the Phanerozoic encompasses the last 542 million years of our planet's existence and is subdivided into four successive *eras*, the first three of long duration and the last (and most recent) a virtual blip in time. They are, respectively, the **Paleozoic** (literally, "ancient animal life"), the **Mesozoic** ("middle animal life" but often referred to in popular accounts as "The Age of Dinosaurs"), the **Cenozoic** ("recent animal life" but popularly "The Age of Mammals"), and the **Quarternary** (the last 10,000 years following the end of the ice ages of the Pleistocene period).

When did animal life first appear? Opinion is divided, with the only general agreement being that it was sometime in the third eon, the Proterozoic. The debate about when in the Proterozoic arises because the two forms of evidence, molecular clock and fossil, initially differed markedly in their results. The first molecular clock evidence, derived from the amount of DNA sequence divergence among different living animal phyla (Box 1.2), placed the origins of animal life as early as 1.2 billion years ago. In contrast, the fossil evidence indicates animals made their first appearance much later, in the final portion of the Proterozoic, the Neoproterozoic, approximately 600 million years ago. Most biologists regard the fossil evidence in this case as the more reliable guide to the time of origin of the animals, while newer molecular clock evidence, corrected for some of the worst vagaries of molecular clocks, places the origins of animal life closer to the more recent date.

The fossils of those early life forms in the Neoproterozoic, however, raise many questions. Collectively, they are termed the **Ediacaran fauna**, named after the hills in southeastern Australia where they were first found, though they are now known from more than thirty locations worldwide. Many are strange-appearing forms indeed and, on the whole, difficult to relate to present-day animal forms. Some circular-shaped ones resemble the umbrella-like medusae of the **Cnidarians,** a phylum whose best-known contemporary members are the corals, sea anenomes, and jellyfish. A comparative handful of others seem to be ancient segmented marine worms (Polychaetes within the phylum Annelida), while a putative mollusk and a few probable arthropods have been identified. In addition to the direct fossil imprints of the Ediacaran fauna, there are *trace fossils*, wiggly paths preserved in the rocks that could only have been made by some form of animal. These could well have been formed in the mud by the slitherings of polychaetes, but some may have been made by other wormlike animals that did not leave any direct fossil remains. For the most part, the late Neoproterozoic stratigraphic record of animal life features emptiness, silence, mystery.

It was with the beginning of the Phanerozoic eon, however, and specifically its first period, the Cambrian, that animal life began to flourish in numbers and diversity. The Cambrian's beginnings are dated to 542 mya and it extended to 488 mya, a span of approximately 54 million years. It is in the *lower* Cambrian strata—namely, the earliest formed in this period—that the first distinctively new animal forms are found. These are tiny but abundant shells that reveal little of the animals they contained. The structure of the animals themselves is unknown, but their complexity was probably similar to that of today's simpler mollusks—clams, oysters, snails. The main significance of these "shelly" fauna is that they provide evidence for a rapid efflorescence of animal life at the start of the Cambrian.[2]

It is with the middle Cambrian, roughly the 20-million-year period stretching from 530 million years to 510 million years ago, that markedly complex life forms are known to have appeared and flourished. It is this period, whose fossil remains from certain fossil-rich sites, *Lagerstaette*, testify to a sudden expansion of the different kinds of complex animals, that is famously known as the *Cambrian explosion*. These animals lacked hard parts—the usual requirement for efficient fossilization—but exceptional preservation conditions allowed the fossilization of many of these soft-bodied animals at these sites, hence providing a good idea of their structures and varieties. It is the diversity of different kinds of animals, many bearing little resemblance to those that came later, and their relatively sudden appearance that give this period its sense of bizarre wonderfulness.

A 20-million-year span may hardly seem the kind of instantaneous event that merits the term *explosion*, yet compared to the preceding 3-billion-year interval in which single-celled organisms were the predominant forms of life, it was a period of rapid and dramatic change in animal forms. Its occurrence was probably made possible by the favorable conjunction of several environmental circumstances involving changes in marine chemical composition. A particularly crucial one was rising levels of oxygen in the atmosphere and ocean surfaces, the cumulative result of nearly 2 billion years of photosynthetic activity of the so-called *blue-green bacteria* (Cyanobacteria). The new animal forms themselves helped establish novel ecosystems in which their own competitive interactions quickened the pace of further diversification. Whatever the precise causes of the Cambrian explosion, the result was the efflorescence of animal types, many showing weird and wonderful morphologies not seen in later fossil beds. Indeed, many of the early to mid-Cambrian animals were initially difficult to place within the classification scheme of living animal phyla, a fact that has led to much debate and controversy. The consensus today is that most were early or *stem* forms of present-day animal phyla, exhibiting some but not all of their defining traits.[3]

Those disputes about the taxonomic affinities of the strange Cambrian animals were part of a larger debate about the relationships, both evolutionary and taxonomic, of the animal phyla extant today. Where connections and putative evolutionary relationships can be reconstructed, such history is most often depicted in the form of a diagram, termed a **phylogenetic tree** for its rough resemblance to a tree, with its progressively bifurcating branches. The best known of these early trees depicting relationships was a hypothetical one drawn by Darwin; it is famous both for its insight that this kind of diagram is a good way to depict evolutionary relationships and for being the only illustration in *The Origin of Species*. (That dearth of illustrations probably reflected Darwin's haste in writing the book, taking less than a year to do so, to establish priority for his ideas vis-à-vis his colleague and competitor, Alfred Russel Wallace). Traditionally, since Darwin's time, such trees of relatedness and descent have been constructed either from fossil evidence, where the fossils could be dated relative to one another, or to inferences of degrees of morphological relatedness of living species from comparative studies, or usually from a mix of the two kinds of evidence. In recent decades, however, as it has become possible to compare DNA sequences, it has become easier to construct probable phylogenetic trees from detailed comparisons of the DNA sequences of living species. (Protein sequences, which reflect DNA sequences, were first used in this way in the 1960s.) Such molecular analyses can, of course, be compared with those from the traditional forms of analysis. A brief, highly simplified description of the principles of **molecular phylogenetics** is given in Box 5.1.[4]

A phylogenetic tree of the Animal Kingdom based on both morphological data and DNA sequence comparative analysis is shown in Figure 5.4. This version divides the animals into three major types based on their overall body symmetry. The earliest animals almost certainly lacked any form of body symmetry and were probably similar to the present-day Porifera (sponges) and Placozoa (even simpler animals), indicated toward the bottom right. The next to arise were the Cnidaria (jellyfish, corals, polyps), mentioned earlier, whose mature forms possess radial (circular) symmetry and were almost certainly represented in the Ediacaran fauna. Grouped with them are the Ctenophora (possibly older), who use tiny cellular hairs termed *cilia* on their cells for propulsion. These two groups make up the Radiata. Still later, and probably evolving from an early cnidarian, appeared the first animals that exhibit bilateral body symmetry, the **Bilateria**, whose various phyla today make up the vast majority of animal types; these animals show their bilaterality either in immature (embryonic or larval) forms, as adults, or both. As previously mentioned, a few recognizable bilaterian-type animal fossils are found in the Ediacaran fossil assemblages at the end of the Neoproterozic, but the great expansion and diversification of the Bilateria took place at the beginning of the Phanerozoic era during the Cambrian. Today, the Bilateria constitute the great majority of animal phyla, consisting of about thirty. Their range of morphologies is considerable: for example, flatworms, mollusks, insects, fish, and humans are all bilaterians. Nevertheless, the Bilateria can be divided into three large constituent groups, each embracing several phyla, hence *supraphyla*, based on shared developmental characteristics between the member phyla. These have been named the **Ecdysozoa**, the **Lophotrochozoa**, and the **Deuterostomia**. The diversification of these three major groupings into their component phyla probably took place in the singular 20-million-year period previously mentioned, between 530 and 510 million years ago, and perhaps much of it to within an even shorter interval, 5 million to 10 million years.[5]

Our focus in this book will be on the vertebrates (or *Craniata*), a subphylum of the **Chordata** (the only one shown in bold in the figure), which itself is a member of the supra phylum the **Deuterostomia**. The Chordata

BOX 5.1 Basic Principles of Molecular Phylogenetics

The basic fact underlying molecular phylogenetics, whose aim is to reveal evolutionary relationships from macromolecular sequences—proteins, RNAs, or DNAs but today primarily DNAs—is that of the molecular clock: each DNA sequence for a specific genetic function is a stable piece of chemical information, but with time and over many generations, it slowly changes with mutational events that accumulate and are passed on. The greater the amount of evolutionary time elapsed, the greater the cumulative amount of change. By comparing the same DNA sequence—for example, the sequence of the same gene from distantly related animal species—we can, with appropriate analytical methods, form a hypothesis as to their evolutionary relatedness.

The procedure can be likened to successive rounds of copying of a manuscript. Imagine an ancient text in an early medieval monastery that is independently copied by three different monks in three different monasteries. If the text is sufficiently long, mistakes in copying are bound to occur, but these will probably be different in the three initial copies. When each of these first copies is subjected to further rounds of copying in each monastery, the original mistakes will probably be perpetuated and new ones will occur. After fifty to one hundred generations of copying, were all the manuscripts to be collected and then placed before an analyst without telling her which manuscript came from which monastery, it should be possible for her to closely compare the texts and determine the three "families" of manuscript, corresponding to the original three copies. Indeed, if early in the history of this copying exercise, one copy was transported to a fourth monastery and the cycle recommenced there, it should be possible to distinguish the copies derived from it as distinct from the other three sets. The analysis should reveal a *phylogenetic tree* for all of the copies. Every phylogenetic tree—whether of manuscripts or DNA molecules—should be seen as a hypothesis but one that can be tested as new data (new sequences) become available.

As with molecular clock analyses, the principle is basically straightforward, while the implementation is complicated, with about four different major strategies possible, each with its variants. An enormously large scientific literature now exists on the different approaches to phylogenetically reconstructing relationships from molecular sequences, and the strengths and shortcomings of each method.[1]

---

1   There is a vast literature on cladistics, but a good place to start is the classic review by Hennig (1966). An authoritative modern text is that of Felsenstein (2004), and an older but useful practical guide to molecular systematics is that of Hall (2001). An excellent if older introduction to the principles and basic strategies of molecular phylogenetic analysis is to be found in Li (1997), chapters 5 and 6.

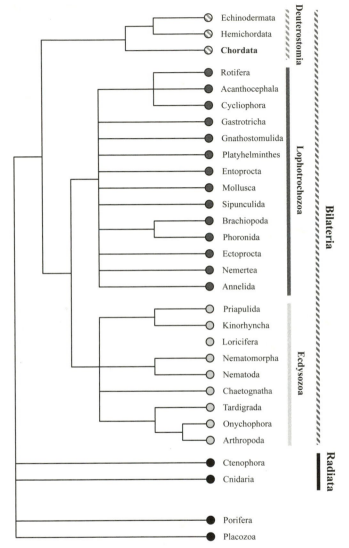

FIGURE 5.4 Phylogenetic tree of the Animal Kingdom. Of the thirty animal phyla indicated here, the focus of this book is solely on one subphylum, the Craniata (Vertebrata) of the Chordata (in bold). The earliest-arising animal phyla were probably the Placozoa, Porifera, and the two phyla of the Radiata, the Ctenophora and the Cnidaria. Some fossil remains of both Porifera and Radiata are found in the late Neoproterozoic. Most animal phyla belong to the complementary supergroup, the Bilateria, all of whose members display bilateral body symmetry at some point in their lives and whose embryos have three germ layers (ectoderm, endoderm, and mesoderm). The Chordata are one phylum in the superphylum Deuterostomia, whose early embryos share two openings, in contrast to those of the two other bilaterian superphyla, the Ecdysozoa and the Lophotrochozoa.

A.

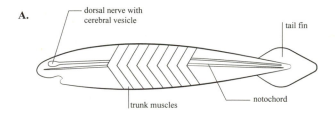

B.

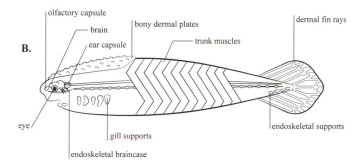

C.

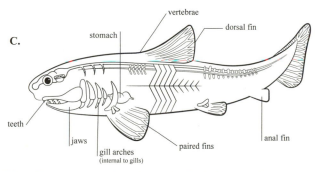

FIGURE 5.5 A. Diagram of a hypothetical early chordate lacking paired sensory apparatus in the head. **B.** Diagram of an early craniate; the striking difference to the other chordate subphyla is in the presence of a true head bearing paired visual organs (indicated) and olfactory organs (not shown). **C.** Picture of a hypothetical early gnathostome possessing jaws. (Modified from Maisey 1996.)

have four diagnostic shared features. The first is a thin, hard element running the length of the animal, termed the *notochord*, from which the phylum derives its name. The second is the dorsal neural tube, which runs parallel to and above the notochord and develops in vertebrate embryos, as we have seen, into the brain and spinal cord. The third is a set of blocks of muscle, *myotomes*, running the length of the body in the larval or adult swimming forms of the chordates. The fourth defining chordate feature is the visually most obvious: a tail that stretches beyond the anal opening, the *postanal tail.* Some of the chordates do not exhibit all these features in all their life stages, but all show these traits at some point in their life histories.

A depiction of a hypothetical early chordate is shown in diagrammatic form in Figure 5.5A. It is envisaged as being similar in structure to today's cephalochordates, another subphylum of the Chordata. Though the prefix *cephalo* in Cephalochordata means "head," the head of a cephalochordate is a modest thing, containing a rudimentary brain (a small bulge at the anterior end of the dorsal nerve cord) and a small mouth and a single, central eyespot for rudimentary visual guidance. There is no face in these animals nor in the other main invertebrate chordate group, the urochordates or *tunicates*, although the larval forms of the latter have larger anterior bulges that can be termed heads.

While the existence of the postanal tail is the easiest identifying external feature of the Chordata, the defining trait of its subphylum, the Vertebrata, or at least its early members, is the presence of a true head. It is that structure that gives the Vertebrates their alternative (and preferable) name, the Craniata, as mentioned previously. Comparing the early chordate diagrammed in Figure 5.4A with that of an early craniate in Figure 5.5B, the differences in relative size and complexity of the anterior ends of the bodies of these animals are obvious. Thus, any attempted understanding of the evolutionary origins of the vertebrates should focus on how the head evolved. That was the starting point of a theory put forward in 1983 by two American zoologists, Glenn Northcutt and Carl Gans. This theory replaced what had been a complete mystery, albeit one largely unrecognized as such, with a concrete idea that revolutionized the subject.[6]

The New Head Hypothesis of Vertebrate Origins

The heads of the simplest living vertebrates, the hagfish and the lampreys, are already far more complex than those of contemporary cephalochordates or urochordate larvae, so it is difficult to envisage how vertebrates could have initially evolved from animals of either type. The "new head hypothesis" of vertebrate origins, as the idea of Northcutt and Gans is known, provides a first-order explanation and is rooted in two central premises.

The first is that the earliest vertebrate-like animals actively sought out their food as opposed to the presumed filter-feeding mode of the adult forms of the earliest chordates. As its name suggests, that latter form of feeding involves passive filtration of water coming through the pharyngeal gill slits to obtain small nutritious particles, whether unicellular organisms or small pieces of organic detritus. A life based on filter feeding does not require much motility or fast movement and, correspondingly, many such feeders, which are found within most marine animal phyla, are either sessile or move relatively slowly. In contrast, craniates actively seek their food, and this requires efficient motility. It seems highly likely, therefore, that fins evolved fairly early in the craniates. Hunting, of course, also requires a mouth suitable for ingesting prey. Such a mouth must be built of strong materials, such as the pharyngeal basket of cartilaginous arches seen in the lamprey, a present-day primitive vertebrate. The greater activity involved in hunting, relative to filter feeding, would also have required more efficient respiration to supply the requisite energy demands, and this would have been facilitated by vascularization of the pharyngeal arches to increase the flow of water through the pharynx to secure more oxygen.

In addition, a hunting animal needs sensory apparatus for efficiently finding prey. Such efficiency requires broad sensory scanning of the environment. For a bilaterally symmetric animal, that requires sensory apparatus on both sides of the head, hence bilaterally symmetric sensory organs, in particular eyes and olfactory apparatus. Thus, if an animal is to be a successful hunter, it needs a face. It is possible that the first craniates were herbivores, grazing on algae, but the earliest unambiguous vertebrate fossils suggest that predation (presumably on more stationary animals) was probably the main source of nutrition.

Finally, and not least, a hunting animal requires a neural system that is complex enough to process the incoming sensory information from the face as well as direct the movements of the animal, especially its mouth, toward the prey. Such a neural-processing system should be fairly close to the sensory organs that supply the information. Such a center is termed a *brain*, and a relatively large brain compared to that of any of the nonvertebrate chordates is a defining feature of vertebrate structure, one that must have been present at the origins of the group. Correspondingly, it is surely not coincidence that the major groups of animals that employ active, locomotion-based predation to obtain food—namely, the arthropods and the vertebrates—have both faces and relatively large and complex central neural-processing centers: brains. Similarly, the only kinds of mollusk that hunt are the octopuses and the squids, highly motile animals, which possess large sophisticated eyes and complex brains, in striking contrast to their more sedate, nonhunting molluskan cousins. The larger brain of an early craniate—that is, relative to a chordate precursor—is indicated in Figure 5.5B.

Of course, no set of complex structural innovations of this sort could have evolved without selective pressures that favored them. These selective pressures, or perhaps one should say selective incentives, for obtaining one's nutrition through predation would have been strong. The opportunities for predation would have multiplied as the potential food sources—numbers of prey species—increased. Given the increases in both animal species and their total biomass during the Cambrian and into the Ordovician, it seems likely that *primary production* of plant material was increasing during both geological periods. Such increases would have favored a flourishing of herbivore species, but those increases, in turn, would have potentiated increases in the numbers (and biomass) of predator species.

While the key role of predation in shaping craniate beginnings was the first element in the new head hypothesis, its second and even more innovative element involved its venture beyond traditional evolutionary concerns of adult structure and selective pressures into the realm of development. The question that Northcutt and Gans posed was, What are the tissue sources of the new structures located in and required for constructing a hunting animal's head? They identified three. The first is mesodermally derived tissue that contributes to the muscles of the pharynx; as such, it makes possible more powerful and active pumping of water for obtaining oxygen. It also contributes to the muscles in the vertebrate trunk.

The other two, however, are innovations unique to the Craniata and essential for developing the unique structures of the craniate head and face. At this point, neural crest cells will be familiar to the reader, but the **neural ectodermal placodes** mentioned earlier, may not be. These consist of bilaterally symmetric thickenings on the outer ectoderm of the head that give rise to the unique paired sensory structures of the vertebrate head—in particular, the eyes, the olfactory apparatus, and the auditory apparatus. The unique structures of the vertebrate head that arise from the neural ectodermal placodes are indicated diagrammatically in Figure 5.6A. Both the neural crest and the neural ectodermal placodes develop from the border area of the neural tube, the neural crest cells developing on the inside (neural side) and the neural placode cells developing only in the cra-

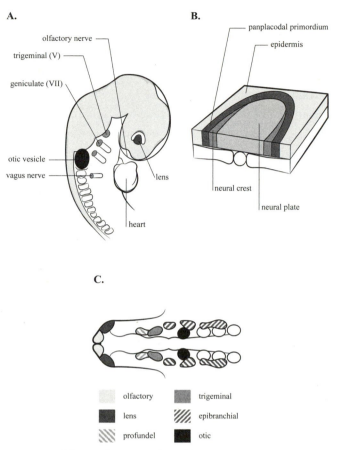

A.

olfactory nerve

trigeminal (V)

geniculate (VII)

otic vesicle

vagus nerve

lens

heart

B.

panplacodal primordium

epidermis

neural crest

neural plate

C.

olfactory

lens

profundel

trigeminal

epibranchial

otic

FIGURE 5.6 **A.** Some of the neural structures derived from those placodes; the roman numerals indicate different cranial nerves. The trigeminal (V) innervates the maxillary and mandibular regions; the geniculate (VII) innervates the mimetic muscles; the olfactory nerve from the olfactory placode; the eye lens from the lens placode; the otic vesicle, precursor to the hearing apparatus, comes from the otic placode; the vagus nerve (X) innervates the art and gut and derives from the vagus placode. **B.** A diagram of the spatial relationship of the panplacodal region (from which the neural ectodermal placodes arise) and the preneural crest region in the embryo. (Modified from Schlosser, 2008.) **C.** Longitudinal horizontal (coronal) section through a later-stage vertebrate embryo showing the bilaterally symmetric placement of the neural ectodermal placodes.

nial region, adjacent to but external to the neural crest cell region of the neural tube. The two respective regions of origin of the neural crest cells and neural ectodermal placodes are diagrammed in Figure 5.6B, and the sites of the different placodes in the early embryo are shown in Figure 5.6C.

The crucial general point with respect to vertebrate origins is that these two kinds of cells are found only in the craniates. Given their essential roles in forming the head, the evolutionary origins of the craniates must be intimately connected to the origins of these cell types. That was the dramatic new element in the theory: that the vertebrate head is not an elaboration of the simple heads of cephalochordate or urochordate-like larva but truly a new head, a novelty generated by evolution. At its inception, that new head would undoubtedly have been both simpler and smaller than the majority of vertebrate head types today, but it was unquestionably more complex than the anterior end of any other type of chordate. Although the new head hypothesis was seen as revolutionary a little more than three decades ago when first proposed, it is taken as obviously true today. Yet it raised new questions at its inception, and they are still debated. We look at those matters next.

Three Unresolved Issues

The first unresolved issue concerns the precise form of the ancestor from which craniates arose. For more than a century, the cephalochordates, the genus *Amphioxus* or, less formally, *lancelets*, had been considered a good model of the probable vertebrate ancestor (Figure 5.5A). DNA phylogenetic analysis, however, has overthrown that conclusion; it is now accepted that the Urochordata are the most closely related group, or **sister group**, in the parlance of the field. The Urochordata or tunicates in their sessile adult forms, however, resemble neither vertebrates nor their larval forms as closely as does the lancelet. At present, one can only infer that the ancestral or *stem* craniate probably did resemble the lancelet even if the present-day cephalochordates are more distant relatives than the urochordates.[7]

The second issue raised by the new head hypothesis concerns the nature of the ancestral cell types that were the forerunners of both the neural crest cells and the neural ectodermal placodes. Since the neural crest cells and placode cells arise, respectively, from the dorsal edge of the neural tube and the ectoderm immediately adjacent to it (Figure 5.5B), both cell types almost certainly originated from cells in comparable positions in the chordate ancestor(s) of the vertebrates. If so, then there should be clues to those relationships from gene-expression patterns. Such clues have been sought—and found.

Recall that the genetic network that underlies neural crest development has been well characterized in terms of its component modules and their constituent genes (Figure 3.3). Those modules consist of groups of interacting genes that, respectively, (1) produce chemical signals for induction of the neural plate; (2) specify the borders of the future neural crest cell territory; (3) determine neural crest properties within the delimited dorsal region, genes known as *neural crest specifiers*, including their ability to migrate within the body; and (4) participate in differentiation of the descendants of the migrated neural crest cells into the particular kinds of cell types that ultimately develop from them. In recent years, tests for expression of the corresponding, or **homologous**, genes in the neural tubes of both the cephalochordate *B. floridensis* and the urochordate *Ciona intestinalis* were made and revealed striking similarities between these two species and craniates in expression pattern in and around the developing neural plate and neural tube, as well as—of course—differences.[8]

In one species of the Urochordata, cells with neural crest-type properties have been found. These cells are found in the larval form of the tunicate *Ecteinascidia turbinate,* their development being followed by injection of cell-tracing molecules into these cells, a marker that allows tracing both these cells and their descendants. These cells, termed *lateral trunk cells*, arise near the dorsal neural tube and migrate ventrally, the first ones from the anterior part of the animal, followed by more posterior ones. These cells give rise to orange-pigmented cells, which are outwardly similar to the vertebrate pigmented cells known as *erythrophores* which, like the better-known pigment-producing melanophores, derive from neural crest cells. In principle, this resemblance between the urochordate pigmented cells and vertebrate erythrophores might be accidental, a so-called **homoplasy**, brought about by coincidentally similar evolutionary paths instead of an evolutionary relationship, but the urochordate cells express several characteristic neural crest specifier and differentiation genes, which strengthens the case for their relatedness to neural crest cells.[9]

In effect, it seems likely that the cell types that evolved into the neural crest cells were located like true vertebrate neural crest cells in the border area of the neural tube in the ancestral chordates. What then was lacking in the ancestral cells of neural crest cells that prevented them from being true neural crest cells? That missing element was probably the molecular machinery that makes motility and cell migration possible in mesenchymal cells. Genes underlying that capacity were probably present in chordates for some limited cellular movements; if so, they could have been employed in a new capacity to endow protoneural crest cells of early chordates in the Cambrian, with true neural crest cell-type properties. Such **gene recruitment** (or *gene co-option*) of preexisting genes for novel developmental functions in new cellular or tissue locations has been a common element in evolution. The epithelial–mesenchymal transition that true neural crest cells undergo and their migration are strikingly reminiscent of metastasizing cancer cells, and it is possible that some of the same genetic machinery is involved by means of a transient reactivation in cancer cells.[10]

In contrast, the cells that give rise to the neural placodes do not undergo an epithelial–mesenchymal transition nor become migratory, although some of the cells they give rise to leave their locations—a process termed *delamination*—and move short distances. Like the neural crest cells, however, the neural placodes give rise to a large diversity of cell types, including different kinds of neural cells and neurosecretory cells. The molecular genetic comparisons between the regions in the neural tube that give rise to neural crest and placodal cells, respectively, indicate that the two cell types share significant molecular-genetic properties as well as exhibit critical differences. Future comparisons to comparable regions in the dorsal regions of neural tubes of nonvertebrate chordates should provide further clues to the evolutionary origins of both the crest and placode cells.[11]

The neural placodes come in pairs, distributed symmetrically on the two sides of the cranium (Figure 5.6C). Given the known role of sonic hedgehog (SHH) in growth and expansion of the face, it may be that in the evolutionary transition to vertebrates an expansion of the SHH domain in the head region contributed to the bilateral symmetry of the neural ectodermal placodes. In vertebrates, the precursor structure for the eyes is a single eye primordium in the early embryo. It splits into two parts, under the growth-and-development promoting influence of SHH, and those daughter elements develop into the lenses of the two eyes. In contrast, cephalochordate larvae possess a single eye primordium that does not split and develops directly into a single eye spot. Comparably, in the development of vertebrate embryos, if there is a strong deficiency of SHH in face development, the midface fails to develop and the primordium does not split, with the development of the single-eye or cyclopia condition, the extreme version of the condition of holoprosencephaly discussed earlier.

Indeed, that anterior expansion of SHH, accompanied by a similar anterior expansion of Fgf8 expression, might have been responsible for another crucial event accompanying or perhaps even driving the development of the new head: the expansion of the anterior neural apparatus into a much larger one to form a true brain (compare Figure 5.5B with 5.5A). This idea is supported by several recent molecular studies comparing gene expression between nonvertebrate chordates on the one hand and lampreys, one of the two most primitive craniates alive today, on the other. This work employed genes whose expression is diagnostic of different regions in the vertebrate brain and indicates that the small cerebral or sensory vesicles of nonvertebrate chordates have diencephalic qualities by these molecular criteria. Thus, formation of the telecephalon might have been the crucial element in the origination of the new head. Additional results support this idea: while high levels of expression of Wnts in the developing brain promote diencephalic development, lower levels accompany telencephalon development, which exhibits strong anteriorized expression of at least one inhibitor of Wnt expression, particularly in the early developing telencephalon. The telencephalon, with its production of SHH and Fgf8, could well in turn have promoted the formation of the facial primordia in those early craniate ancestors, just as it does today in living vertebrates.[12]

A third general question about neural crest cell origins is, How did these cells acquire their range of developmental and differentiation capabilities? A particular puzzle concerns their acquisition of the capacity to give rise to skeletal-type elements. Such features, including the bones of the face, generally arise from so-called connective tissues that derive from embryonic mesoderm. Neural crest cells, however, develop from the neural tube, whose origin is the primary embryonic ectoderm. In effect, neural crest cells seem to have acquired some mesodermal properties in their evolutionary origins. In principle, this could have involved gene-recruitment events from mesoderm to neural crest cells. One specific possibility involves the recruitment of specific genes for making cartilage. Traditionally, it has been believed that no nonvertebrate chordates make cartilage, the original vertebral skeletal material. However, because the genes of the genetic regulatory network (GRN) responsible for cartilage production in neural crest cell derivatives are now known, it has been possible to look for them in simpler chordates. One study found that all the major genes of the cartilage GRN are present in the amphioxus genome and are expressed in various tissues—in particular, mesodermal tissues. The authors concluded that the cartilage GRN of neural crest cells had been assembled in a series of gene recruitment events to protoneural crest cells during the origins of the craniates. A more recent study, however, has reported that cartilage does exist, though only transiently, in the oral skeleton of amphioxus. This implies that cephalochordates at least possess a cartilage GRN, one that could have been recruited to protoneural crest cells in early craniate evolution. Crucial transcriptional regulators of the cartilage GRN in vertebrates are in the *SoxE* gene family, and these same investigators report that vertebrate neural crest *SoxE* genes have acquired distinctive *cis*-regulatory enhancers for turning or keeping this gene "on" in neural crest. Such persistent activation of *SoxE* activities could keep the cartilage GRN switched on in those neural crest cell lineages that produce cartilage such as the facial prominences.[13]

Despite the current gaps in information about the origins of the craniates, it is even clearer today than it was thirty years ago that the new head hypothesis was a major breakthrough in understanding the origins of the vertebrates. Yet while the earliest craniates were undoubtedly far more efficient searchers for food than their chordate precursors, they were still relatively limited in their feeding capacities. What they lacked were efficient, movable jaws; these evolved in the next major stage in vertebrate evolution. The advent of jaws may seem a rather trivial structural change compared to the development of a head and a face, but it was actually highly significant. It was probably the powerful increase in feeding efficiency made possible by jaws that, eventually if indirectly, made possible the rise of vertebrates as dominant animals in the food chain. Had vertebrates remained jawless, the world would almost certainly be an entirely different place today, not least in the fact that we would not be here to observe and alter it.

The Advent of Jaws: From Agnathans to the First Gnathostomes

Up to this point, we have discussed the origins of the vertebrates without dealing with their probable time of origin. As will be discussed in the next chapter, precisely dating the time of origin of any major new animal group is impossible, and the difficulties become progressively greater the further back in time we go. Nevertheless, the fossil record hints at the approximate time of origins of the craniates. Thus, two putative early species approximately an inch long have been tentatively identified from "upper" Cambrian strata in China and date to about 500 to 490 mya. Their classification depends on their possession of somewhat enlarged heads and some anterior ventral fretwork interpreted as pharyngeal baskets, a vertebrate-specific neural-crest–derived characteristic. The interpretation of these fossils as the earliest craniates is not universally accepted, however, and some paleontologists regard them as noncraniate chordates, differing only by degree from some of the clearly identified ancient chordates.[14]

By the early Ordovician, however, roughly 480 mya to 470 mya, there are fossils of larger and unambiguous jawless fish, this period marking the latest dates for the origins of the vertebrates. By the late Ordovician, circa 445 mya, there were several different kinds of **agnathans**, many reaching more than a foot in length, some slightly over two feet, and abundant in the seas of that time. Though they lacked an internal bony skeleton (possessing almost certainly a cartilaginous one), their fossils are well preserved due to their possession of large plates of mineralized, skeletal material—*dermal armor*—that overlay their skin and conspicuous hard, bony shield-like heads. Though no longer recognized as a unified taxonomic category, these fish were long termed *ostracoderms* (literally, "bony skins").

The first jawed animals or **gnathostomes** apparently did not arise until the late Silurian or early Devonian, hence in the period 425 mya to 410 mya, or roughly 75 to 90 million years after the first craniates appeared. A picture of a hypothetical—and idealized—early gnathostome is shown in Figure 5.5C. The early gnathostomes were undoubtedly more complex in multiple aspects of their body architecture than the first craniates (Figure 5.5B) but the critical change was the addition of jaws. The first gnathostomes were probably small, like the early chordates and first craniates, but like the abundant agnathans, the earliest gnathostomes to leave fossils were medium-sized and also had dermal armor, permitting ready fossilization. In addition, and perhaps unlike the very earliest gnathostomes, they had internal bony skeletons. Once they appear in the fossil record, however, their fossils become abundant in later strata, indicating that there must have been a strong selective advantage in having jaws, permitting their rapid expansion and diversification. By the late Devonian, circa 370 mya to 359 mya, the gnathostomes had largely displaced the agnathans to become the predominant craniate type in the oceans. Many of

these species were large too, up to six feet in length, perhaps an indication that jaws permitted more efficient nutrient intake and consequently
increased body size.

The nature of the biological innovations that led to the first gnathostomes is becoming clearer in light of recent comparative molecular studies.
The strategy is to compare the embryonic development of the mouth area
in the embryos of lampreys, a living agnathan, hence a stand-in for the
putative agnathan ancestor, with that of two present-day gnathostomes,
the chick and the mouse. By looking at the structural changes in the jaws
that must have taken place in conjunction with examining the patterns of
expression of potentially important genes, it is possible to infer what gene
changes in development might have led to the evolution of jaws in the agnathan ancestors of the gnathostomes. The strategy relies on the large
degree of conservation of gene-expression activities among different vertebrates; differences in such expression can often be informative about
developmental differences.

A diagram of the lamprey head is shown in Figure 5.7. In contrast to
the gnathostomes, its mouth has limited mobility and consists of upper
and lower "lips," these consisting of hard skeletal material that is unlike
mammalian fleshy lips. The lamprey uses its lips to make contact with prey
fish and then suction to fasten on to prey flesh. Then its sawlike teeth just
inside the lips start to work on the victim. Just behind the teeth lies a series of U-shaped cartilaginous bars, continuous on the ventral side and
open dorsally, the so-called pharyngeal bars or *branchial basket*. As in gnathostomes, these structures are products of cells derived from neural crest
cells, wrapped in endodermal coats, and are numbered from the front
(anterior), the first being termed the *mandibular arch*.

There are two broad views about how jaws originated, but both share
the premise that the mandibular arch of the agnathans was the precursor
of the mobile lower jaw of the gnathostomes. The classical view is that the
mandibular arch as a whole possessed some sort of genetic-molecular
property that provided it with a potential to form a joint in its middle, allowing its lower half to extend forward and thus become the basis of the
lower jaw. One such distinguishing feature is already known: recall that
the mandibular arch differs from its more posterior branchial arches in
mammalian and bird embryos in that it is the only one that does not express *Hox* genes in its development. Whatever the first event that created

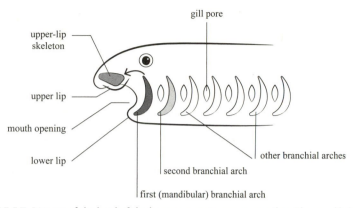

FIGURE 5.7 Diagram of the head of the lamprey, a contemporary agnathan. The mandibular arch of the lamprey is considered to be the precursor of the gnathostome lower jaw, but its relationship to the upper jaw is controversial; see text for description of the debate. (Modified from Mallatt 2008.)

a protojoint, the lower half of the newly jointed mandibular arch could then have extended forward, perhaps involving a sequence of mutations of individually small effect, each promoting a quantum of forward growth to become the cartilaginous precursor of the lower jaw. In this hypothesis, the upper half of the lamprey mouth also extended forward in a comparable fashion to become the upper (maxillary) jaw in the gnathostome lineage, being driven perhaps by selection for a stronger bite involving both upper and lower jaws. In this scheme of jaw evolution, the upper jaw of gnathostomes is simply a modified upper lip of ancestral agnathans.

In one of several alternative scenarios, the emphasis is on the formation of the joint within the mandibular arch. It focuses specifically on the genetic changes that might have generated a joint and relies on comparison of different gene-expression patterns in lamprey versus gnathostome embryos. In these schemes, the key events are shifts in gene-expression patterns that create a joint in the mandibulary arch. Subsequently, there would have been forward growth of both upper and lower halves to become, respectively, the upper and lower jaws. In this view, the forward-growing upper half (the future maxillary) would have fully replaced the upper lip of the lamprey-like ancestor.[15]

Specific molecular evidence supports the second theory. In particular, there are shifts in expression of signaling proteins—specifically, FGF8 and some bone morphogenetic proteins—between lamprey and gnathostome embryos, that may be involved in forward-directed outgrowth of the mandibular arch. In addition, two transcription-factor genes that are involved in joint formation in other points of the gnathostome have been found to be expressed in the early mandibular arch in gnathostome embryos but not in that of lamprey embryos. The first is *Bapx,* a known regulator of joint formation in vertebrate embryos; the second is a critical gene for joint formation that *Bapx* regulates, *Gdf5.* The complete set of findings suggest that these genes would not by themselves have directly specified the key new joint but would have done so in combination with other gene-expression differences along the dorso-ventral (top-bottom) axis along the ancestral mandibular bar. In so doing, they would have laid the conditions for development of the initial joint in the mandibular bar.[16]

If a movable lower jaw did originate in this way, the process would have involved the recruitment of expression or one or two genes to a new site in the developing embryo. This particular event must have been an unusual and unlikely one given the existence of a 75- to 90-million-year gap between the first agnathans and the first gnathostomes. The DNA sequence evidence from living species indicates that all gnathostomes derive from a single ancestral species; in the language of evolutionary biology, they are a **monophyletic** group. The evolution of jaws was thus not a highly probable event—in which case, there might have been several independent originations—but a fortuitous and improbable one.

With the coming of and successful establishment of the gnathostomes, the basic structure of the vertebrate face was complete. The evolution of jaws opened up a new world of feeding possibilities and consequently evolutionary potential. In the terminology of evolutionary biologists, it was a **key innovation**—namely, an evolutionary change that facilitates a major adaptive change that in turn permits a major expansion and diversification of the organisms bearing this new feature, an **adaptive radiation**. In the case of jaw evolution, its success is evidenced in the fact that the agnathans, a highly diverse and successful group of animals, began to fade away as the gnathostomes increasingly dominated the seas of the world. In the evolutionary history of the craniates, the advent of the jaws was only second in importance to that of the initiation of the vertebrate head itself. In a sense, perhaps, it was more important. Had craniates never moved beyond the agnathan "grade," they would probably not have become the top predators, dominating the food chains in both the seas and on land.

Their success is further testified to by the retention of the basic gnathostome "ground plan" for head and face architecture in all subsequent vertebrate evolution.

Conserving the Basic Vertebrate Face Plan While Evolving Significant Changes in the Details

It is striking how little the vertebrate face has changed in general structure from the first full-fledged gnathostomes of the Devonian period to those of today—namely, all contemporary vertebrates apart from the handful of agnathan species. Indeed, that *conservation* of craniofacial form was one of W. K. Gregory's major points in *Our Face from Fish to Man,* in which he emphasized humankind's evolutionary connectedness to the animal world. Relative to the dramatic changes in physiology, metabolism, and overall body structure that took place as vertebrates colonized the land, the basic structure of the craniate head has remained remarkably stable over 400 million years of evolution, which has also been noted by others.[17]

That strong conservation of craniate head structure was accompanied by a vast number of changes in the details of the head in all vertebrate lineages (fish, amphibians, reptiles, birds, mammals). Despite Gregory's ringing affirmation of how much had stayed the same in vertebrate heads, his book was actually devoted to describing all the interesting differences that had taken place in the evolution of those lineages from the early gnathostomes to humans. These changes included simplification of the development of the skull, in both neurocranium and jaws, via fusion of smaller bones into larger ones; development of the mammalian hearing apparatus from changes connected to the changes in jaw structure, in particular the movement into the head of two of the smaller jaw joint bones; and in the mammals specifically, an increase in the complexity and diversity of the teeth, the latter resulting in a tremendous variety of mammalian dental patterns for different diets. A wealth of information is available about all these changes, but in order not to lose sight of our subject—the evolutionary track that led to the human face—what follows is a sketch of the principal evolutionary changes along that path.

From Marine Gnathostomes to the Tetrapods to the First Mammals

In older textbooks on biology and evolution, the great transitional event in vertebrate evolution—from life in the sea to life on the land—was often described as the "conquest of the land." It required two essential major modifications of vertebrate physiology and anatomy; these adoptions led to many more modifications. First, there was a change of respiration from extracting oxygen from water via the gills, supported and moved by the pharyngeal basket, to breathing air directly. The second involved the evolution of limbs from fins to permit movement on land. Neither transition was simple or rapid, but the eventual outcome was the appearance of the first land-dwelling animals—namely, the earliest amphibians—by the late Devonion period, roughly 360 million years ago.

The phrase "conquest of the land," however, evokes an image of fish-like creatures massing in large numbers and making an assault on the new territory, scoring a decisive if not necessarily quick victory. As with military conquests, there is the implication that somehow the new territory was a particularly desirable new field to conquer. The reality, however, was surely different: the land was a hostile environment for newcomers from the sea, although by the Devonian, there was extensive plant cover and insect life where the land met the sea, making it more hospitable than it had been previously. Altogether, the sequence of steps that led to the evolution of land-dwelling vertebrates seems to have taken place over an interval of 20 million to 30 million years. Furthermore, judging from the fossil skeletons, it was initiated by only a handful of species out of many thousands of types of marine-dwelling gnathostomes. Indeed, there was nothing about the general marine environment that would have spurred its vertebrate inhabitants to flee. Water was, and remains, a perfectly good environment for those vertebrate species that can extract oxygen from it directly: approximately half of all living vertebrate species today—about 25,000—are fish, although a disproportionate number are freshwater species, perhaps because separated freshwater environments are more conducive to evolutionary diversification than the marine environment.

Thus, the move to land by the relatively small number of animal lineages that made it was certainly a pioneering move, but if we let ourselves anthropomorphize slightly, it was a forced and reluctant one in which the pioneers were making the best of difficult circumstances. They found themselves in marginal environments, coastal or estuarine, marshy habitats, where locomotion over the substratum in the water was more efficient with limblike appendages than fins. The fossils testify to the relative gradualness and complexity of the evolution of this new form of movement. Furthermore, the adaptation was almost certainly driven by the need to survive in situations where the watery environment would temporarily disappear. The lungfishes, for example, can survive for long periods by burying themselves in mud and undergoing metabolic arrest there, illustrating one solution to the problem of such marginal environments. The real ancestors of the first land dwellers, however, were almost certainly the *sarcoptyrigian* or lobe-finned fishes, of which the coelacanth, a species thought to be extinct for more than 50 million years but discovered as a living species in 1938, is a good example. Their appendages were not cartilaginous fins but fleshy, with muscles, permitting movement on solid substrates such as the muddy bottoms of estuaries. It was these fishes that, again, if we permit a touch of anthropomorphism, are the heroes of the story, however slow and tentative their first (literal) steps into the new environment were.

The long sequence of evolutionary events by which vertebrates came to inhabit the land, however, involved two major and distinct stages. The first, just mentioned, involved the initial invasion of the land by the sarcoptyrigian lineage, leading first to protoamphibians and then to true amphibians. It started in the mid-Devonian, probably about 390 mya to 385 mya, and was complete by sometime early in the following period, the Carboniferous, perhaps 355 mya to 350 mya. Certainly, amphibians were increasingly common during the Carboniferous. Those early land-living pioneer vertebrates were the founding group of what are designated today the **tetrapods** (literally, "four-footed" animals).[18]

In addition to the evolution of true limbs from the sarcoptyrigian lobe-fins to permit efficient movement on land, the transition also involved the evolution of true lungs. The sarcoptyrigian pioneers almost certainly had rudimentary lungs, just as the aptly named lungfishes do, and early amphibians probably had some capacity for taking in oxygen through their skin as modern amphibians do. Yet, despite limbs and lungs, the first amphibians were tied to regions that possess ample freshwater to lay their eggs, as the great majority of living amphibian species are today. To ensure survival of an adequate number, large numbers of eggs are produced and laid in water; this means that the newly hatching juveniles cannot be too complicated in their structure but sufficiently complex to be free-living, swimming entities—namely, larvae (tadpoles).[19]

A subsequent evolutionary change, however, is what liberated one line of the vertebrates from those constraints and made possible the real conquest of the land. This entailed a key innovation that sounds relatively simple and trivial: the development of a new membrane surrounding the egg, an **amnion**. It is from the possession of that structure that all land-living vertebrates—reptiles, birds, and mammals—take their designation as *amniotes*. The amnion simultaneously reduces the rate of evaporation of water from the egg and permits the diffusion of oxygen, directly from the atmosphere, into the interior of the egg to permit the development of the embryo. This permits the construction of larger eggs that, in turn, can support longer and more extended patterns of embryonic development to yield young that are more complex on hatching than the young of earlier vertebrate forms. Most significantly, however, it obviates the need to lay eggs in water. The first amniotes to leave unambiguous traces of their existence in the fossil record lived in the mid- to late Carboniferous period, approximately 320 million years ago, as judged from the fossil evidence. The true time of origin of the first amniotes, however, may have been several to many millions of years earlier.

Within a relatively short period of time after the amniotes arose, they diverged into four branches, of which the two most important in terms of later flourishing were the **diapsids** and the **synapsids**. The feature that distinguishes these two groups seems a minor one: a difference in the number of holes or *fenestrae* (literally, windows) in the temporal (lateral posterior) region of the skull (Figure 5.8). The diapsids are characterized by two such temporal fenestrae on each side, an upper and a lower one, while the synapsids have only one, the lower one (Figure 5.8). Initially, this difference might have had relatively minor functional significance to jaw strength and diet (see the following), yet these two lineages would come to have distinctly different evolutionary fates. The early diapsids would eventually give rise to the dinosaurs and all present-day reptiles—namely, crocodiles, snakes, lizards, and turtles—and ultimately the birds (the only living descendants of the dinosaurs), while the earliest synapsids, initially called *mammal-like reptiles,* founded the lineage that would eventually give rise to the immediate progenitors of the mammals and from them the mammals themselves.[20]

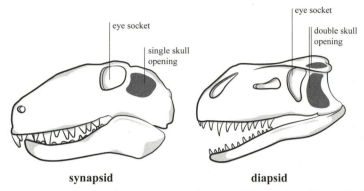

FIGURE 5.8 A comparison of representative early synapsid (A) and diapsid (B) skulls. The former group, possessing a single temporal fenestra, eventually gave rise to the mammals. The latter includes both the extinct dinosaurs (the sauropsids) and their living descendants, the birds, as well as contemporary reptile groups (snakes, lizards, crocodilians, and turtles).

During the period following this split, the late Paleozoic era, the synapsids came to be far more abundant and diverse than the diapsids. The dramatic geological and climatic events circa 252 mya that brought the Paleozoic era and its final geological period, the Permian, to a close, however, produced a mass extinction of both land and marine animals and plants. Indeed, it was the most severe of all the mass extinctions, leaving the land and seas depauperate of animal life for 5 million years. One of its consequences was the elimination of most species of the once-dominant synapsids. The ensuing period, the Triassic, would see a reversal of the relative fortunes of the two groups. During this period (251 mya to 201 mya), it was the diapsids in the form of the **sauropsid** reptiles, most notably the dinosaurs, that became the more diverse and dominant land animals, with the synapsids reduced to minority status and surviving with a relatively small number of lineages. Nevertheless, those survivors carried the seed of the future, giving rise to lineages that would eventually yield the first true mammals. The Triassic featured the first truly complex land ecosystems, with highly diverse vertebrate forms and abundant insect life.[21]

From Early Synapsids to the True Mammals

How did the initial rather reptilian-looking early synapsids give rise to the mammals? It clearly involved many intricate and complex changes and was a correspondingly lengthy process, taking place over approximately 150 million to 180 million years. Its intricacy is indicated by the list of mammalian traits that came into being during this long period: (1) a simplification in the number of bones, both in the head and in the trunk regions; (2) the displacement of two previously positioned bones of the lower jaw, necessary for the articulation of upper and lower jaws, to a more interior position, where they came to function as bones of the middle ear; (3) a corresponding posterior and dorsal expansion of the anteriormost lower jawbone, the dentary, to become the entire lower jaw bone with new connections posterior to the upper jaw to form the jaw joint; (4) development of a secondary palate, separating breathing and eating functions and facilitating manipulation of food within the mouth; (5) a stable internally temperature-regulated metabolism, **homeothermy**; (6) a more erect gait suitable to faster movements for both chasing prey and escaping predators, a shift made possible by homeothermic metabolism; (7) a fur coat or *pelage* along with face whiskers; (8) nourishment supplied directly from the mother's body during embryonic and fetal development (seen today even in the two kinds of egg-laying mammals, **monotremes**, whose females donate uterine fluid to the developing embryos through the porous, leathery egg shell); and (9) a special form of nourishment for the newborn young involving milk.

This set of traits looks initially like a confusing jumble of characteristics, with little connection to one another and with most having no obvious relevance to the evolution of the face. In fact, each one is related to some or all the others in its functional consequences, and all had consequences for the evolution of the early mammalian face, though some quite indirectly. A full account is unnecessary here, but a short explanation may be informative.[22]

As mentioned, the defining skull feature of the synapsids are the paired, single low-temporal fenestrae on each side of the skull and their robust lower bony arches. Those lower ridges permitted a new lower jaw (mandibular) muscle attachment, which in turn made stronger jaw action possible; with it came more possibilities for expanded feeding opportunities. That strengthened cranial support for jaw action might have, in turn, created a potential for upper cranium expansion, a requirement and forerunner for the enlarged mammalian brain. The development of the temporal fenestra, with its lower shelf, thus permitted a growth of opportunities for preying on other animals, in effect an expanded carnivorous diet. Many of the early synapsids, however, were plant eaters.

Whatever its precise functional significance, the single temporal fenestra of the synapsids is what permits a fairly secure dating of the origins of the synapsids. The first unambiguous synapsid fossils have been dated to the mid-Carboniferous, approximately 320 million to 310 million years ago. These early synapsids, termed *pelycosaurs*, were (outwardly) lizardlike creatures showing the characteristic reptilian sprawling gait. Yet some of the earliest pelycosaurs show the first signs of one essential and defining mammalian trait—differentiation of the dentition into different tooth types. Those initial differences were in tooth size, with most teeth in these animals crudely resembling canines, the sharp piercing teeth used to seize prey and deliver the initial flesh-piercing bites. Though the first pelycosaurs were small animals, certain lines achieved moderate sizes, up to two meters or more, such as the iconic early mammal-like reptile, the sail-backed *Dimetrodon* familiar to many museum visitors, though often taken to be an early dinosaur (see **Plate 6**). As tooth complexity evolved further during subsequent synapsid diversification, molars and premolars appeared, permitting more thorough mastication of food, especially of plant material. This, in turn, permitted a further expansion of dietary items and consequently the additional evolutionary diversification of the synapsids themselves, primarily into **herbivores**. During the Permian (290 million to 252 million years ago), some achieved great size.

The more mammalian-like **therapsids** arose during the late Permian, though without displacing the pelycosaurs. Therapsid fossils display limbs placed more directly under the body, producing a more erect posture and a faster gait. Such traits are indicative of both a higher rate of metabolism and an ability to control that metabolism to a set level, the property of homeothermy. The therapsids also showed more advanced, mammalian-like dentition and, correspondingly, faces that looked more mammal-like in their jaws. Judging from their skeletons, the therapsids probably had mammalian-like metabolism and might well have had the beginnings of fur coats.

These features would become even more apparent in the **cynodonts**, a group of therapsids that first appeared in the late Permian but would achieve their real flourishing early in the subsequent era, the Mesozoic, specifically during the mid-Triassic, before they were to a large degree displaced from prominence in the food chain by sauropsids, specifically the early dinosaur species. The cynodont lineage gradually evolved into a range of smaller forms during the Triassic as the sauropsids became the dominant land animals. A cynodont is shown in **Plate 7.** The cynodonts were the direct precursors of indisputable mammalian-like species, *mammaliaforms,* that arose during the Jurassic, probably around 190 million years ago. The mammaliaforms were the immediate precursors of the earliest true mammals, so-called **stem group** mammals.

What did these mammaliaforms look like? The long-held belief about mammals during the dinosaurian era, based on the relative handful of fossils that had been identified up until the 1970s, was that they were small, probably insectivore-like (moles, shrews) creatures, figuratively cowering in the shadows of the various large dinosaurs and probably most being nocturnal, with the darkness of night providing cover. In recent decades, however, a plethora of new fossil findings has overthrown this view. The indisputable mammaliaform species that existed in the mid- to late-Jurassic embraced many different types with morphologies resembling true mammals, which arose from them later. Judging from their skeletal features, in particular their teeth and limbs, these various types included carnivores, beaverlike animals, tree climbers (scansorial forms), small insectivorous forms (as in the traditional view), digging animals, and gliding (flying-squirrel–like) forms. Many of these early lines died out without leaving recognizable descendants, but there were further rounds of diversification, with those lineages again often coming to an end without issue. Evidently, there was great diversification in the late Jurassic and Cretaceous of many different types of mammaliaforms, despite the ecological dominance of the dinosaurs. Nor were all of them small; one beaverlike animal was relatively large, perhaps weighing 10 kilograms or so.[23]

A critical question, however, concerns the evolutionary relationships between the late Cretaceous mammals and modern mammals. Present-day mammals come in three main groups: the monotremes, the marsupials, and the placentals. The monotremes, consisting of one mammalian order, are the smallest group, consisting today of only two kinds, both native to Australia: the platypus and the echidna. They are also the most primitive, laying eggs like the earlier synapsids. The second and third groups, the **marsupials** and **placentals**, respectively, are more advanced, arose later, and are grouped as the *Therians* (beasts). They also look more typically mammalian in their faces, as helped by the presence of external ears or *pinnae,* which frame their faces. The marsupials, also making up only one mammalian order, bear live young but at a very immature state such that late fetal development takes place while the young nurse within a protective pouch on the mother. There are considerably more marsupial species today than monotremes but, within the mammals as a whole, they are still a minority group, and the vast majority of species are native to one continent: Australia. The placental mammals, in contrast, are the largest, most diverse, and most widespread group of mammalian species, with a truly global distribution. Their young develop through the fetal stage in a true uterus, with nutrition provided to the growing embryo and fetus by a special organ, the placenta, hence the name of the group. This mode of development allows the development of young before birth to considerably more advanced stages than those of marsupial young in their mother's pouches. The three major divisions of mammals probably originated in the Jurassic period, and in the temporal sequence of monotremes, marsupials, and placentals, but the precise dates of origin of these lineages remain controversial.[24]

Because our focus is human evolution, we will concentrate on the placental mammals. Altogether, the current taxonomy has eighteen **orders** of placental mammals, of which the Primates—our mammalian mother group—are one. The placental mammalian phylogenetic tree as reconstructed from many molecular studies is shown in Figure 5.9. It reveals four major branches termed **superordinal groups**, each consisting of a group of more closely related mammalian orders. These four superorders have been named the **Afrotheria**, the **Laurasiatheria**, the **Euarchontoglira**, and the **Xenarthra**. Their deduced phylogenetic relationships are indicated in the figure. The greatest number fall into the Laurasia and the Euarchontoglira, which are more closely related to each other than to the other two superorders and are sometimes grouped into a still larger superorder, the *Boreotheria*. Perhaps the most intriguing fact about this DNA-based phylogenetic tree is how poorly its main divisions match the older traditional tree that had been derived primarily from general morphological resemblances. The most startling departure from the anatomical picture is provided by the placental superorder the **Afrotheria**, which embraces such morphologically diverse animals as elephants, hyraxes (which resemble large rodents but possess disconcertingly menacing

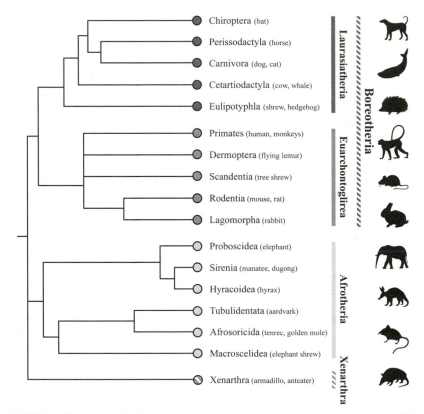

FIGURE 5.9 Phylogeny of the placental mammals. From molecular phylogenetic analysis, the original eighteen Linnaean orders of placental mammals—those whose young receive extended nutrition as embryos and fetuses via a placenta—are now regarded as falling in to four *superorders*: the Laurasia, Euarchontaglira, Afrotheria, and Xenartha. The Laurasia and Euarchontoglira have been more recently grouped in the superorder Boreotheria. The time of origin of the modern or *crown group* orders of the Placentalia, however, whether before the end-Cretaceous extinction or following it, is a matter of debate as discussed in the text. (Modified from Nishihara, Marayama, and Okada 2009.)

canines), manatees (sea cows), and aardvarks. Despite the seeming improbability that such different-looking animals are closely related, there is little doubt that the new placental mammalian phylogenetic tree is correct. The Afrotheria illustrate how evolution can mold closely related lineages of animals along strikingly different paths of morphological change.

The placental mammal phylogenetic tree raises some intriguing evolutionary questions. In particular, how many, if any, of these modern or **crown group** mammalian orders can be traced directly to the late Cretaceous mammals? (A crown group consists of all the modern representatives of the taxon, defined by a set of key traits, plus their immediate ancestral groups. Earlier forms that lack one or more of those key traits would be considered *stem* forms.) Or did the origination and diversification of the modern placental mammalian orders take place only after the large land-based (nonavian) dinosaurs had perished during the end-Cretaceous mass extinction? The fossil evidence shows that mammalian diversity expanded enormously in the early Cenozoic era (the "Age of Mammals") after the demise of the nonavian dinosaurs, but it has long been ambiguous as to when the modern mammalian orders—those indicated in the figure—originated and began to flourish.

Traditionally, there have been three answers to this question. First, that many of the main placental mammalian lineages extend fairly deep into the Cretaceous but for tens of millions of years there was little morphological diversification among them—probably most being small, long-snouted insectivorous species. The new abundant fossil evidence previously discussed effectively refutes that idea of little diversification, which in turn makes it less likely that crown group mammals had roots deep into the Cretaceous among fairly similar-looking ancestors. Second, that most of the mammalian placental orders extend from the Cretaceous but only from its final stages, perhaps 10 million to 20 million years before its end, at 65 mya. Molecular clock evidence, corrected as much as possible for its many anomalies, provides the chief evidence for this. Third, that all eighteen modern placental mammalian orders arose just after the Cretaceous, at the beginning of the Cenozoic era, in the first period of that era, the Paleocene.

The consensus view today is that the second answer is correct: most of the placental mammalian orders originated in the late Cretaceous, but the flowering of diversity within the individual mammalian orders took place in the Cenozoic era after the giant sauropods had vanished. The assessment is based on both fossil and molecular clock evidence, the former indicating that many mammalian groups survived the end-Cretaceous extinction despite massive losses of many species and even higher-level taxons (genera, families).[25]

A dissenting view, however, was put forward by a group of eminent paleontologists in 2013. They claimed that the evidence supports the idea that all modern placental groups trace to the very beginnings of the Cenozoic, in the Paleocene, and specifically to one ancestral species at that time. The analysis involved a thorough retrospective reconstruction of mammalian history from comparisons of living mammalian species as well as fossil evidence of extinct mammals, comparing skeletal and soft tissue traits (the latter solely from living species, of course), trait by trait, and establishing likely phylogenetic relationships step-by-step, progressively back into the past. They used molecular phylogenetic evidence to establish the likely evolutionary relationships but did not include molecular clock evidence for dating, the latter provided solely from the fossils. The analysis permitted a tentative reconstruction of the morphology of the putative ancestor species of the modern placentals; it is shown in **Plate 8.** In this reconstruction, the ancestral placental was probably insectivorous, scansorial (tree climbing), and small, weighing no more than 250 grams and quite possibly much less.[26]

This issue remains controversial. Apparently, there are no fully accepted mammalian crown group fossils that date to before the end-Cretaceous extinction. On the other hand, the absence of fossil evidence has long been known to be problematical for asserting the lack of existence of particular animal types. Furthermore, the weight of the genetic-molecular evidence favors mammalian crown group roots in the late Cretaceous. An uncertainty in evaluating the latter is how much that massive extinction would skew back-extrapolations based on molecular data because much of the mammalian biota, as well as the dinosaurs, were wiped out and the early Cenozoic brought an efflorescence of mammalian diversification.[27]

Evolving Mammalian Characteristics: Four Traits Relevant to the Face

Altogether, the evolutionary journey from the first synapsids to the first true mammals took place over an interval of probably no more than—and possibly less than—150 million years, from roughly 310 mya to 160 mya. In that period, the faces of the animals of this lineage changed dramatically from the dinosaurian-like visage of the pelycosaurs to the fur-covered, mammalian muzzle with differentiated teeth, and erect-eared early marsupials and placentals. Fortunately, we can now probe beneath the morphological changes and begin to identify the specific genetic and molecular alterations that made these traits possible. Here let us consider four general mammalian traits that affect the face and what has been learned about their developmental and genetic foundations.

Mammalian-Type Teeth

At first, we might not regard teeth as a facial feature since they are not visible when the mouth is closed. They are, however, part of the mouth, and the mouth is definitely part of the face. Furthermore, when the lips are parted—as a smile, a shout, a threat, or simply in talking—the teeth visibly and immediately become a perceived part of the face. In humans, they influence one's perception of the person: beautiful teeth always enhance attractiveness, snaggly ones have the opposite effect, and toothlessness, even with the mouth closed, alters the whole appearance of a face, making it seem aged. Were someone's smile to reveal prominent canines, as seen in many mammals—and fictional human vampires—one would recoil in fright. Teeth also influence sound production in speech, further influencing social interactions mediated via the face. Thus, teeth are certainly part of the face, even if concealed most of the time. Of course, the development of the teeth is intimately connected to that of the jaws. Evolution of the jaws both influences the kinds of teeth and their arrangements and is, in turn, influenced by the evolution of the teeth themselves. As we have seen, one of the signature characteristics of synapsids is different types of teeth; in general, mammals possess four distinct kinds of teeth—incisors, canines, premolars, and molars—with each species having a characteristic number of each type on each half of the upper and lower jaws, its *dental formula*. Such tooth diversity was not found in the sauropsids or their descendants.

One crucially important set of genes for the development of the teeth and jaws are members of the *Distal-less* (*Dlx*) gene family, as shown with mice that have had different *Dlx* genes experimentally inactivated. (These genes, it will be recalled, are also important in limb development.) Despite the existence of specific *Dlx* genes and specific types of teeth, however, there is no simple 1:1 relationship between the activities of specific *Dlx* genes and particular tooth types. Instead, it appears that different kinds of teeth reflect the expression of different combinations of *Dlx* genes in the developing jaws, with different patterns and degrees of spatial overlap. The ultimate consequence is the production of different tooth types at specific positions. This relationship has been termed the *Dlx code*.

How might the *Dlx* code produce different kinds of teeth? The central consideration is that each *Dlx* gene encodes a transcription factor. Thus, in principle, each factor is capable of turning on the expression of numerous worker genes. If the different *Dlx* genes encode transcription factors that differ, even slightly in the sets of worker genes they control, then different combinations of the small set of *Dlx* genes will yield the expression of different sets of worker genes. This supposition is reasonable: the gene members of transcription gene families always display some differences in gene sequence, and some of those sequence alterations might affect their activities, affecting which worker genes are turned on. The activities of the *Dlx* genes would first lead to shaping the jaw itself and then to the development of each kind of tooth *germ*, either indirectly via the jaw effects or directly. The latter's development will include not only effects on its general shape and size but also, for the molars and premolars, the detailed patterns of the cusps of the teeth. The idea also provides a clue as to how different tooth patterns might have emerged in evolution: all that would be required are spatial shifts in the regions of expression, their *domains*, of different members of the *Dlx* gene family (or of other genes that are involved). Genetic changes in the *cis*-controlling sequences for these genes would produce such changes while changes in the coding sequences of the DLX proteins themselves could contribute further to subtle changes in the development of individual tooth types.[28]

Fur (Hair)

The striking and instantly perceived general difference between most mammals and reptiles is the fur coats of the former. Hair and fur are exclusively mammalian properties. Yet full fur coats did not suddenly emerge with the mammals. Rather, the origination of individual hair shafts was the fundamental innovation and these would have been relatively sparse initially; full pelages arose much later. That putative difference in hair density between the first synapsids bearing them and contemporary mammals suggests that hair did not function primarily as an insulating device as it does today but must have been selected for other reasons such as tactile sensing—for example, to detect solid objects while moving in the dark.

The starting point for thinking about how hairs might have originated is that they are largely composed of one subclass of the group of proteins termed **keratins**, resilient fibrous proteins that are found not only in hair but also in all protective animal surface structures, including skin. These protective keratins are evolutionary derivatives of the original keratin molecules of eukaryotic cells, where they make up one element of the cytoskeleton, one of the kinds of intermediate filament proteins. The keratin genes today consist of a large gene family that can be divided roughly into those that are *soft* and *hard*, respectively. It is the hard keratins that give strength and shielding to both skin and the various structures that animals bear on the outside of their bodies: scales, feathers, hooves, claws, vibrissae (face bristles), and, not least, body hair. The toughness and resilience of these elements is due to the combined effects of their keratins, to other keratin-associated proteins that bind to them, and still other proteins that are not in direct association with the keratins but part of the same structures. The hard keratins, however, are the core proteins that give epidermis and hair their strength. Thus, the evolution of hair involved the specific redeployment of molecules, originally employed in the cytoskeleton, in constructing the epidermis and later the hair shaft.

Two ideas have been offered to explain how hair originated. The choice between them depends largely on the kind of skin on which they initially appeared: was it scale-covered skin as seen in fishes and present-day reptiles, or was it more like the smoother, scale-free skin of present-day amphibians? If mammalian-type hairs developed from heavily scaled skin, then it probably did so as part of the reduction in number and size of scales, with the early hairs first appearing at the junctions or *hinges* between scales. Indeed, groups of three hairs are found at such hinge regions in the small number of mammals that bear scales, as seen, for example in rat tails and armadillo and pangolin skin. Alternatively, if the first hair-bearing synapsids had smoother, less scaly skin, hairs probably evolved from secretory glands in that skin, probably sebaceous glands for the secretion of oils and mucous. In such an evolutionary transition, the changes would have involved a switch in the material being secreted from oily substances to keratinaceous materials. (Both models, however, predict an initially sparse hair coat.)

No definitive evidence decides the issue, but the few fossilized remains of early synapsid skins suggest they had leathery skin with few scales, a finding that favors the second hypothesis. Furthermore, hairs are produced as initial internal secretions from small glandlike structures related to the sebaceous glands, gelling into hard outward-growing shafts. It is not difficult to imagine the evolutionary conversion of glands used for producing one kind of product to the synthesis of another kind of structure such as keratin-rich hair. (If so, the small number of mammalian examples of scales associated with three-hair structures should be seen

as a secondary and late evolutionary innovation.) In this evolutionary scenario, new variants of keratins, keratin-associated proteins, and still other
proteins would have arisen later and become specially adapted to hair
production.[29]

The special relevance of the evolution of hair and fur coats to humans,
especially the human face, concerns what may be considered the reversal of these events, the progressive loss of body hair in the evolution
of humans. The initial loss was probably of facial hair in the anthropoid
primates, early in primate evolution, as discussed in the next chapter.
The reasons are unknown but may relate to the process of sexual selection in which certain traits in one sex enhance appeal for prospective
mates. (Later, those traits may evolve to be present in both sexes.) One
consequence, however, of losing facial fur would have been the creation
of a face whose facial expressions could be read more easily because it
lacked a fur covering. Thus, face-mediated social interactions would have
been enhanced by the loss of facial hair. Then during the evolution of
the hominins, there would have been a further loss of (most) body hair.
We will return to this subject in Chapter 8.

Facial (Mimetic) Muscles

All vertebrates have muscles in their faces anchored to their skulls. They
serve to open, close, and narrow the apertures surrounding the eyes, nostrils, and mouth. These *deep* facial muscles are a commonly shared, hence
conserved set of muscles in the vertebrates. Mammals, however, in contrast
to all other vertebrates, also have a second and more superficially located
set of muscles, the facial or *mimetic* muscles. As noted earlier, these were
the focus of Sir Charles Bell's work in the early nineteenth century. Although rather neglected in most treatments of mammalian evolution, the
mimetic muscles are a unique and general mammalian trait. They are essential for subtle eye, lip, and brow movements and consequentially for
the whole range of facial expressions seen only in mammals. The original
functions of the eye movements mediated by certain of the mimetic muscles almost certainly had as their primary functions the sighting of prey
or predators, while the lip movements that only mammals possess were
probably originally for the unique mammalian trait of nursing and probably also for manipulating food items. Those abilities, however, were evidently later co-opted during evolution into the making of facial expressions

for the purposes of communication, especially in primates. Consistent with that idea is the fact that the mimetic muscles reach their most extensive development in the primates, the most sociable of mammals. Every mammalian species has a characteristic number and set of mimetic muscles. In humans, there are twenty-one kinds of mimetic muscle, most of which are paired, with the members of each pair placed symmetrically opposite each other on the two sides of the face. Uniquely, all but two of the mimetic muscles attach to the dermis of the skin; no other muscles in the mammalian body, or indeed any in nonmammalian vertebrates, attach to the skin.

The ultimate developmental source of the mimetic muscles is the neural crest. Neural crest cells that have migrated to the second branchial arch give rise to both the cells that will generate the mandibular facial primordia and cells that will become migrating mesenchymal cells. These cells move out and come to just beneath the dermis in several streams—in humans, there are four principal streams—and to specific sites, starting about weeks eight to nine of embryonic development, when there is already a rudimentary face. The streams move to distinct locations, splitting off groups of cells at different locations to form the individual facial muscles. The cells in each such muscle founder group undergo further cell divisions to generate a long laminated structure consisting of muscle precursor cells (myoblasts). These, in turn, transform into fused premuscle structures termed *myotubes* and then later fully formed muscles that attach to the dermis. A diagram of the migrating streams in the human fetus is shown in Figure 5.10.

The existence of the facial muscles as a distinct mammalian attribute illustrates the continuing ability of the neural crest to evolve—specifically, its ability to evolve new derivatives long after it originated. It was suggested earlier that the different derivatives of the neural crest almost certainly did not arise simultaneously in the beginnings of craniate evolution when the neural crest itself originated but probably in a sequence of steps. Such innovation would have required only slight genetic modification of regulatory processes within these multipotential stem cells to yield new cell-type derivatives. The fact that the mimetic muscles are a mammalian specialty shows that the neural crest retained its ability to evolve new cell types long after their origination. The mammalian mimetic muscles presumably arose about 300 million years after other basic neural crest cell derivatives.

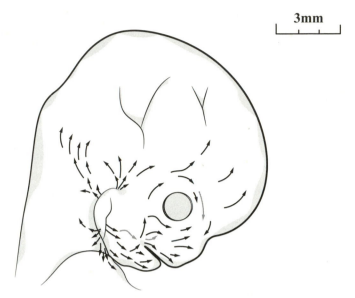

FIGURE 5.10 Pathways of migration of mesenchymal cells in the human embryo that will eventually form the facial (mimetic) muscles. (Modified from Gasser 1967.)

The precise genetic basis of the evolution of the mimetic muscles is unknown, but it is not difficult to imagine how it might have taken place. One element was probably the capture of the basic muscle-cell–making machinery by a new genetic regulatory gene in the region of the developing neck, allowing new muscle production to originate there. A parallel event probably underlies the relationship between the skeletal musculature of the trunk and the deep craniofacial muscles previously mentioned. While these two sets of muscles are highly similar in structure, they differ in their upstream regulators. Presumably, the deep craniofacial muscles of verte-brates evolved after the trunk muscles and their evolution involved capture by new regulator genes expressed in the developing face. For the mimetic muscles, however, there would almost certainly have been addi-tional genetic changes that allowed these new muscles to form attach-ments to the underlying dermis. This would presumably have entailed shifts in the battery of worker genes to allow attachment of the developing muscles to the underside of the dermis. Altogether, the actual evolutionary tinkering of the underlying genetic networks, both upstream and down-stream of the conserved modules, would have been relatively modest, but the consequences in terms of form and function were significant: the development of a new class of muscles that ultimately permitted a major unfolding of facial expressive capacities.[30]

Nursing and the Evolution of the Mammalian Muzzle

Nursing of neonates with maternally supplied milk is one of the two defining traits of mammals (the other being their distinctive jaw joint). Indeed, mammals were named in honor of their unique possession of mammary glands, whose function is to produce milk for nourishing newborn young. All mammals, including the egg-laying mammals or monotremes— platypuses and echidnas—have mammary glands that produce milk for their young. (The monotremes lack nipples, however, and the mother secretes milk through a specialized area of the skin on her underbelly.)

The way in which milk production or *lactation* evolved is an intriguing question, one that vexed Darwin and was seized on by one of his main critics, St. George Mivart, as a process whose gradual evolution, according to Mivart, was impossible to conceive. Mivart argued, how could animals develop, step by step something that only had value when it was fully developed? A present-day answer to the problem, however, is now emerging: it seems probable that lactation arose in synapsids initially not to provide nutrients but to supply liquid and antimicrobial substances via skin glands to the somewhat porous leathery eggs they laid. With time, the composition of the secreted material changed to or began to include that of more strictly nutritional elements, some of which at least were related in structure to some of the antimicrobial substances.[31]

We may ask, how does this phenomenon relate to the evolution of the mammalian face? Milk, after all, is supplied by mammary glands on the mother's body, well away from her face. The potential connection to the evolution of the face is through the physical requirements of the young to nurse. Long muzzles, in principle, should tend to make for less efficient nursing. The latter is aided by close face-to-nipple contact, which is precisely what is possible with the foreshortened muzzles of nursing young. In most placental muzzles, only after nursing is complete does full muzzle elongation takes place. (Some grazing animals are born with muzzles and begin grazing relatively early, but their nursing capability may have evolved in connection with elongated nipples of the mothers or other means to cope with the problem.) With humans, of course, there is no muzzle, hence no muzzle elongation. This helps make possible the often prolonged nursing of young humans, even into the second and third years of life in many older and more traditional societies, including hunter-gatherer tribes. (The functions of muzzles and the features of human life that render them unnecessary for us will be discussed in the next chapter.)[32]

On to the Primates...

In this chapter, we have taken the story of the evolution of the face from its first appearance in the earliest vertebrates, around 500 million years ago during the late Cambrian or early Ordovician, to the emergence of the first modern (crown group) placental mammals and their diversification into the four major supraordinal groupings, one of these being the **Euarchontaglires**, which includes the Primates. In the next chapter, we will trace and discuss the events that led from a seemingly unpromising primate ancestor, perhaps smaller than a mouse and possessing a rather generic mammalian face, to our species with its highly distinctive one.

six

HISTORY OF THE FACE II: FROM EARLY PRIMATES TO MODERN HUMANS

Introduction: Diversity of the Primates—and its Modest Roots

The human species is unique in the Animal Kingdom in terms of what it has created collectively and in what its members can do individually. No other animal species on Earth comes close to our capabilities or accomplishments—remaking the ecology of the planet (not necessarily for the better), going to the moon, and probing the deepest secrets of the universe and of living things. In retrospect, it is hardly surprising that many of Darwin's contemporaries found his idea of evolution as applied to humans to be beyond belief. Yet by all biological criteria, we are bona fide primates and what makes us special must have, in some way, grown out of our "primate-ness." (That statement may strike the reader as either obvious or outrageous, but this chapter will try to show that it is neither banal nor absurd.) Hence, I begin this chapter, devoted to the evolutionary journey in which humans evolved from earlier primates, with a look at the order of **Primates** as a whole.

The scientific study of the primates is itself of interest, not least because it begins with a nice irony: humans were first grouped with the monkeys and apes not by Charles Darwin but by someone who did not believe in evolution: Linnaeus. As with all his classifications, it was based entirely on morphological resemblances. He first proposed this grouping in 1735 in the first edition of *Systema Naturae,* but he only coined the term *primates* (signifying the "first" in importance) in 1758, in the definitive tenth edition. That was just over a century before Darwin published *The Origin of Species.*

Despite the honor Linnaeus thus bestowed on the primates in naming them, he could not have appreciated how unusual they are given the relatively small number of species known in his time. Comprising 440 known species today, or about 8 percent of all living mammals, the primates are not only one of the most species-rich mammalian orders but also one of the most diverse, whether that diversity is judged in terms of size, morphology, color patterns, or behaviors. In size, for example, they range from the tiny mouse lemurs of Madagascar, whose adults weigh only two ounces, to lowland gorillas, weighing up to nearly 500 pounds. In coloration, primate fur coats run from various shades of brown to black-and-white to bright yellows and even orange, and the colors of their naked skin can be particularly showy; the gaudy red and blue facial skin colors of the male mandrill are unmatched among mammals. The extent of their pelages is also quite varied, from the thick coats of the beautiful white-and-black Colobus monkeys to a nearly furless species, ourselves. Modes of movement range from brachiation (swinging by forelimbs from tree to tree) to quadrumanous ("four-handed") species (such as the orangutan, who moves on trees by limb-over-limb grasping movements with hind as well as forelimbs), to terrestrial knuckle-walking gorillas and chimps, to the bipedal movement of humans. In dietary preferences, most primate species are primarily frugivorous, but there is a range of diets and some are highly specialized. In degrees of sociality, the range is also great, extending from Bornean orangutans who lead solitary lives to several monkey species that live in large troupes up to the hundreds. High sociability is the general rule for primates, however, to a degree unmatched by other mammals, and we *Homo sapiens* are perhaps the most sociable of all: each of us has the potential, and often the willingness, to interact socially with any other newly met member of our species, a most unusual trait.[1]

What features unify such a large and diverse group, or, more precisely, what shared features (synapomorphies) define a primate? These traits are few and modest: a *postorbital* bar, namely, a bony ring that surrounds the eye; nails instead of claws on the digits; an opposable first digit on the hands (the *pollex* or the thumb in humans) as well as for most primates—humans are a key exception—on the feet. These features must have been possessed by the ancestors of the group, which, as we shall see shortly, were probably small, tree-living, insect-eating animals. Those modest origins would have given little hint of the kind of diversity seen in today's primates, let alone the properties of *Homo sapiens*.[2]

We will now, however, leave the vibrant, multifarious world of living primate species to concentrate on the past and specifically on the path that led to humans, the long hominocentric lineage path (LHLP). Fortunately, a surprising amount about this history is now known. In particular, many of the specific changes in the final part of the sequence, when the homi-nins branched off from the chimpanzee lineage, have been well character-ized from fossils. Furthermore, it is now possible to make deductions about both the genetic changes that underlay the evolutionary ones and the selective factors that fostered the latter. This chapter will provide a concise history of the sequence of events in primate evolution that led to humans and, specifically, to the human face. As will become apparent in the later part of this account, the evolutionary history of the human face is entan-gled with that of human mental traits and behaviors, a set of connections that will be explored more fully in the next chapter.

The time frame of the story is important. While one line of evidence, to be discussed shortly, indicates that the primates originated in the late Cretaceous, clearly most of their history unfolded afterward during the Cenozoic era. Hence, in this recounting there will be much necessary reference to events occurring at different periods of the Cenozoic, which is diagrammed in Figure 6.1. The traditional scheme divides it into six periods—the Paleocene, Eocene, Oligocene, Miocene, Pliocene, and Pleis-tocene. The more current division simplifies it into two, the Paleogene and the Neogene (also shown); for this account, the older version will be used because it provides a finer-scale temporal division of the history.

Stem Groups Versus Crown Groups: Tracing Primate Beginnings

In thinking about the evolutionary history of the primates, we would like to know when they first appeared. This is an important question for any group of animals of interest but is always difficult to answer. There are two reasons for that difficulty. The first is the obvious, predictable and near-universal one: the lack of a complete or even near-complete, well-delineated sequence of fossils. We have seen the difficulty in the debate about the ori-gins of placental mammals, but it is just as problematical for the early fossil record of the primates, which is poor. The second problem is sub-tler but just as important; it hinges on how one defines the group of interest. Any living taxon of animals is defined by a distinctive set of traits—anatomical, physiological, biochemical, behavioral—yet those features typically will have arisen not simultaneously but in a temporal sequence. Thus, each change initially often involves one trait at a time, a

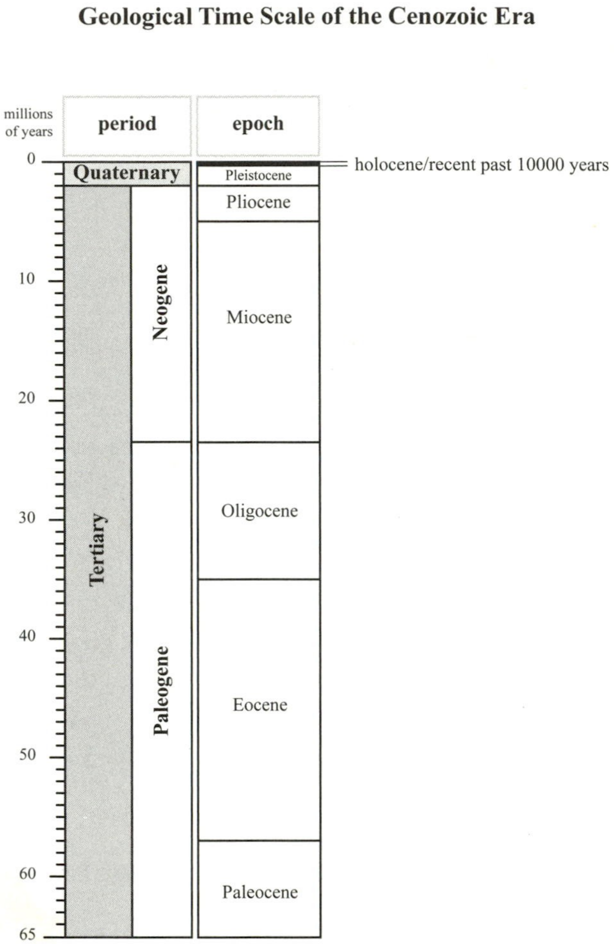

Geological Time Scale of the Cenozoic Era

FIGURE 6.1 The geological time scale of the Cenozoic (Tertiary) era, from 65 mya to 10,000 years ago, during which most of primate evolution unfolded. It has traditionally been divided into five geological periods: the Paleocene, Eocene, Oligocene, Miocene, and Pliocene (right). More recently it is described as involving two: the Paleogene and the Neogene (left).

pattern of trait acquisition termed **mosaic evolution**. In effect, the modern taxon only comes into existence at the end of that series of trait acquisitions. Indeed, many fossils bear witness to mosaic evolution, showing a mixture of traits, some belonging to one present-day taxonomic group, others to another. The classic example is the birdlike *Archaeopteryx,* for whom the term *mosaic evolution* was coined. *Archeopteryx* possessed wings and feathers, but it lacked a "wishbone" and possessed teeth and a long bony tail, features not found in modern birds. Another previously encountered example of mosaic evolution is that of the so-called true mammals, which were derived from the **mammaliaform synapsids**. In fact, the evolution of all large taxonomic groups entails sequences of change involving multiple traits, hence it is always mosaic in nature.

Nevertheless, it is often possible to have an inkling about the beginnings of a group when an identified fossil marks an early point in the divergence of that group to become a distinct lineage. That animal will have possessed one or more but not all of the modern taxon's defining traits. Such early precursors are referred to as **stem groups**. An example is, again, *Archaeopteryx,* which is regarded as a stem group member of the birds, the class Aves. In contrast, a later group that has acquired the set of key characteristics that define the modern taxon is referred to as a **crown group**, and it includes all the species with those traits, both extinct and living, including its original ancestral species. Thus, modern passerine birds and all their ancestors possessing their distinguishing traits are a crown group within the Aves.

By the methods of **cladistics** (Figure 1.4, Box 1.3), we can often reconstruct the evolutionary history of the living crown group members of a group. That history, however, usually provides little insight into the earliest history of the taxon, which involved its stem group members. For that, we need fossils, but for the primates, there are no fossils of universally agreed stem group members. The earliest recognizable fossil primates date to about 56 million years ago and the beginning of the Eocene. These are considered *euprimates* or crown group members and were members of the genus *Teilhardina,* resembling today's marmosets (though smaller). Another group of animals closely related to the primates, the **plesiadiforms**, existed earlier. Previously thought to be the primate stem group, they are now regarded as a probable **sister group** of the early stem primates. They were arboreal, roughly resembling long-limbed squirrels, and left fossils dating to 63–62 mya, thus in the early Paleocene.[3]

✓ There are no known fossils of primate-like creatures before this point, but indications are that there must have been earlier, stem group forms well before. In particular, various early molecular clock estimates placed the primates as diverging from other, related mammalian lineages (of the **Euarchonta**) perhaps close to 100 million to 90 million years ago during the Cretaceous, the final period of the Age of Dinosaurs. Molecular clock estimates, however, as we have noted, can be problematic, and many paleontologists questioned such an early origin of the primates. Of course, if the placental mammals as a whole only originated after the end-Cretaceous extinction, as claimed in the big, fossil-based analysis of 2013 discussed in the previous chapter, then the primates obviously would also have originated after that event. Still, newer molecular clock data, which have been put through various analytical corrections, persist in placing the origins of the primates in the Cretaceous, albeit with a considerable range of dates.

One way to resolve the discrepancy between those estimates and the paucity of fossil data of primate-like animals in this period involves a method of indirect inference utilizing known patterns of diversification of a group, the average lifetimes of its component species, and estimates of loss of fossil information. With reasonable values for those parameters, starting with information on current primate species abundances, the analysis can reach back into the past to arrive at estimates for species numbers at successively earlier times. Needless to say, the analyses are both sophisticated and dependent on various assumptions that create a degree of uncertainty. A chief difficulty concerns the validity of back-extrapolating rates through the end-Cretaceous extinction, a major singularity if there ever was one. Nevertheless, different studies with somewhat different data sets have achieved surprisingly good agreement for primate origins at around 85 mya to 80 mya, hence well within the final part of the Cretaceous era. Though not universally accepted, this conclusion is the consensus today: the primates originated in the late Cretaceous while their crown group diversification only began strongly in the Eocene.[4]

Such findings, of course, do not reveal what the ancestral primates looked like. Recent studies, however, of the variable rates of the molecular clock have been used in an imaginative way to reconstruct some features of the stem primates—in particular, their likely size(s). These analyses have revealed that the changes in rate of DNA sequence alteration in the primates correlate with certain life-history characteristics of the animals themselves and thus provide clues to those features. Specifically, there is a rough inverse correlation between the rate of ticking of the molecular clock, on the one hand, and, on the other, body size, absolute endocranial (skull) volumes, and relative endocranial volumes. Thus, the larger the animal and its endocranium and the more long-lived it is, the slower its molecular clock runs. (This was initially suspected decades ago for apes and hominins, and was originally tagged the "hominoid slow-down.") Some possible explanations of these relationships between DNA sequence change and life-history properties are given in Box 6.1.[5]

Whatever the precise explanation, the crucial conclusion is that increases in all three properties—body size, absolute endocranial skull volume, and relative endocranial volume—took place in the major primate lineages. (There have also been decreases within some lineages, but the general trend was toward increases.) As estimated from the fossils, the rates of change in these physical features can then be used *in* reverse via plots of rates of molecular change against the measured body and cranial measurements to estimate the molecular rates and extents of change in the earliest stages of primate evolution. This analysis, in turn, makes possible estimates of the body size of the ancestral or stem primate. The findings indicate that this animal had an approximate body weight of 55 grams (or about 2 ounces) and a small endocranial volume of about 2.3 cubic centimeters and thus on the order of the smallest living primate, the mouse lemur. This is also close to the estimated size of an early primate relative, *Ignacius,* a member of the *plesiadapiforms* (a sister group of the primates) previously mentioned.[6]

This is but one way to reconstruct the ancestral primate, and not all experts agree. Furthermore, it does not provide a detailed picture of the way the ancestor actually looked. A different form of analysis based on deduced morphological features, the phylogeny of the primates, and the assumption that the earliest primates had nails, not claws, making grasping with the digits easier, has produced a picture of a somewhat larger animal, perhaps 200 to 450 grams, that perhaps looks rather like a contemporary dwarf lemur. That reconstruction is shown in **Plate 9**.[7]

The primates thus probably started with tiny to small ancestors, but with time and divergence over many different lineages they generally increased in size. (Within each main lineage, however, there were lines exhibiting decreases in size.) One of those major lineages was the hominoids, the lineage that includes both the apes and the hominins. We shall return to the hominoids, but first let us examine the main features of primate evolution.

BOX 6.1 Rate Variation in the Molecular Clock as a Function of Life-History Variables

The fundamental premise of the molecular clock is that it "ticks"—creating base-pair substitutions—at essentially a constant average rate (for a particular gene), independent of the lineage of the organism (see Box 1.2). Yet in the fifty years since the idea of the molecular clock was proposed, it has become clear that the premise is false. In particular, small-bodied animals tend to have faster rates in their molecular clocks than do large-bodied ones. An example: in the hominoids (the apes plus humans), the molecular clock runs slower than in many other, smaller mammals, generating the now generally accepted *hominoid slowdown* of the molecular clock. Such slower-than-expected rates can lead to underestimates of the time elapsed since a divergence of two lineages from a common ancestor; in contrast, faster-than-expected rates can lead to artificially lengthened estimates. The inverse relationship between body size and molecular clock rate cannot reflect direct causation but must be a second-order reflection of several life-history traits that are more directly related to mutation rate. These include generation time, life-span duration, and number of offspring. Thus, smaller animals reproduce earlier, live shorter lives, and have greater fecundity—and exhibit faster clocks—than larger animals.

The question is why? The standard answer is that the rate of molecular clock ticking should be a function of the number of DNA replications over time, and shorter-lived animals that reproduce more frequently will generally have greater numbers of DNA replications per unit of time (years or decades). The correlations between numbers of replications and molecular clock ticking, however, are far from linear, suggesting that that is not the whole explanation. The other reason often invoked is that smaller animals generally have higher rates of basal metabolism (BRM), and such higher rates should generate more oxygen radicals, which can cause mutations; hence a higher BRM should correlate with more mutations and a faster molecular clock. Larger-brained animals, however, tend to have higher BRMs, even when normalized for body size, so this idea predicts a faster clock for animals with larger brains, which is the exact reverse of what is observed. Thus, at present, the reason why larger endocranial volumes (as a measure of brain size), both absolute and relative, should be correlated with slower-running molecular clocks is still unknown. One possibility, as yet untested, is that because larger-brained animals have larger bodies, hence more cells, there has been selection for more accurate DNA replication and thus lower mutation rates to maintain the integrity of cells and tissues.[1]

---

1   See Steiper and Seifert (2012) for fuller account

The Big Division in Early Primate Evolution and Two Ways to Think about It

As reconstructed from fossil evidence and comparative anatomy and DNA sequences, the phylogenetic tree of the primates is diagrammed in Figure 6.2. It is now universally agreed that this group divided into two major subgroups early in its history. That division, however, can be described in two alternative ways as indicated in the figure that reflect two different views of the nature of the evolutionary process.[8]

The older, classical form of the division is that between **prosimians** (*premonkeys* in effect), and the anthropoid primates (the *manlike* primates) or *simians*, an older term embracing the monkeys and the apes. This division accords with presumed evolutionary *grade*, namely, the relative level of biological complexity. In this view, the living prosimians—the lorises, lemurs, and tarsiers—are the more primitive living primates. While they share the defining traits of primates, they are closer in morphology, especially their heads and faces, to the other orders within the Euarchontoglires [the tree shrews, flying lemurs, rodents, and lagomorphs (rabbits)], as would be expected if they are closer to ancestral primates. In contrast, the anthropoid primates have diverged further from those ancestral stocks and are more humanlike, especially in their faces, teeth, and head shapes. This group includes the monkeys—both major groups, New World and Old World monkeys—the apes, and humans. In such traditional gradistic thinking, the anthropoid primates are "higher" than the prosimians, concordant with a pre-evolutionary view of animal life, the "Great Chain of Being."[9]

The second way of splitting the primates is not in terms of their relative degrees of advancement but their relatedness as deduced by cladistics. In this view, two of the prosimian groups—lemurs and lorises—make up a category of related primates termed the **strepsirrhines** or literally "curly noses" (a reference to the curly, comma-shaped nostrils) while all the remaining primates constitute the **haplorhines** (literally, "simple-nosed"). That difference in nostril shapes, however, is probably less biologically important than another nasal characteristic, the respective degree of "wetness" of the noses of the two groups. The strepsirrhines, in common with many mammals (for example dogs, mice, seals, deer), have moist rhinaria, namely, the area of the nose immediately surrounding

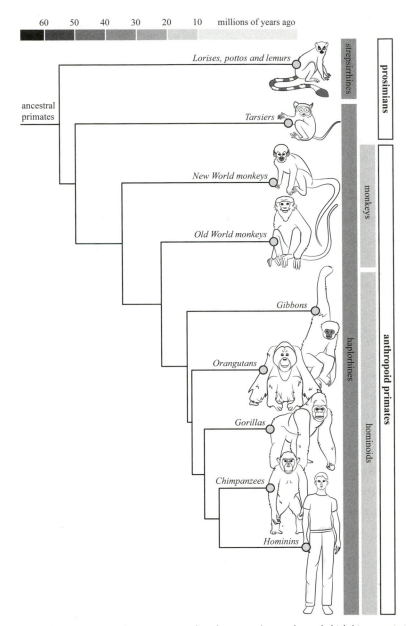

FIGURE 6.2 The primate phylogenetic tree and its alternative divisions by grade (right) into prosimians and anthropoid primates and by cladistics (left) into strepsirrhines and haplorhines. The two classifications are largely congruent except for the tarsiers, classed as prosimians but with strong haplorrhine affinities. The LHLP runs through the haplorhines as described in the text.

the nostrils; the anthropoid primates, in contrast, possess dry noses as seen in monkeys, apes, and humans. The significance of this difference is that animals with moist rhinaria rely more on olfaction than vision. Fittingly, most of the prosimians are primarily active at night, when vision is less useful and olfaction and touch (primarily via whiskers) much more so. In contrast, the anthropoid primates rely far more on vision and, accordingly, are mostly diurnal animals.

Another difference between the strepsirrhines and haplorhines concerns the upper lip. The strepsirrhines have a "split" upper lip, with the division running from the middle of the upper lip to the bottom of the nose, again like dogs and cats, while the haplorhines have a unified upper lip. The latter property gives the upper lip more mobility for eating, making additional sounds, and, not least, making facial expressions involving the mouth.[10]

In practice, these two classifications of the primates, the traditional *gradistic* one and the more modern evolutionary or *cladistic* version, largely overlap, although there is one exception, the tarsiers, which are distinguished by their especially large eyes. They were traditionally classified as prosimians but, when examined for their individual traits, show a mix of strepsirrhine and haplorhine characteristics. Though a nocturnal species, they almost certainly evolved from a diurnal one, their prominent eyes presumably reflecting selection for better vision in the dim light conditions of dusk and dawn. The tarsiers probably diverged from an ancestral group that also gave rise to the rest of the anthropoid primates; if so, they would be a sister group to the anthropoid primates.

The emergence of the anthropoid primates was a crucial stage in laying the foundations for the human face. Three basic anthropoid traits are of special relevance in this regard and presumably characterized the anthropoid stem group: (1) a more convergent placing of the eyes, facilitating acute close-up vision; (2) the loss of fur from the face, permitting facial expressions to be more easily seen; and (3) some reduction of the muzzle, with corresponding changes in the teeth and lips, affecting eating, vocalizing, and the making of facial expressions. Much of the ultimate distinctiveness of the human face thus involved elaboration of the traits that early in the haplorhine lineage set them apart from the strepsirrhines.[11]

When did the stem anthropoid primates arise? Although the first unambiguous anthropoid-like fossils from North Africa date to about

37 million years ago, some of the Asian candidate fossils for anthropoid status are considerably older, supporting a latest time origin of this group at the start of the Eocene, 55 million to 56 million years ago. In particular, there is *Teilhardina,* mentioned previously, which was possibly a stem haplorhine. The high diversification of the anthropoid primates, however, apparently began 15 million to 20 million years later during the late Eocene, judging from recovered fossils.[12]

From Early Anthropoid Primates to First Hominins: an Overview

Although the anthropoid primates are a monophyletic group arising from a single ancestral group and sharing many features, they divide into two major subdivisions, the Platyrrhini and the Catarrhini. The divergence between these two major clades is generally taken to be about 35 mya to 45 mya, as anthropoid primate diversification was accelerating, and thus perhaps as much as 30 million years after the strepsirrhine–haplorhine split.[13]

As with the basic primate division into strepsirrhines and haplorhines, the names of the two anthropoid taxons denote a distinctive nasal characteristic. *Platyrrhine* literally means "flat-nosed," but the striking characteristic of the **platyrrhine primate** nose is not its flatness per se but its flaring nostrils; in contrast, the Catarrhini have closer-placed, downward-pointing nostrils. Another difference is the number of premolars, with the platyrrhines having three instead of two on each side of the mouth, top and bottom. In addition, many platyrrhines have prehensile tails by which they can hang from trees, a trait not seen in any **catarrhine primates**. A further difference is physiological: the catarrhines have full color (*trichromatic*) vision, while nearly all the platyrrhines have only partial or *dichromatic* color vision, and the owl monkeys lack it entirely, the *monochromatic* condition. (There is also one group of platyrrhines that have trichromatic vision, but they evolved it independently.) The better color vision of the catarrhines undoubtedly had initial adaptive advantages though, as we shall see, the precise nature of those advantages has been debated.

A final and crucial distinction is geographic: contemporary platyrrhines all are native to Central and South America and are thus referred to as the New World monkeys, while the catarrhines are all native to Africa or Asia.

Given the many resemblances between the New World monkeys (platyr-rhines) and the Old World monkeys, one of the two major branches of the catarrhines, it is likely that the immediate ancestral anthropoid stock for both groups broadly resembled monkey species. (The way in which the platyrrhines came to inhabit the New World is an interesting question because, at the time of their separation from the catarrhines, there were no land bridges between the Americas and other continents—but that is a separate matter.) The other major catarrhine group is the apes, or Hominoidea, which includes the hominins, and they arose well after the platyrrhine–catarrhine split. The fossil evidence as to where these Old World monkey–ape lineages first separated, whether in Africa or Asia, is ambiguous, but the weight of the evidence favors Africa. It is the ape lineage that is of crucial relevance to our story and to which we now turn.

The apes, or Hominoidae, are distinguished from the Old World monkeys by several characteristics: their lack of tails, highly flexible elbow and wrist joints, the presence of an appendix, and a facial similarity that is hard to characterize but is distinct from that of monkeys and involves larger and more human-like faces. The oldest known fossil apes date to 23 million years ago in the early Miocene (23 mya to 25 mya) in Africa and comprise the genus *Proconsul,* which consists of four species. In size, *Proconsul* was that of a large monkey; in its bodily characteristics, it had a mixture of monkey and ape traits. A reconstructed version of what *Proconsul* looked like is shown in **Plate 10.**[14]

Another early ape-like species, placed in a second genus, *Rangwapithecus,* was also African and dates to the same period. From the structure of their teeth, the diet of these earliest-known ape species is believed to have been principally tough, fibrous fruits. Furthermore, their limbs indicate a life lived primarily in the trees, probably semitropical forest. By the mid-Miocene, 17 mya to 15 mya, other ape species were living in what is now Europe and the Middle East. This geographical spread of these early ape species indicates the first of the "out-of-Africa" migrations of Hominoid primates. In both Africa and later in Europe, there was a branching out and diversification of the apes, though eventually all of the European species died out.

Unfortunately, existing fossils do not illuminate the evolutionary pathway from these Miocene apes to modern apes. The living apes, however, divide into two groups, the lesser apes and the great apes, and the molecular clock evidence indicates that divergence probably took place in the mid-Miocene, again probably in Africa, with some species later spreading into Asia. The lesser apes, most often designated as the gibbons, are long-armed wholly arboreal species whose faces are more monkeylike than those of the living great apes or the extinct apes **(Plate 2)**. those facial changes were probably later and hence "derived" traits. The gibbons, however, are not our concern, since the LHLP leading to humans runs through the great apes, whose present-day members are the orangutans, gorillas, and chimpanzees. From molecular clock analyses, the orangutans split off early from the other great ape lineages, about 16 mya to 15 mya in the mid-Miocene, while the divergence of gorilla ancestors from chimpanzee ancestors occurred a few million years later. From the origins of the apes, around the beginning of the Miocene, circa 23 mya, to the appearance of the hominins, 6 million to 7 million years ago, the faces of most of these species, despite changes in habitat and diet among the different lineages, did not deviate strongly from the apelike visage of *Proconsul* and that genus's successors.[15]

Evolutionary changes in the face were also not a marked feature during the first few million years of the hominin lineage. The more striking alterations were in the trunk of the body, especially those that led to bipedal movement and the consequent freeing of the hands from involvement with body locomotion. It is to the hominin lineage that we now turn, beginning with a brief history of the discoveries that inform our current understanding.

From First Hominins to *Homo sapiens*:
I. History of the Search for Human Origins

Of all of Darwin's hypotheses, the idea that human beings had evolved from ape-like creatures was undoubtedly the most contentious in nineteenth-century Britain, though hardly the most fundamental. Was it credible, many wondered, that the most beautiful women and the handsomest men in society were ultimately descended from ape-like creatures? Could it really be that a species whose capabilities and accomplishments included Copernican cosmology and Newtonian physics, the weaving of exquisite silk scarves, the invention of steam engines, the sculptures of

Michelangelo, the paintings of Leonardo da Vinci and Rembrandt, and the music of Bach, Mozart, and Beethoven had evolved from creatures who lacked even simple speech? Even if such were possible, was it conceivable that it took place by the Darwinian mode of natural selection of small variations, each of small effect? It seemed preposterous. Whatever the facial and bodily resemblances between apes and humans that might be conceded, the mental distance between these two groups looked unbridgeable. Why should any sensible Victorian believe that humans had originated from ape-like creatures?

Nor was it just members of the public who found the idea unbelievable. When faced with the apparent enormous difference in mental capacities between apes and men, Darwin's codiscoverer of the principle of natural selection, Alfred Russel Wallace, blinked. Wallace believed, as did Darwin, and in sharp contrast to most Europeans of the time, that nonwhite racial groups in Africa, South America, and Asia had essentially the same mental capacities as Europeans, though Europeans (of that time) had, self-evidently to the Victorians, developed that potential further. It followed that the earliest humans would have had similar mental potential to modern Europeans. Wallace could not understand how natural selection could have created the mental capacities of modern man, which went far beyond anything needed for survival when modern humans first appeared. A basic tenet of Darwinian theory was that selection delivers more or less what is required, not vastly more. The reason is, simply, that new properties entail "costs"—in energy, developmental time, physiology, or other respects—and natural selection presumably would work against any such extra costs. Hence, in Wallace's view that natural selection could not have yielded humankind's extraordinary gifts when the environment demanded so much less, Wallace was more Darwinian than Darwin. Though not conventionally religious, he opted in the late 1860s to believe that divine creation was responsible for humans' unique mental capabilities. His conclusion caused Darwin anguish, which he expressed in a letter to Wallace. To deny the efficacy of natural selection to the point of invoking divine intervention to explain the source of a trait was, in his eyes, to "murder" the theory. (His objection to Charles Bell's theocentric view of the origin of the human mimetic muscles was essentially the same.)

Darwin himself skirted the issue of human origins when he first published his evolutionary theory 1859. Only in a later edition of *The Origin of Species* did he mention this issue and there devoted precisely one sentence to it, promising his readers a detailed account later. His evasion was deliberate. In part, he must have been deterred by the paucity of fossil evidence that could support the claim of human ancestry from ape-like animals. Furthermore, he knew that the proposition of human origination from ape-like ancestors would be painful to many members of the public, especially those raised in traditional Christian belief, and he did not want discussion of his theory of evolution turned into a debate about human origins. In the event, that nearly happened anyway.[16]

When Darwin finally marshalled his arguments and tackled the subject of human evolution in *The Descent of Man* in 1871, the fossil evidence was not much better than it had been twelve years earlier. True, there were Neanderthal skeletons and skulls, the first coming to light in Germany in 1856, in the valley of the Neander River (hence, the name), three years before *The Origin* was published, and by the 1870s an even larger set had been found. Although not fossils of modern humans, they were far closer in form to humans than to any ape species. Furthermore, later measurements would reveal that they even had cranial capacities that were slightly larger than those of modern humans. Thus, Neanderthals, though later caricatured in the popular press as ape-like, were actually too close to existing humans to be a convincing bridge between apes and humans. (One interpretation had them as degenerated humans.)

The fossil evidence was not substantially stronger by the end of the nineteenth century. This period in the history of evolutionary biology was a low point for Darwin's theory and is often glossed over in popular histories of science. Although the agency of evolution in creating the diversity of living things was by then widely accepted in the scientific community, Darwin's claim that natural selection, especially the idea of natural selection by means of small differences, had been the chief driver of evolution was unpersuasive to many. To the critics, natural selection acting on small changes did not seem a powerful enough mechanism to account for the creation of complex new characteristics of organisms (see note 8 and Box 6.2). By the late nineteenth century, many scientists viewed Darwin's theory of evolution by natural selection as flawed and inadequate.

The popular opposition to Darwinism, however, was not about such relative subtleties. It was grounded in a frank disbelief that all living things from amoeba to ape to humans, were related in the way that Darwin

BOX 6.2 A Key Issue in Darwinian Evolution:
Mutations of Small versus Large Effect

The central issue for Darwin and his critics was the relative importance in evolution of *variations* (mutations) that involved small versus large effects. Of those who accepted the reality of evolution, Darwin was almost alone in believing that it proceeded entirely by the accretion within populations of variations of small visible effect. His reasoning was based entirely on the analogy to artificial selection, where breeders of new strains of animals or plants did exactly that, selecting minute differences and amplifying them over time and through multiple generations, continuously selecting for those differences. The importance he placed on this fact is indicated by the fact that he begins *The Origin of Species* with this subject. He argued from this that natural selection would have the same potential efficacy with variations of small effect. In contrast, virtually all the other important figures, including Thomas Henry Huxley and St. George Mivart, felt that only mutations creating noticeable large effects must be the actual source material for natural evolution. Such visible, dramatic changes were termed *saltations*, and most nineteenth-century biologists who believed in the reality of evolution were *saltationists*. They argued that the tiny differences Darwin argued for would have insufficient traction for natural selection to act on—in contrast to the situation in which breeders were deliberately applying such selection—so they could not believe that this mechanism really had the potency to create truly novel features. Because Darwin had so firmly nailed his colors to the mast on the importance of variations of individually slight effect, while most scientists found it unpersuasive, his theory that natural selection acted primarily on such variations was seen as fatally flawed by many of his colleagues. While he had succeeded in convincing most scientists that evolution was a reality and must be the explanation of most perhaps all differences in species, he failed to convince them that his theory explained how evolution worked. It would be about sixty years before Darwin's position began to be vindicated within the new field of population genetics. The rough solution to the paradox of how bold new properties can be generated by the accretion of small effects is tied up with the phenomenon of *emergence* as described in the text.[1]

---

1   The history of Darwinism's decline in influence over more than sixty years—from the publication of *The Origin of Species* to the early 1930s—is given in Peter Bowler's *The Eclipse of Darwinism* (1983). For a relatively recent evaluation of the question of the phenotypic size effects of mutations on which natural selection can act, see Orr (1998).

claimed. The specific idea that human beings had descended from ape-like creatures simply added insult to incredulity. This resistance to Darwinism was buttressed in the case of human evolution by the absence of fossil evidence linking humans and ape-like animals, a gap immortalized by the term *the missing link*. The absence of such evidence was all too

readily taken as evidence for the absence of the proposed evolutionary connection. Such an inference is, of course, a classic mistake in logic, but it
was widespread and contributed to the scepticism of many people.[17]

As we know today, the early Darwinian story of human evolution lacked
not a single so-called missing link but many. The first to be discovered
were those of the species (or species group) now known as *Homo erectus.*
The initial skull was discovered in 1891 by Eugene Dubois (1858–1940),
a visionary Dutch amateur. He had set out to find fossils of human
precursors in Southeast Asia, specifically Indonesia, in the belief that
modern humans had originated in Asia, the predominant scientific view at
the time. Those skulls were distinctly more primitive than those of the
Neanderthals. Sadly for Dubois, his interpretation was widely disputed
and discounted by most scientists for several decades, hence it had little
immediate influence in the field itself or on public perceptions. Dubois
himself died a disappointed man several decades later, even doubting
the significance of his discoveries and suggesting that perhaps the fossils
were after all an extinct large gibbonlike primate. Actually, he had been on
the right track as shown later when *H. erectus* fossils were found in China
("Peking man") in the late 1920s; by the late 1930s those fossils and
Dubois's discoveries were taken seriously. By then, however, Darwinian
evolution had resurrected itself as a scientific theory, one to which the
scientific community was giving increasing support after decades of
vigorous dispute.[18]

It was, however, the rich lode of discoveries of fossil hominins in
southern and eastern Africa that came to fill the evidentiary void between
ape-like and human-like creatures. The first of these fossils was found by
workers at a quarry in South Africa in 1924 and then soon brought to a
young Australian scientist, Raymond Dart, who had recently arrived at
the University of the Witwatersrand from a post in London. It was the skull
of an infant hominoid primate. Dart could see from the characteristics of
its teeth and the position of its foramen magnum (the opening in the skull
for the spinal cord) that it was neither ape nor modern human; he named
it *Australopithecus africanus.* Dubbed by the press the "Tuang child" after
its site of discovery, Dart published its first description in the premier scientific journal, *Nature,* in February 1925. Though his conclusion that it
represented a primitive ape-like ancestor of humans was controversial, its
potential importance was recognized. In the same week as the publication

of the *Nature* article, one major South African newspaper hailed it as the possible long-sought missing link between ape-like creatures and humans. By the late 1940s, among **paleoanthropologists**, there was general acceptance that the Taung child skull represented a possible human precursor species. *A. africanus* is shown in **Plate 11**.

Over the next six decades, but accelerating from the early 1970s, larger numbers of hominin fossils of different ages and different degrees of completeness were unearthed in Africa. They included new and different species of *Australopithecus* and new members of the genus *Homo,* both earlier and later than *Homo erectus.* As a result, the major gaps between the lineage of the great apes and the first hominins, the lineage of the australopithecines and *Homo erectus,* and finally the lineage of *Homo erectus* and *Homo sapiens* were filled in.

From First Hominins to *Homo sapiens*:
II. The Inferred Phylogenetic Pattern

A diagram of the time of appearance and approximate duration of some of the better-known species of hominin, as deduced from their fossils, is shown in Figure 6.3. The picture is provisional, of course, and will almost certainly be revised as new fossils are unearthed. Altogether, more than 26 possible hominin species have been proposed, but some of the species designations, which are often based on a small number of bone fragments, will not survive. Nevertheless, it provides a rough guide to the evolutionary history of the hominins. The great majority of the fossils come from the southern and eastern parts of Africa, from what is present-day South Africa at the southern tip of the continent to Ethiopia at its northeast corner via Malawi, Tanzania, and Kenya. One critical fossil, a skull, which comes from Chad in the northwestern Sahara, is the earliest. Despite those early hominin fossils from western Africa, it is not clear where precisely in Africa the hominin lineage originated.[19]

The sequence of hominin types can be divided roughly into three groups that correspond to three discernible stages in their evolution. The first and least well-characterized were ape-like early hominins with genus designations of *Sahelanothropus, Orrorin,* and *Ardipethicus,* all found within the period of 7 mya to 4.5 mya. They possessed distinctive but

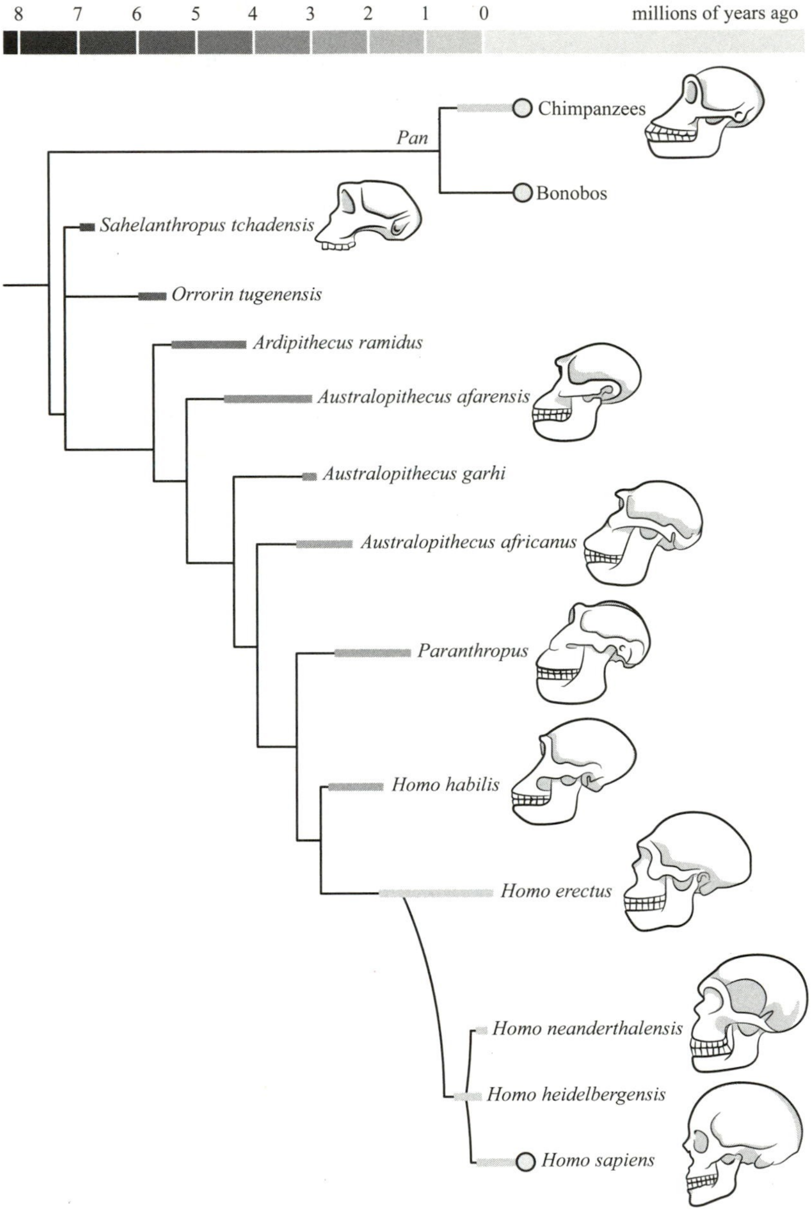

FIGURE 6.3 Estimated periods and durations of existence of the better-characterized hominin species based on fossil evidence. Estimated divergence points and even various aspects of the phylogeny should be regarded as provisional. Nevertheless, it provides a good present-day picture of the hominin phylogenetic tree and dates of the major branch points.

subtle cranial trait differences from earlier apes and the placement of the foramen magnum in *Sahelanthropus* indicates that it held its head vertically. They may therefore have been bipedal to a degree, but they lived in forests and were probably largely arboreal.

The second stage consists of the australopithecine hominins, which are now considered to comprise four species of the genus *Australopithecus* and at least two of the genus *Paranthropus,* the latter distinguished primarily by stronger, more robust skeletal features, both in the head and the body, and slightly larger brain sizes. More ape-like than human in their faces, yet less so than the earlier species, they resembled chimpanzees in both body and brain size. Yet they differed from earlier hominins as judged from bone structure and preserved footprints in (probably) being the first fully bipedal primates. The earliest of the australopithecines, *A. ramidus* and *A. anamensis,* appeared 4.5 million to 4.0 million years ago and the last, *P. boisei,* became extinct about 1 million years ago.

The third stage in hominin evolution began with the advent of the genus *Homo,* which overlapped that of the australopithecines in both time and space. In the now traditional accounting of the history of *Homo,* the earliest definable species types are those classified as *H. rudolfensis* and *H. habilis,* and their beginnings date to approximately 2.3 mya to 2.1 mya. *H. rudolfensis* had a somewhat larger cranial size than *habilis* and a flatter, taller, and more vertical face. The morphological differences between these earliest *Homo* species and their presumed ancestors, australopithecine hominins, were relatively minor and primarily involved somewhat larger body size and cranial capacity. The trait that was initially used to designate these two species as *Homo,* however, was not anatomical but behavioral: stone toolmaking. This was deduced from the findings of primitive stone implements for cutting meat and associated with the earliest fossils of *H. habilis.* A reconstruction of *Homo habilis* is shown in **Plate 12**.[20]

The ensuing changes in the evolution of *Homo,* however, show an accelerated rate of change, particularly in head and face morphology, body shape, and brain size. Recall that increasing body size is a general trend in primate evolution, as it is within the mammals as a whole (though not without exceptions and reversals). *Homo erectus* typifies this trend; this species possessed larger and more human-like heads, faces, and bodies. Its body shape, in particular, showing a barrel chest rather than the conical chest shape of the earlier species and of apes, is distinctly more human in morphology. That probably, in turn, signifies a change in diet, with a reduced dependence on vegetation, which requires longer digestion time, and a heavier reliance on animal protein. A reconstructed *H. erectus* is shown in **Plate 13**.

The first *H. erectus* fossils date to about 1.9 mya in eastern Africa, yet apparently this species, unlike the preceding hominin species, was soon on the move and over great distances. Judging from fossil finds in Asia, *H. erectus* had spread to Asia within a few tens of thousands of years, reaching Indonesia by 1.66 mya. This was, apparently, the first great "out-of-Africa" movement of a hominin species and was quite possibly succeeded by one or more migrations by *H. erectus* or similar species. (The last major hominin migration involved our own species, *H. sapiens,* which began approximately 70,0000 to 60,000 years ago, as will be described in Chapter 8.) A striking feature of different *H. erectus* populations was the range of body sizes within those populations. Such variation probably reflects the property of **developmental plasticity** and presumably, in this case, the effects of variations in environment and food supply. Some of that developmental change presumably later became "fixed" as genetic change, leading to genetic differentiation of some populations, possibly forming new species. Today, indeed, *H. erectus* is viewed as a *species cluster* rather than a single species.[21]

H. erectus was almost certainly the precursor of several later hominin species, although the dividing lines between different species are often not clear, given the variability within populations, overlaps in trait ranges, and geographical location between different populations. Nevertheless, presumptive later species identifications are possible. In particular, one of two species, *H. ergaster* and *H. heidelbergensis,* was the probable immediate precursor species of *H. neanderthalis,* the Neanderthals, who first appeared in Europe between 400,000 and 200,000 years ago. (See **Plate 14**.) Correspondingly, the ancestor of our species, *H. sapiens,* was probably either *H. heidelbergensis,* whose fossils have been found in both Africa and Europe, or possibly another, wholly African species, *H. rhodensis* (but see endnotes 19 and 22).

The question of when "modern" *Homo sapiens* arose from previous "archaic" *Homo* species *(H. heidelbergensis, H. rhodensis)* is, like all such time-of-origin questions, impossible to answer precisely, but here the fossil findings and the molecular clock evidence converge to a large degree, placing the date at about 200,000 years ago (as will be described in more detail in Chapter 8). If so, the history of our species falls into approximately only the last 3 percent to 4 percent of hominin history.[22]

How does the emergence of the human face fit into this global picture of hominin evolution? As reconstructed from fossil skulls, the major changes in face structure took place essentially coincident with the rise of the genus *Homo* during the past 2 million years. These reconstructions have been carried out by the methods of forensic science as based on a detailed knowledge of human anatomy. The contours of the skull determine the patterns of the overlying muscles in characteristic and thus predictable ways. These, in turn, dictate the placement of other soft tissues. From these forensic reconstructions, what the faces themselves looked like in the flesh can be deduced. Several have been shown in **Plates 9-14**. Even in this highly abbreviated pictorial series of hominin evolution, one can see the approach to modern human faces, despite the necessary caveat that one should not view it as a strictly linear progression.[23]

Probing the Genetic Foundations of Two Distinctive Human Facial Traits: The Loss of the Muzzle and the Gain of a Forehead

When the face of the apelike australopithicines is compared with that of modern humans, the human face is quickly perceived to be both shorter and more vertical. The human head is rounder, the face is topped by a proper forehead, and the muzzle is gone, with the jaws essentially level with the plane of the midface. These gross shifts are, of course, correlated with many others, but at the visual level the immediate impression is that human faces differ from ape-like faces by the gain of a forehead and the loss of the muzzle. To understand how this transformation took place, we need to know something about the genetic changes that were involved and the ways that natural selection promoted these changes. Here we will look at the first aspect, the genetic changes, beginning with the reduction of the jaws.

On Losing the Muzzle
Humans lack something that most mammals possess: a muzzle. A muzzle is simply the forward-projecting part of the face that contains the jaws. The primary function of forward-projecting jaws is simple: to grab food, often

living prey. While many mammals have muzzles without being hunters or carnivores, they are descended from ancestral species that hunted and used their mouths to seize other animals, whether insects or vertebrates. (In herbivorous mammals, the muzzle serves comparably for seizing plant material while also protecting the eyes from leaves or twigs by keeping them a safe distance from the food being consumed.) Conversely, the reduction and loss of a muzzle signifies the ability to obtain food without grabbing it with one's month. (Other mammals that use their forepaws or hands for obtaining food—for example, marmots and raccoons—in principle can dispense with a muzzle, but these forepaw adaptations are presumably relatively recent.) Smaller jaws also comport with dietary items that need less chewing, as seen in highly frugivorous primates, namely, most of the haplorrhines.[24]

The earliest primates had typical mammalian muzzles, as do contemporary strepsirrhine primates. Muzzle reduction along the LHLP began with the anthropoid primates, as can be seen in the fossils of ancient haplorhines as well as the great majority of living anthropoids. The baboons, with their prominent muzzles **(Plate 15)**, are an exception, but that trait almost certainly involved an evolutionary reacquisition within the anthropoid primates. The general reduction in muzzle development among the anthropoid primates testifies to the increasing importance of the forepaws or hands in obtaining and handling food, while the redevelopment of the muzzle in baboons was presumably related to other unknown selection pressures.

Nevertheless, the great apes, who primarily use their hands to put food in their months, have muzzles, albeit reduced ones, as did the earliest hominins. Perhaps these minimalist muzzles are a neutral trait, involving neither a special adaptive function nor involving a cost in their possession. What is certain is that during hominin evolution along the hominocentric path, the muzzle was reduced to the vanishing point, a feature that contributes markedly to the distinctiveness of the human face. While we have jaws and teeth, which are as essential to us as they were to our muzzle-bearing ancestors, our jaws do not project. As a side effect, this permits a greater range of expression with our mouths, particularly given the absence of fur on our faces, another ancient anthropoid primate feature. In considering the genetic and developmental changes that made possible the reduction and eventual loss of the muzzle (while permitting retention of jaws and teeth), the basic facts about jaw development (Chapters 2 and 3) must be relevant.

In particular, remember that the upper and lower jaws develop from the paired maxillary and mandibular facial primordia, respectively. As we saw, their formation requires sonic hedgehog (SHH) signaling to promote cell division in and hence growth of these facial prominences. The greater the SHH signaling, the wider the face. Humans with wider than normal faces are termed **hyperteloric**, and the evidence from animal experimental systems indicates that it is a consequence of excess SHH signaling. In addition, as we saw for the cichlids, mutants in the SHH signaling pathway can affect the particular shape and extent of projection of the jaws. It is but a small extrapolation to conclude that the extent of the muzzle is probably, in part at least, a function of the strength or duration (or both) of SHH signaling during development of the maxillary and mandibular facial primordia. As we have seen, a second set of vital molecular components in mammalian jaw development are the bone morphogenetic proteins (BMPs). Finally, fibroblast growth factor (FGF) signaling is important; given the close connections between FGF signaling and SHH signaling, the two are probably linked in promoting jaw outgrowth. There is, indeed, some evidence, in chicken embryos that FGF8 and SHH work synergistically in beak development to promote growth. In contrast, the Wnt signaling pathway—also important in face development—seems to be involved primarily with setting certain patterning details of the upper jaw and midface rather than muzzle extent.[25]

These signaling pathways are crucial for muzzle formation during fetal development in mammals. A possible corollary is that a muzzleless mammalian species such as ours may exhibit reduced signaling in one or more of these pathways in the maxillary and mandibulary facial prominences following prominence fusion and completion of the face. Of course, the face grows and expands between early fetal development and birth. What does not happen in humans, however, is distal extension of the jaws. Cell proliferation and growth of the face take place during this period but in breadth and depth, not in the antero-posterior axis, the dimension that would generate a muzzle.[26]

For other mammals, however, what happens in fetal development is not the whole story. Newborns of many mammalian species show reduced muzzles relative to adults. Their heads are rounder and their faces flatter than the adults of their species, creating a stronger resemblance to human newborns than their adults possess (something that doubtless activates humans' parental responses to young mammals). This flatter face facilitates nursing and may have been selected specifically in connection with it as discussed previously. The consequence is that most muzzle outgrowth in many mammals happens postnatally during juvenile development in those species. If SHH, BMP4, and FGF8 signaling is what regulates the extent of muzzle outgrowth, then it presumably involves some kind of setting of postnatal growth in the mid- or late-gestation stages of fetal development. But, of course, other pathways and processes may be involved.

On Gaining a Forehead

In Chapter 3, we examined the basis of another distinctive human feature, the forehead, which contributes to the verticality of the human face. We saw how its existence reflects in part the large increase in brain size that accompanied the evolution of our species. The trajectory of brain size increase in hominin evolution is shown in Figure 6.4. Relative to body size, the human brain is roughly three times larger than that of living chimpanzees or extinct australopithicines. As discussed previously, increased human brain size was not the only factor producing the rounder head and more vertical face, but it was a major factor.

Large human brain size ultimately reflects a longer and more extensive period of production of neural progenitor cells within the cerebral cortex relative to other animals. More progenitor cells lead to more radial units that, in turn, produce a larger cortical sheet, which is accommodated within the human skull by extensive folding to produce the characteristic sculpted shape of the human brain with its gyres and sulci. The disproportionately large human brain prompts questions about both the genetic developmental mechanisms and the selective reasons that favored it. Here we will look briefly at the possible genetic foundations of this increase in neural progenitor cells, and later we will examine the possible adaptive functions of a larger brain. As will be seen, the frontal part of the brain, the prefrontal cortex, which lies just behind the forehead, houses some of the special neural circuitry responsible for the distinctive mental properties of human beings, including those that involve complex decision-making. The enlargement of the brain, including the prefrontal cortex, is

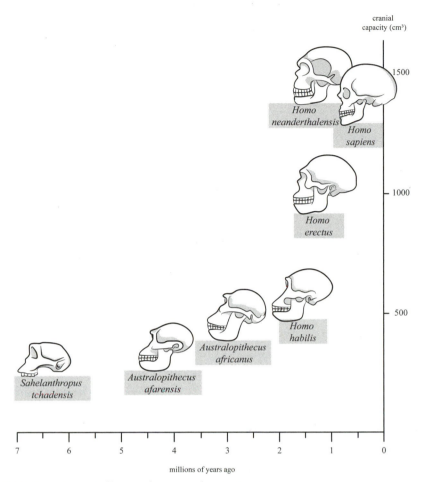

cranial
capacity (cm³)

1500

*Homo
neanderthalensis*

*Homo
sapiens*

1000

*Homo
erectus*

500

*Homo
habilis*

*Australopithecus
africanus*

*Australopithecus
afarensis*

*Sahelanthropus
tchadensis*

7   6   5   4   3   2   1   0

millions of years ago

FIGURE 6.4 Diagram of hominim brain size enlargement over time.

therefore relevant to the face in two general ways, both in its shape (morphology) and in what may be regarded as its behaviors.

In light of the previous discussion about the relationship between neural
progenitor cell production and growth of the cerebral cortex, it must be
the case that the evolution of a large brain is related to the capacity to generate more neural progenitor cells. Correspondingly, it might be suspected
to be related to the activities of the *microcephalic* genes, whose wild-type
activities are essential for neural progenitor production. That possibility
has been considered and investigated. Indeed, early comparative studies
of one of these genes, *ASPM,* revealed evidence of its rapid DNA sequence

evolution in the primates. That increase, however, is believed to have begun much earlier in the hominoid lineage, hence long before the major expansion of the human brain. Yet a more fundamental objection to trying to explain increase in brain size in terms of the microcephalic genes is that it is difficult to imagine how any changes in these genes— which function simply to ensure accurate symmetrical divisions—would *extend the number* of symmetrical divisions and lead to more neural progenitors.[27]

What seems to be needed is that these genes stay on over a longer period during human cerebral cortex development than during brain formation in other primates. Extended activities in time would permit the extra production of neural progenitor cells. If so, the questions concern matters of gene regulation. What are the regulatory genes that are upstream of the microcephalic genes and might their expression be extended in humans, relative to other primates, to give greater neural progenitor cell production?

One possibility involves the expressions of specific transcription factor genes in different regions of the early developing cerebral cortex. In the early embryo, it will be recalled, the telencephalon is specified in part by inhibition of Wnt activities, which, in turn, inhibits development of diencephalon-like properties. The dorsal telencephalon, in turn, which will give rise to the cerebral cortex, soon begins to express four different transcription factors in the regions that will give rise to the major regions of the mature cerebral cortex (the frontal, parietal, temporal, and occipital lobes). This process of area delimitation is termed **arealization** and is initiated by the production of signaling molecules, the same ones we have already encountered in face development (FGFs, SHH, Wnts, BMPs) from specific locations in the newly formed brain. These molecules diffuse through regions of the developing brain, creating overlapping gradients, and those combinations of concentrations in specific regions evoke unique gene-expression patterns of downstream genes. Each transcription factor required for arealization is necessary to promote the production of secondary neural progenitor cells within the area it helps specify; mutations that inactivate these genes greatly reduce neurogenesis in the region in which each is expressed. Conversely, enhancement of neurogenesis in the developing cerebral cortex in human evolution could have been through temporally extended expression of these four transcriptional regulatory genes.[28]

In addition, other genes promote neurogenesis in the developing cerebral cortex. In primate brains, the subventricular layer of the neocortex is expanded relative to the ventricular layer. In itself, this serves to expand the pool of secondary neuroprogenitor cells, presumably in all regions of the developing cerebral cortex. One gene required for this expansion is called *FoxG1* (formerly *Brain Factor-1*). A mouse mutant that is defective in *FoxG1* shows greatly reduced cerebral cortices. Another gene expressed in the developing brain and implicated in expansion of the human brain is *NBPF*. This gene includes multiple copies of a DNA sequence that encodes a special delimited **protein domain** termed *DUF1220*. In humans and chimpanzees, the number of DUF1220 repetitions within *NBPF* is greatly expanded relative to other primates and other mammals, leading to corresponding numbers of the encoded domain in the protein itself. Thus, humans have 272 of these domains in *NBPF*, chimpanzees only 125, and the mouse lemur and the mouse only two and one, respectively. Several additional findings involving human genetic variants implicate the number of DUF1220 domains in *NBPF* as related to human brain size (and to neural functioning). It is unknown whether these two neurogenesis-promoting genes (*FoxG1* and *NRBF*) or others yet to be discovered are upstream or downstream of the arealizing genes, though probably downstream. The key point is that specific genes and processes that might have been involved in the evolutionary expansion of the human brain are now known.[29]

On Darwinian Incrementalism and the Question of Whether Big Qualitative Changes Can Result from Accumulated Small Steps

Both the reduction of the jaws and the growth of the brain during hominin evolution can be conceived of—and almost certainly involved—a series of individually but relatively small changes and effets. This pattern may be termed *Darwinian incrementalism*. Indeed, in human evolution, *all* of the major physical or morphological changes in shape of the hominin face, head, and trunk—from the australopithecines to *Homo sapiens*—are most readily interpreted as involving incremental changes of differential growth, both positive and negative with respect to australopithecine body proportions. Saltations need not be invoked.[30]

Darwin's posthumous vindication on this point, that major morphological changes can result from accumulated small ones, is little noted today. One reason is that his position is often misstated as the claim that evolution is "gradual," meaning slow. Today, we know that evolutionary change can proceed quickly. Indeed, Darwin knew this too; he realized that evolution could and did proceed at different rates. His position was thus a statement not about rates but of pattern: that evolution involves *a* sequence of change by relatively small steps, though the tempo of those steps may vary. The hominin fossil record, despite its existing gaps, essentially bears him out. The increase of brain size in the hominins, as deduced from fossil skulls, provides a good illustration within the more general case of human evolution. That increase was initially slow, indeed gradual in the usual sense, but in the final 2 million to 1 million years it accelerated . The key point is that it can be seen as a trend without large discontinuities. Not all hominin lineages experienced it—the australopithicines maintained a fairly constant brain size over 2 million to 3 million years, and one *Homo erectus*-derived population experienced a reduction in brain size—but the overall trend is real.[31]

This example, however, raises two questions: the first, a specific question about hominin evolution, and the second, a general one about Darwin's incrementalist mode of evolutionary change. The specific issue concerns the selective pressures that must have driven the increase in hominin brain size. There must have been some such pressures because brains are expensive in terms of energy, hence in terms of nutritional intake. Making up about 2 percent of human body weight, the adult human brain consumes about 20 percent of daily energy use. To evolve bigger brains is therefore costly, and there must have been some offsetting benefits that outweighed the costs. This is one of the big questions in human evolution, and we shall return to it and examine the possibilities.

The more profound question is the one that vexed Darwin and his critics: how can incremental increases in a particular property—brain size in this instance—translate into large *qualitative* differences—mental qualities in this case? It is clear, for example, human brains not only became bigger but also capable of new tasks. Decades of testing of the cognitive abilities of chimpanzees have indicated that there are some genuine and inherent differences in mental capacity between chimps and humans. For instance, while they can learn to use colored shapes as symbols for words, they learn far fewer and do so more slowly than human children. One returns to Wallace's doubt: was natural selection really capable of creating truly novel human mental capabilities by Darwinian incrementalism?

The simple if insufficient answer is that there are often thresholds or tipping points in many processes involving quantitative change when the next increment of change triggers something new. This is true of many physical conditions, biological phenomena, and psychological properties. A simple example of such is the freezing of ice: progressively cool water and it remains liquid until the temperature reaches 0 °C, at which point the water becomes ice. A catchall term for such changes is *emergence*. New properties that could not be predicted—and often cannot be explained— suddenly manifest themselves or *emerge*. Such is almost certainly the case for various mental properties in relationship to increasing brain size that appeared during human evolution. In the nineteenth century, had either Darwin or Wallace made this argument, it would undoubtedly have seemed like begging the question to many of their critics, though Darwin's disciple, T. H. Huxley, understood it. (As an example, Huxley cited the properties of liquid water itself as an unpredictable, indeed emergent, property from its atomic constituents, oxygen and hydrogen). Today, the phenomenon of emergence is a commonplace observation. Nevertheless, each case of emergent properties demands an explanation or at least an attempt to reach one.[32]

As a test case of the ability of Darwinian incrementalism to create a striking new property, let us take a particularly difficult one: the origins of language and speech. These are capacities unique to our species, although some forms of sophisticated oral communication in songbirds and a few mammals, particularly whales, are known. Can a Darwinian framework accommodate the birth of language? As discussed below, we will see that the answer is probably "yes." That excursion, however, will bring us back to the subject of the face since the evolution of facial expressive capacities was almost certainly connected to the emergence of language.

Gestures and Facial Expressions: The Beginnings of Language?

Language is not only the most distinctive of human behavioral traits but also the most complex, requiring a unique vocal apparatus and elaborate and intricate brain neural circuitry, which is still poorly understood. In fact, the subject of the origins of language has been one of the most contentious and difficult of all scientific issues.

Perhaps surprisingly, an appropriate historical point in considering the birth of language is the human hand. In *The Descent of Man,* Darwin assigned a key role to the development of the human hand in promoting the growth of human intelligence (and brain size). He argued that three special human traits—bipedalism, increasingly sophisticated stone tool manufacture and use, and brain size—had all been closely linked. The thesis was that bipedalism "freed" the hands from any role in body movement. In Darwin's view, this soon led to their employment in the making of stone tools. The use and demands of that technology, in turn, would have quickly prompted further changes that required higher cognitive ability and therefore larger brains. It was not a poor argument, but the now-reconstructed pattern of brain size increase over 4 million years has undermined it. First, there was a lag of about 2 million years between the onset of australopithecine bipedalism and the use of stone tools. Second, the most rapid increase in brain size began well after the first stone tools, more than a million years (paralleled by a similar lag in their technological improvement). This is not the pattern that Darwin suggested but rather a temporally extended case of mosaic evolution.[33]

Although Darwin's scenario was wrong, his identification of the human hand as a key innovation in human mental evolution—and the evolution of human mental abilities—was probably right. The aspect of its use that he missed was its expressive capacity.

Human hands carry out three basic functions: (1) grasping objects, (2) manipulating objects with precision, and (3) signaling to communicate wishes, intentions, and meanings to others. Furthermore, this was probably the temporal sequence in which those abilities arose. In particular, grasping and manipulating objects are ancient primate capabilities and would have surely been characteristic of stem primates. Grasping by fingers without claws is requisite for primate tree climbing and depends on primates' opposable first digit, which permits efficient movement along tree branches. Similarly, the ability of primates to use their hands to manipulate objects such as food items and sticks was also undoubtedly an early primate trait.

In contrast, signaling with the hands is seen widely only among the hominoid primates and almost certainly is a far more recent and restricted primate ability. Apart from begging for food with an outstretched arm and open palm, which is seen in some monkeys, nonhominoid primates do not use hand gestures. Unlike grasping and manipulating objects, traits that are useful for dealing with the physical world, the purpose of signaling with the hands is social—the communication of intentions and requests between individuals and hence the transmission of simple ideas. Among humans, some examples are: pointing the index figure to indicate "look there"; Churchill's hand gesture with raised index and middle finger and the other fingers folded to symbolize "V" for victory; a wagging finger in front of a child's face to signify "you should not do that!"; a circling index finger close to and pointed toward the temple, meaning "he's crazy"; opening and closing fingers in an open palm to signify "come here." Although the repertoire of hand gestures in apes had earlier been thought to be somewhat limited compared to humans, recent work indicates quite an extensive set of gestures for communicating in wild chimpanzees. The implication is that early hominins also would almost certainly have had extensive gestural capability.[34]

Humans, however, have taken this ability even further. In particular, the development of various sign languages, which can have just as great a grammatical and syntactical complexity as any spoken language, suggests there is nothing inherently limited in the use of manual gestures to convey meaning. Furthermore, the autonomous development of a complete and complex sign language by a group of Nicaraguan deaf children indicates that the ability to build highly complex gestural communicative systems is an inherent human propensity.[35]

Sign languages, however, do not rely solely on hand gestures; they also involve facial gestures—namely, facial expressions. Thus, hands and faces are intimately linked in gestural communication systems. Furthermore, it is not just humans who display this linkage: in the great apes, face-to-face interactions also involve both hands and faces. Previously, I described facial expressions as constituting a form of backup communication system for speech in which the play of expressions reinforces, modifies, or sometimes (unintentionally) contradicts the verbal message. Even if one does not hear the spoken words, the facial expressions themselves can convey much about the speaker's emotional state and sometimes his or her intentions. Indeed, a large range of relatively subtle human expressions are used in combination with speech; in effect, they provide the emotional subtext of the spoken words, which is often more important than the words themselves.

Given the ancient hominoid roots of gestural expression, it is reasonable to think that a gestural communication system was not a supplementary communication system but initially the main one, preceding full human speech and language, though undoubtedly accompanied by some vocalizations. This idea has roots in the eighteenth-century writings of the Abbé de Condillac (1746), but it has been recently revived and elaborated in light of recent knowledge as the "hand-to-mouth" theory of language evolution.[36]

There is, however, a major problem with the idea that purely gestural language preceded spoken language: the nature of the transition from gestures to speech. It is extremely difficult to imagine how a purely gestural system could have evolved directly into a speech-based one. It seems far more probable that language evolved as part of a system employing both physical gestures and increasing vocalization. Some of those gestures, in fact, could have involved hand-to-mouth motions. For example, it could have been hand motions to the mouth, in bringing or signaling the bringing of food items to the mouth, that constituted the first hand-mouth gestures, and these would have gradually expanded within the anthropoid primates to include facial and articulatory gestures involving parts of the mouth. Later in hominin evolution, this could have expanded to more gestures involving the hands and, more frequently, accompanying specific vocalizations that reinforced the meaning of the gestures. With increasing evolution of human sound-production capabilities, the range of sounds—and eventually words—would have expanded and been incorporated in the system, eventually displacing gestures as the primary conveyors of meaning. In effect, gestures and vocal communication would have coevolved.[37]

The preceding scenario provides a rough idea of how a gestural system might have evolved into a speech-based communication, but it leaves many of the crucial details untouched. Three issues in particular should be noted. First, it neglects the evolution of the human vocal apparatus itself and its capacity for making the full range of basic sounds, **phonemes**, that is characteristic of human speech. This ability is lacking in our primate cousins and cannot be taught to young chimpanzees or bonobos, though valiant attempts have been made. The human vocal system involves both the relatively low position of the larynx in the throat, which is necessary to produce many of these sounds, and various aspects of the geometry of the oral cavity that seem to be specific to *H. sapiens.* That oral cavity configuration may not have existed even in archaic human species such as *H. erectus* or *H. heidelbergensis,* though that conclusion is controversial. Many of the necessary evolutionary changes in the vocal apparatus would have involved relatively subtle developmental shifts in the tongue and parts of the oral cavity but would have markedly increased the range of possible sounds.[38]

A second and more profound puzzle concerns the changes in brain neural circuitry that made both language and articulate speech possible. The acquisition of language and the evolution from a primarily visual-gestural system to a vocal-sound based system must have involved significant "rewiring in the brain" and undoubtedly expansion of the some of the underlying neural infrastructure. Embedded within this general phenomenon of brain reconfiguration and expansion, however, is the most mysterious aspect: the evolution of the mental capacity for symbolic representation, in particular the use of words as symbols via the complex medium of syntax. Chimpanzees and bonobos have some capacity for symbolic thinking, which can be brought out and developed in training by humans, but the human ability in this respect is orders of magnitude greater and must be considered a qualitative difference, not a mere quantitative one. The evolution of this capacity must have involved elaborate rewiring and expansion of neural circuitry in several brain areas. We will, however, defer discussion of that phenomenon until the next chapter.[39]

The third issue may be considered a special case of the second (rewiring and expansion of brain circuitry to create new functions), but it merits separate attention—namely, how facial expressions during speech are linked to the spoken words so that the expressions and words together convey the information. As we have seen, speech involves facial movements that are experienced by the listener as gestures; the ability to lip-read depends on the fact that every spoken word is associated with and requires a characteristic mouth configuration. Neural circuitry for producing sequences of sounds that are word strings is essential, but precisely how speech capacity is neurally linked to facial expressive capacity, which in turn is linked with emotional states, is unknown and has received little attention. We will also return to this issue in the next chapter.

Yet however much remains obscure about the origins of human language and speech capacities, their evolutionary roots have become clearer in recent years. In particular, the roots of facial expression in conjunction with speech may be glimpsed in anthropoid primates, especially Old World monkeys and apes. Those connections lie in three rhythmic facial expressions these animals display: lip smacking, tongue smacking, and teeth chattering. In particular, lip smacking, which is used in various social situations, including face-to-face interaction, has the same rhythmicity in rhesus macaque monkeys (about six per second) as their vocalizations, although these take place separately. The same rhythmicity is seen in human speech in the production of spoken syllables. It seems probable, therefore, that the evolution of speech in hominin evolution involved a "marrying up," via neural rewiring, of the mimetic muscles and the vocalization machinery involved in speech. A possible precedent exists: male gelada baboons have their own version of lip smacking, "wobbles," which they employ with vocalizations in courting females. No other baboons have been seen exhibiting this behavior, hence it must reflect an evolutionary acquisition of this ability in this baboon species, one involving changes in neural circuitry and perhaps parallel to some that occurred in our hominin ancestors as speech capacity evolved.[40]

Diet and Sociality as Profound Shapers of Primate Faces

In this chapter, I have sketched some of the important events in the evolution of the distinctive features of the human face beginning with the earliest primates, tiny arboreal creatures, but accelerating with the appearance of the anthropoid apes, experiencing further changes with the appearance of the early hominins, and concluding with the evolution of modern *Homo sapiens*. Those events transformed what was initially a fairly standard mammalian face—one covered with fur, dominated by a muzzle for grasping food, possessing no or minimal forehead, and with eyes positioned fairly laterally—to a furless, muzzleless face topped by a real forehead fronting a large brain and close-set, forward-positioned eyes. The whole transition from earliest anthropoid primates to us took place over approximately a 60-million-year period, but the key changes occurred in two relatively short periods. The first, which established the basic anthropoid primate face, probably took place within the first 5 million to 10 million years of the appearance of the primates. That face lacked fur, had a reduced muzzle, and had relatively closely spaced eyes. In appearance, these were still far from human but were more human-like than the faces of the earliest primates (or present-day strepsirrhines). The next major changes, which transformed generic anthropoid facial features into truly human ones, occurred 50 million or more years later and primarily within the last 2 million years; it was these that created the human face as we recognize it today.

Both sets of changes must have been driven by natural selection. Much about these selective forces is unknown, but some tentative conclusions are possible. First, all of the changes in primate evolution took place in what had become a relatively permissive external environment for their survival and evolution; it was free of the large sauropods that had dominated the food chain for more than 160 million years and whose presence undoubtedly slowed and thus limited mammalian diversification. The elimination of that particular threat at the beginning of the Paleocene gave mammals a chance to evolve to larger sizes, a trend seen in many mammalian groups, including the primates, as well as to diversify strongly. In the anthropoid primates, the increase in body size led to bigger heads with larger faces. One consequence would have been bigger eyes, which are better at close-up visual resolution and therefore useful for several things, including face-to-face social encounters.

Many evolutionary changes in the primates also involved the mouth, throat, jaws, and teeth, hence the entire apparatus for eating. Those divergences almost certainly reflected alterations in diet since different food sources require different modes of chewing. Those changes in diet, in turn, were consequent on alterations in climate. When climate changes, the vegetation changes; when vegetation changes, the herbivores that live on it must evolve new dietary habits (or die out); when the herbivore populations change, the omnivores and carnivores that live on the herbivores must also evolve or perish. The requisite evolutionary changes in jaws and teeth at each of these transitions must have come about through relatively subtle modifications of the developmental processes that create them, hence ones that should be relatively easy (not improbable) in genetic and developmental terms.

Thus, climate change inevitably generates selective pressures for structural change in animals, especially via changes in nutritional sources. The Cenozoic era, the period in which most of primate evolution unfolded, was notable for extreme climate changes compared to the preceding 185 million years of the Mesozoic. One such extreme was a remarkable period, approximately 55 million years ago, named the Paleocene-Eocene Thermal Maximum (PETM), in which Earth's average temperature shot up a remarkable 10 °C to create an ice-free planet and a period in which a great deal of mammalian diversification took place. The phenomenon was apparently produced by a huge increase in carbon dioxide into the atmosphere from sources unknown. Although the PETM lasted a relatively brief 150,000 years, it almost certainly and strongly affected mammalian evolution and that of all other complex living forms. It was during this period and probably within the space of only 25,000 years that *Teilhardina,* one of the earliest true primates, spread from Asia to Africa and Europe, and from Europe by then-existing land bridges to North America, displaying what may be a basic primate propensity toward range expansion when the conditions are right. We saw another example previously, in the spreading out of *Homo erectus* populations in the early Pleistocene, and in Chapter 8 we will examine our own species' rapid range expansion during the late Pleistocene.[41]

At the other end of extreme climate change, there were the ice ages of the Pleistocene, occurring during the last 2.5 million years of Earth's history, a period of enormous climatic variability, that undoubtedly influenced human evolution in multiple ways. In between the PETM and the Pleistocene ice ages were many less dramatic but highly significant events such as the long periods of cooling and drying out of Africa, which reduced forests and jungles and created savannahs and deserts. During these changes, African animals, including our primate ancestors, often moved into new territories and there either evolved (into new species) or simply went extinct. The change in diet from largely fruit and other plant matter, which the early forest-dwelling hominins survived on, to the more meat-based diet of later species of *Homo* on the savannahs, was undoubtedly driven by the consequences of climate change.[42]

All long-term changes of diet would have been accompanied by selective pressure for changes in the teeth and jaws. The reduction in muzzle size that the anthropoid primates experienced was almost certainly partly driven by dietary change and permitted by their ability to handle food items with their hands. This would have been particularly important in the picking of fruit, which probably constituted a major part of the nutrition of these early primates, judging from the teeth in early fossil skulls. Muzzles for seizing food would not have been necessary. Changes in teeth, in particular reduction of the canines in many species and elaboration of molars and premolars for processing plant material, would have been additional favored changes. All the evidence thus suggests that dietary changes were a major source of selective pressures that shaped primate faces over the past 55 million years or so. In this, there was continuity with the whole period of vertebrate evolution in which nutritional needs and changing sources of food played a major part in shaping jaws and hence faces.

There is, however, another set of evolutionary pressures that undoubtedly molded the evolution of the primates: social ones. Of all the major groups of mammals, the primates are the most consistently and extensively social. These social interactions go far beyond the minimal sociality required to find a mate or the relatively impersonal social environments of the large herbivores that graze in large groups for mutual protection. For nearly all living species of primate, and undoubtedly for most of those that have gone extinct, the life of the individual is heavily conditioned by being part of a social group. Take most individual primates out of their groups, force them to live in isolation, and in general most do not survive long. This is obviously true for humans, but it is also true for most primates. With apologies to John Donne for de-emphasizing precisely what he deliberately tried to stress, one can say that truly no primate is an island.

How would selection for traits that promote social functioning take place? Traditional evolutionary theory emphasizes that natural selection can only act on genetic changes that confer benefits to the individual (and its descendants). It also argues quite reasonably that selection cannot be for changes that only confer some future benefit for the group. Yet any changes that promote better social functioning in highly social animals must also create here-and-now advantages for the group and simultaneously favor the survival ability of individuals within that group. This process may be termed **social selection**. Correspondingly, such changes should be selected over time. Social interactions that involve sharing and cooperation and thus promote group and individual survival should not be seen as altruism, the process of extending help to others, often at a cost to oneself. Instead, such social interactions should instead be seen as a low-cost means of enhancing group survival that indirectly provide benefits to individuals.[43]

Social selection should not only promote better social interactions but also influence their kinds, depending on the characteristics of the species involved and the environments in which they live. The loss of facial fur is an example. Initially, this may have been driven by sexual selection, namely, the selection for a trait found in one sex that enhances the possibility of obtaining a mate of the other sex—which is, undeniably, one major form of social interaction—but as furless faces came to be present in both sexes, it probably acquired a wider function. (How an initially sexually selected trait, arising in one sex, might come to be present in both sexes will be discussed in Chapter 8.) What might that new role have been? I suggest that that function was the comparative ease of reading facial expressions. It would have been selected and spread because such expressiveness would have contributed to social cohesion, thus favoring survival of the individuals possessing this property.

A similar argument pertains to a distinctive human facial trait involving the eyes. We are the only mammal to have our colored irises set within a larger white area, the sclera, in such a way that the sclera shows. (Rarely, individual chimpanzees show this trait.) This serves to let fellow humans see our direction of gaze. This is a social trait and one whose selective advantage is again only explicable in terms of aiding social interactions. Another form of gaze ascertainment, though not involving the sclera, has been found in wolves, one of the most intelligent and socially interactive canids.[44]

Finally, there is our ability to use language. As we have seen, this trait may have originated in part with the ability to make gestures with the hand and the face. Although it could be argued, albeit implausibly, that traits such as hairless faces and the whites of the eyes might have arisen as neutral traits from **genetic drift**—possessing neither cost nor benefit—it is impossible to make that claim for language, the ultimate social trait. Language is only useful if one is interacting with someone else; it has no survival value to the isolated individual, and individuals who lack language as an instrument of thought and communication are as lost socially as if they were alone in an immense desert.

Thus, with the beginnings of the anthropoid primates, there began to be a shift in the selective forces that shaped the face. Up to that point and throughout most of vertebrate history, the principal factors that had influenced the structure of faces had been nutritional ones. Finding and ingesting the right kinds of food was the predominant selective force in shaping the evolution of vertebrate faces, in particular the detailed morphology of the jaws and teeth and the characteristics of the sensory apparatus. With the advent of the anthropoid primates, a new set of forces—social interactions—came into play. Diet and nutrition remained crucial for survival, of course, but sociality became a strong shaping force in the evolutionary molding of the face.

Accordingly, the focus of this account will shift from matters of external morphology to internal factors, specifically to mental processes and the brain as the ultimate source of changing mentalities and behaviors. In the next chapter, we will explore those matters directly and discuss how the brain and face evolved together to create what is today the human face.

— seven —

BRAIN AND FACE COEVOLUTION: RECOGNIZING, READING, AND MAKING FACES

Introduction: The Intimate Relationships between the Brain and the Face

By now, it will be apparent that the evolutionary history of the face is neither a small nor neatly delimited subject but one that spreads out and connects with many others. These include matters as diverse as diet, nutrition, and feeding behaviors; the evolution of hair; the requirements of suckling; brain size and complexity; modes of communication (both visual and oral); and more broadly, the social interactions of at least highly sociable primates. This web of relationships between the face, the body, and behaviors, however, is hardly surprising. After all, the face is intimately connected to the biology and activities of the individual in numerous ways, and its evolutionary history is bound to reflect those relationships.

Nevertheless, there is one set of links between the face and one other part of the body that is especially important and thus deserves particular attention: those connecting the face and the brain. These ties are indeed so numerous and close that it is inconceivable that either the brain or face could have evolved without influencing the evolution of the other. The term for such mutual evolutionary dependence is, as noted previously, **coevolution**, and this chapter will explore the ways in which the brain and face co-evolved.

One class of these relationships has already been examined: the developmental interactions between brain and face that take place during early embryogenesis. These are crucially important; without the developing brain and the chemical signals it produces, the face would not form. Yet the far more extensive and longer-lasting set of interactions between brain and face are of a different character; they involve behavior. Of these, the earliest in evolution concerned the pursuit of food. The sensory organs of the face transmit information about possible food sources to the brain, which then sends appropriate signals to the animal's body that result in some action to obtain that nourishment. This most basic of face–brain relationships pertains to all vertebrates throughout their lives and has existed from the beginnings of their history.

Here, however, we will concentrate on a more recent and more intricate set of relationships between the face and brain, those involving social interactions. These seem to be primarily a mammalian phenomenon rather than a more general vertebrate feature, but they are especially important in the primates and probably, most of all, in humans specifically. Indeed, brain–face connections are at the heart of our emotional and social lives—how we express ourselves to one another—and retain that centrality throughout the individual's life, from cradle to grave. Three aspects of these relationships that are crucial to human sociality will be the focus of this chapter: recognizing the individual faces of others, making facial expressions, and reading the facial expressions of others.

To understand these matters, however, we need two kinds of background information. The first concerns the ways in which the human brain as a whole is organized in order to function. The second involves the nature of vision and, specifically, primate visual capacity. Let us look at brain organization first.

A Brief Circumnavigation of the Human Brain

If the face functions as the sensory headquarters of an animal, the brain can be regarded as the body's general executive branch, combining the roles of central intelligence agency and top command center. It is within the brain that all sensory messages from outside the animal are received and then decoded and from which all instructions for the response to that information are issued. The brain is thus the seat of all understanding (*cognition*)—from instantaneous and hardly conscious to fully cognizant—and of the actions that flow from that comprehension. A key part of that understanding stems from the emotions generated within the brain that guide internal and external responses to external events. These responses often include physiological responses of the internal organs and, more visibly, speech and body movement. Thus, the emotions are an essential *evaluative* element in cognition.[1]

These basic brain functions are, of course, possessed by all other mammals—indeed, by all vertebrates. The complexity of the human brain, however, and the cognitive, emotional, and behavioral responses that flow from its activity are unparalleled. That complexity requires two things: (1) an enormous degree of internal structural specialization in the form of regional areas or domains crucially dedicated to performing particular functions and (2) an intricate *wiring* system between those domains that integrates all the information into a larger functional system. This allows either rapid unconscious assessment or slightly slower conscious evaluation followed by generation of the appropriate responses. Much of the basic functional regions of the human brain are largely the same as in other mammals, though often larger, but the wiring, especially involving the cerebral cortex, must be considerably more complex than in animals that lack the complexity of our cognition and behavior.

To understand how the brain works, we must first know in broad terms how its parts are spatially organized with respect to one another—its *functional geography*. Recall that in its earliest stages of development, the brain has a tripartite structure that first becomes visible in the early embryo. It consists of: the *prosencephalon* (forebrain) at the anterior end, the *mesencephalon* (midbrain) just behind the prosencephalon, and the *rhombencephalon* (hindbrain), which lies just behind the mesencephalon (see Figure 2.7). During subsequent embryonic and fetal development, these three divisions give rise to the main physical compartments of the fully developed brain as shown in Figure 7.1.

The fully developed human brain is dominated in size and shape by structures derived from the forebrain, which will be the focus of our attention, and the hindbrain, both of which show much expansion and regional diversification during embryonic and fetal development. In contrast, the midbrain of the early embryo undergoes less growth and does not segment into major structural divisions. It serves principally to relay information about certain aspects of vision and hearing to other parts of the brain. For vision, its role as relay station is primarily to help direct gaze at objects within the field of vision; in general, for perception, it is secondary to the **thalamus**, a forebrain-derived structure. The hindbrain, which is vital to much of the automatic physiological functioning of the organism, gives rise to the *brain stem*, which connects the brain to the spinal cord. It consists of two relatively large structures, the **pons** and

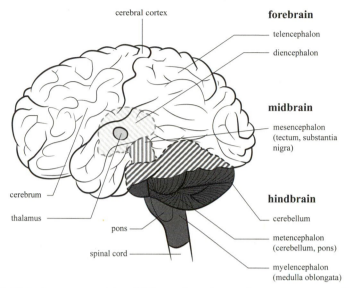

FIGURE 7.1 Midsection view of the adult human brain showing the main components of the developed forebrain, midbrain, and hindbrain. For the forebrain, the main component of the mature telencephalon is the cerebrum; that of the diencephalon is the thalamus, which lies centrally and underneath the cerebrum. The principal derivative of the midbrain is the tectum, a routing station for various sensory signals. The main parts of the mature hindbrain are the cerebellum and the brain stem, which is largely concerned with the routing of motor (movement) information.

the **cerebellum**, which are both concerned with relaying sensory information from the body (trunk)—that is everything posterior to the head—to the neocortex and instructions for movement or **motor responses** back to the body. The hindbrain's special relevance to the face is that it contains a special nerve center, cranial nerve VII, which controls facial movements and hence facial expressions. (Recall that, like all the cranial nerves, that nerve derives from a pair of neural placodes in the early embryo; see Figure 5.6).

It is the derivatives of the two parts of the forebrain, however, the telencephalon and the diencephalon, that are most directly and extensively involved in brain–face interactions. The telencephalon gives rise to the cerebral cortex, which constitutes the main mass of the outer brain, as well as the **basal forebrain nuclei**. The latter are situated centrally within the brain just beneath the cerebral cortex. Made up of globular-shaped structures that are densely packed with neurons, each devoted to a particular function, the basal ganglia function to relay information for the execution of various commands from the cerebral cortex to produce various body movements. The diencephalon, which lies directly behind the telencephalon in the early embryo, gives rise to the thalamus, which consists of two dense nodules, each the approximate size of a walnut, that lie in close association with the basal forebrain nuclei just underneath the cerebral cortex. The thalamus is a complex relay center: it transmits sensory information from the various sense organs (except for olfactory signals) to the cerebral cortex, and it relays neural signals between several different parts of the cerebral cortex. It is thus an essential transit station for helping to mediate many behaviors. It has also been implicated in sleep, maintenance of consciousness, and memory.

The largest and most distinctive structure in the brain is one we have already looked at when discussing the evolution of brain size: the cerebral cortex, namely, the highly folded, multilayered sheet of cells that forms the outer surface of the human brain, providing both its special bulk and remarkable appearance. It is traditionally referred to as the *neocortex*, with the prefix *neo* denoting its relative evolutionary newness relative to the rest of the vertebrate brain. In more recent literature, it is often designated the **isocortex**, but I will use the earlier term interchangeably with **cerebral cortex**. With its distinctive six-layer structure, it is unique to the mammals, as noted previously. In other vertebrates, the telencephalon gives rise to a structurally different and often smaller cortex that is characterized by a three-layer structure. That comparative structural simplicity in other vertebrates, however, does not necessarily entail either functional or behavioral simplicity. Some bird species evidently have complex mental operations as judged from their behaviors, and they possess their own equivalent of the mammalian cerebral cortex. That structure, however, arose independently in evolution because the birds evolved not within the synapsid branch of the vertebrates but from one lineage of the early theropods.

There are two general ways in which the geography of the human cerebral cortex can be described, one purely physical-structural, the other functional. The first is the simplest: it involves dividing the cerebral cortex into broad geographic regions that consist of one or more distinct major *lobes* as seen from the top and sides. This is shown in Figure 7.2 in lateral and medial views. Given the outward bilateral symmetry of the brain, each kind of lobe is found on both the right and left sides. The terminology gov-

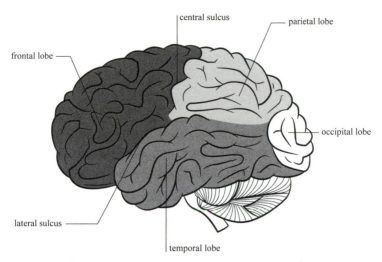

FIGURE 7.2 The different lobes of the cerebrum. The positions and relative sizes of the frontal, parietal, temporal, and occipital lobes are indicated. Two major sulci—the lateral sulcus, which separates the frontal and temporal lobes, and the central sulcus, which separates frontal and parietal lobes—are also shown.

erning this description of brain layout will be familiar from the previous discussion of the neurocranium, the skull (Figure 2.9), whose divisions take their names from the regions of the brain that underlie them. With one exception, the sutures seen in the skull overlie the corresponding grooves, the so-called sulci, which lie between the major brain lobes. (The exception is the one between the frontal and parietal lobes, on both sides, where the boundaries in the brain are slightly posterior to the corresponding skull sutures.) Thus, the most anterior region of the brain consists of the frontal lobes, which lie beneath the frontal bones of the skull, while the brain regions located just behind the frontal lobes are the parietal lobes, and the more ventral and lateral sections are the temporal lobes. Finally, the most posterior parts are the occipital lobes.

This broad spatial division of the brain into distinct lobes is helpful for general orientation but does not identify their differential functional specializations. That *functional geography* of the brain has been elucidated by a combination of anatomical and electrophysiological studies of both normal brains and those bearing specific injuries or *lesions*. Particular brain regions characteristically light up under particular electrophysiological scanning procedures such as **functional magnetic resonance imaging** (fMRI) and positron emission tomography in conjunction with

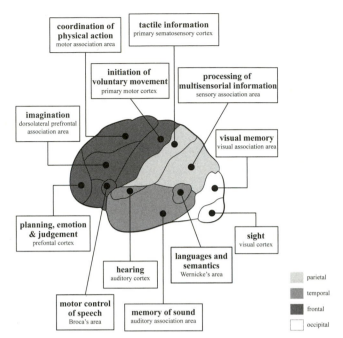

FIGURE 7.3 Functional divisions of the brain as deduced from analyses of injured brains and electro-physiological measurements. The frontal lobe is shown in dark gray, the temporal lobe in medium gray, the parietal lobe in light gray, and the occipital lobe as unshaded. The various areas for initial processing of sensory information (hearing, vision, touch) are as indicated. Broca's area and Wernicke's area are of major importance in language and speech though they have now been joined by many others. The frontal lobes are the principal locus of integration of the different kinds of information and for decisions (executive actions).

particular mental activities, while localized brain lesions often reduce or abolish a particular ability.

There are two major forms of such functional geography. The first is still at a fairly gross spatial scale but effectively indicates which regions of the cerebral cortex, within different lobes, are involved in and required for particular roles in either processing different kinds of sensory data or organizing movement (motor) responses, or in the various processes of higher cognition that involve processing that information and assigning meaning to it. It is shown in Figure 7.3, which indicates the names and locations of various areas and their specific functions. Several of these have major, if not always direct, roles in facial behavior, and we will come back to them. The evolutionary reasons why the brain has the structure it does lie deep in mammalian history. We will not delve into them here, but several good treatments have explored that history.[2]

The diagram of functional specialization in the figure provides useful information but requires a cautionary note: the regional assignments of function should *not* be taken to mean that these areas are in *exclusive* control of those respective functions. Indeed, such a highly modular view of brain function tended to dominate thinking about the brain until at least the late 1990s. Yet that view involved overinterpreting the findings on which it was based. Those findings involved two methods: (1) in normal brains, scanning for elevated localized brain activity by electrophysiological methods in conjunction with specific mental activities, and (2) in patients with brain injuries, correlating losses of particular mental functions with particular localized sites of brain damage. Electrophysiological detection of high brain activity in a particular region associated with a particular mental activity only indicates that region is activated in association with that activity, not that it is essential for it, though that is often the case. More important, such regions of brain activity should not be interpreted as the only ones involved in that particular mental function. In the lesion studies, the loss of a key mental operation with certain localized brain regions indicates that that area is necessary for that activity but not that it is sufficient for it. Nearly all the ostensibly localized mental functions indicated in Figure 7.3 involve inputs from other regions and some kind of automatic, sequential computation of those inputs, depending on which ones they are and their strengths. Understanding the rules of these processes is one of the great challenges in the neurosciences.

Nevertheless, both the electophysiological observations and the lesion studies have identified some especially important localized brain functions. A particularly important set of functions for human higher cognition and decision making concerns the frontal lobes—in particular, their most anterior section, the **prefrontal cortex** region. The prefrontal cortex not only performs vital integration of different and diverse kinds of information—it is sometimes referred to as the *association cortex* to denote its broad integrative functions—but also is the site of conscious decisions for action, the so-called executive functions. Many of the distinctive aspects of human mental activity almost certainly involved changes in the prefrontal cortex, consisting of new linkages of previously disparate functions to yield new mental capacities. Yet many other areas of the brain are also intimately involved in each of those functions, and brain–face co-evolution would also have involved those areas; the brain is far less modular than the diagram makes it look and far more *connectionist* and *computational*.

If the relatively broad regional assignments shown in the figure are the first form of brain functional geography, then the second is at a finer scale and is based on differences in local folding pattern, cellular composition, and structure. These are so-called cytoarchitectural differences between neighboring small regions of the brain—namely, their distinctive cellular and structural properties—whose functional roles were later assigned by a variety of methods. The scheme divides the brain into fifty-two numbered **Brodmann's areas**, collectively named after the neuroanatomist Korbinian Brodmann (1868–1918), who developed the method and first published his resulting map of the cerebral cortex in 1909. This form of division of the brain is diagrammed in Figure 7.4, which provides (top) both an external, lateral view of the brain as seen from the left and a medial view (bottom). Because of the physical bilateral symmetry of the brain, each Brodmann area is represented twice in the brain, once within each hemisphere. Yet the members of each bilaterally symmetrical pair can possess somewhat different functional attributes; these contribute to the long-noted functional differences between the left brain and the right brain. Those differences are particularly marked with respect to language, but they also come into this story because of some important left-right differences in face recognition.

The Critical Importance of Vision for the Anthropoid Primates

To monitor the outside world, all mammals rely on their five senses—visual, auditory, olfactory, taste, touch—but each species tends to rely most heavily on one, though sometimes with major assistance from a second. Many mammals, particularly nocturnal ones, depend primarily on their olfactory sense assisted by auditory cues. If the earliest placental mammals were small insectivorous nocturnal, tree-living animals, they would have been of this kind. Many rodents, however, especially those that burrow or are primarily nocturnal, possess highly sensitive face whiskers (*vibrissae*) and depend strongly on somatosensory sensation via the touch signals conveyed by those whiskers, exploring their immediately environment tactilely as they move forward but using olfaction as their sec-

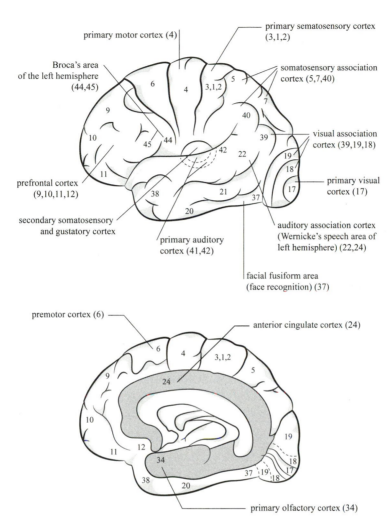

FIGURE 7.4 Brodmann's areas on the left hemisphere, laterally (top) and more centrally (bottom). Particular important areas for language and speech are Broca's area (44, 45), Wernicke's area (22), the primary auditory cortex (41, 42), and the prefrontal cortex (9, 10, and 11). The Brodmann's areas especially important for face recognition are the facial fusiform gyrus (37) and the visual association cortex (39, 19, and 18). The anterior cingulate cortex, lying centrally and within the cortex, plays an important role in nonvoluntary vocalizations in both nonhuman primates and humans, while the prefrontal cortex is important in human speech, cognitive integration, and executive decisions that lead to specific actions.

ondary system. Similarly, the platypus, one of the two kinds of egg-laying mammals, takes in a disproportionate amount of information about its environment through its highly touch-sensitive bill as it searches underwater, probing the mud for food items. Other mammals rely predominantly on auditory information, bats being the obvious example. Primates, in contrast to all of these, are overwhelmingly reliant on vision, being for the most part diurnal animals dependent on sunlight to assess the environment. Even the few nocturnal primates—the galagos, bushbabies, lorises, all prosiminians—that would seem to have less use for vision than diurnal ones have enlarged eyes that pick up visual signals in dim light.

The spatial organization of the cerebral cortex, with different areas devoted to initial processing of differing sensory signals, reflects those specializations. In particular, there is a trade-off between olfaction and vision in terms of development and size of the respective sensory areas. Thus, olfactorily dependent mammals have larger olfactory signal-processing regions than visual ones while animals that are heavily dependent on vision have far more of the cortex devoted to visual processing than to olfactory cue evaluation. This allocation of cortical space is particularly apparent in the case of the anthropoid primates. Thus, rhesus macaques, the principal anthropoid primate used in basic research, have approximately 50 percent of their cortex devoted to the processing of visual information, a far higher percentage than nonprimate mammals who rely less on vision.[3]

Such differences in cortical organization, however, go beyond disparities in relative extents of the cortical areas assigned to the different sensory modalities. Differences within each region can also reflect the different kinds of specific input for each such modality. This is particular apparent in *S1*, the somatosensory cortical region, where parts of the body whose touch sensitivity matters particularly are represented disproportionately in size and intricacy. Thus, in rodents who largely feel their way through life in the dark via their vibrissae, the region within *S1* devoted to processing information from the vibrissae specifically is greatly enlarged relative to others, while in humans the part of the *S1* devoted to hands is disproportionately large. Some of the differences in space allocation in *S1*, according to the relative importance of different body areas for touch sensation, are shown in Figure 7.5. These differences in brain region allocation largely reflect differing extents of innervation for touch receptors in different parts of the body. The more neurons conveying sensory information from a particular part of the body, the greater will be the amount of cortical space devoted to it. For highly vision-dependent animals such as primates, the intricacy of processing visual information is reflected not just in the amount of cortical space devoted to it but also

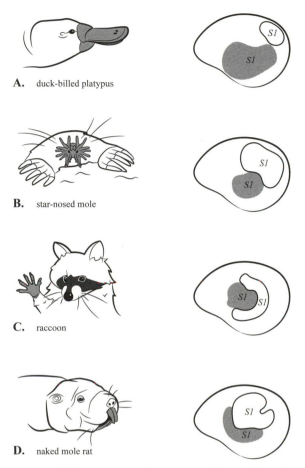

A.   duck-billed platypus

B.   star-nosed mole

C.   raccoon

D.   naked mole rat

FIGURE 7.5 Comparison of different extents of sensory cortical areas within the somatosensory cortex (*S1*), with gray areas in the diagram indicating enlarged areas associated with particularly important touch-sensory functions in the four animals shown. Thus, for example, a disproportionate amount of the *S1* of the platypus is devoted to sensations picked up by its bill in probing muddy stream bottoms, while the part of the raccoon's *S1* devoted to the hands, which it uses extensively, is enlarged relative to most other members of Carnivora. In humans (though not shown), the hand similarly occupies a large part of the *S1*. (Modified from Krubitzer, 2009.)

in the internal structure of these regions and the complexity of their links. In humans, there are about thirty distinct cortical regions devoted to processing visual signals in various ways and then passing that information on to other parts of the brain for eventual integration.

In particular, the processing of visual information involves an initial splitting of the elements of an image (for instance, location, details, movement) to different areas of the brain equipped to handle those specific features and then the reintegration of the information in specific brain places, particularly in the prefrontal cortex. Such reintegrative stations include both special areas in the temporal and occipital lobes for particular objects such as faces and so-called cortical association areas in the prefrontal cortex for the final integration of the visual information with other kinds, including auditory information and memory. Those steps are then often followed by transmission of further signals to the motor cortex in the frontal lobes, which, in turn, sends signals via the hindbrain to trigger appropriate movements in response.

The system is diagrammed as a flowchart in Figure 7.6. The sequence begins, of course, with the eye, and specifically the retina, the sensory sheet at the back of the eye that records the incoming light signals and registers the initial image of what the animal sees at a particular moment. The visual image is transmitted via a special nerve, the optic nerve, to the first receiving station in the brain, a part of the thalamus called the **lateral geniculate nucleus** (LGN). It is within the LGN that the first major sorting of kinds of visual information takes place. That sorting concerns the properties of movement, position, and identity of specific elements of the image transmitted from the retina, and it involves neurons that are specially and differentially activated by those respective properties within parts of the image. These neurons are grouped in distinctive layers, which gives the LGN a visibly striated character when stained appropriately.[4]

The three main classes of neuron in the LGN, each making up distinct layers and possessing special sensitivities for different aspects of the image are the *magnocellular* (M), the *parvocellular* (P), and the *koniocellular* (K). With some oversimplification, it can be said that the M and K neurons are primarily concerned, respectively, with location and movement detection and the P layer neurons more with edges and identity of objects. Both the M and P layers send their information initially to the primary visual area, V1, at the back of the brain, the occipital region. In addition, the M layers communicate with a nearby dorsally located secondary visual area, V4, while the P layers connect to a third visual area, V2, which is located more ventrally. (The K layer cells have more complicated connectivity, which we can ignore for the moment.) Depending on whether the information originated in the M or P layers, it is then differentially dispatched via so-called dorsal and ventral streams (Figure 7.6), the former to specific areas more dorsally in the parietal lobes, the latter more ventrally to the tem-

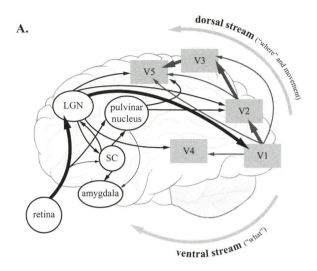

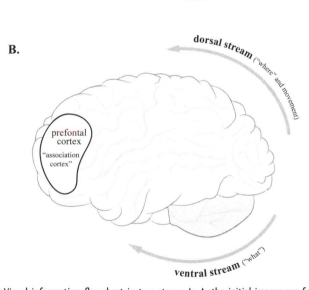

FIGURE 7.6 Visual information flowchart in two stages. In A, the initial images are formed on the retina, then transmitted to the lateral geniculate nucleus (LGN), where there is an initial segregation of elements associated with movement and location, on the one hand, and object identity on the other. This information is transmitted to V1 in the occipital lobe and then transmitted to other visual areas, via the dorsal and ventral streams. (In this diagram, only relative positions, not actual locations, are shown.) B indicates crudely the downstream integration of the information, from the dorsal and ventral streams, in the prefrontal cortex, although there have been multiple intermediary steps of integration as well.

poral lobe for assessment. This form of processing complex information via dorsal and ventral streams of different kinds of information is not unique to vision, however; it is also seen in the general processing of auditory (sound) information and apparently also in the processing of language information specifically. This general two-stream modality of processing information is described in more detail in Box 7.1. Reintegration of all the visual information from the two streams takes place in the prefrontal cortex, as indicated schematically in Figure 7.6.

Let us now consider vision in the anthropoid primates generally and how it probably influenced their evolution. Three significant visual features of these animals are (1) closely set eyes relative to presumptive ancestors and prosimians, (2) acute vision for near objects, and (3) excellent (trichromatic) color vision. The first two—namely, relatively closely spaced eyes and accurate vision of near objects—are related. Closely set eyes permits highly convergent vision of the visual fields provided by the two eyes, the property of **stereopsis**, and it permits accurate in-depth, three-dimensional vision of near objects. The field of visual overlap between such eyes can be as great as 120° (out of 360°, the maximum for wrap-around vision). It was undoubtedly the capacity for stereoscopic vision that provided the initial selective advantage for closer-set eyes during the period in which the anthropoid primates first evolved. In turn, good stereopsis might select for additional space in the brain to be devoted it. In general, there is a good correlation between the degree of stereopsis and the absolute extent and proportion of the cortex devoted to processing visual information. It seems likely, therefore, that both the advantages and demands of stereopsis contributed initially to the evolution of larger brains within the anthropoid primates relative to the prosimians.[5]

The question provoked by these considerations concerns the specific adaptive advantages that such stereopsis might confer. The three main contending ideas stress, respectively, advantages for foraging, spotting dangerous predators, and social interactions. Clearly, good vision in general is useful for all three activities, although the nature of present-day society has reduced the need for the first two. What was the most probable *initial* advantage? It seems unlikely that acute close-up vision for social encounters would have been a strong advantage in the earliest stages of anthropoid primate evolution. Indeed, many primates, including some of

BOX 7.1. Processing Visual, Auditory, and Language
Information via Dorsal and Ventral Streams

Visual information is inherently complex and, correspondingly, the ability of the brain to decode it must be intricate. This usually involves identifying objects within a scene, placing them with respect to one another and to the viewer, and, if any element is in motion, detecting that motion and its direction. Accordingly, the discovery that there are two processing *streams* operating relatively dorsally and ventrally, respectively, initially parsing each image into its different aspects, was a major step in understanding how that complexity is handled. Yet vision is not unique in involving such complexity and a similar division of labor via two streams in its processing. A similar pattern of dorsal and ventral streams is used to process general auditory information (sound) and, in connection with the auditory system, a separate but also two-stream system for language. For the first two, visual and auditory information, the dorsal stream is devoted to establishing aspects of location ("where") and the ventral stream for matters of identity ("what"). For language, however, the division is somewhat different: the ventral stream is involved in decoding the meaning of the sounds (words), while the dorsal stream is devoted to controlling the movements of the mouth and larynx to generate them. For all two-stream systems, the different parts of each such two-track system are, of course, different between the three systems, yet all involve an initial processing step toward the back of the brain followed by intermediary steps that take place in slightly more anterior regions, which in turn are followed by integration in the prefrontal cortex. In contrast to these three dual-stream systems, the relatively simpler sensory information involved in olfaction and taste, which involve primarily matters of identity ("What smell?" "What taste?"), do not involve such intricate neural processing. Nor does somatosensory information (touch), whose primary information—location on the body—is recorded within the somatosensory cortex's map of the outer body.

   While the evidence for a two-stream system for visual, auditory, and language information is strong, in which the information from the two streams is integrated within some central associative and executive centers in the prefrontal cortex, it is not clear that in each of these systems the two streams are wholly separate. For all three, there is evidence there are connecting and integrative steps between them during the initial processing steps.[1]

---

1   See Cloutman (2012) for a review on dual-stream neural processing in the visual, auditory, and language systems.

the platyrrine primates, though highly sociable, do not spend a lot of time inspecting each other's faces. If, therefore, we put that function aside, that leaves the *better foraging* and the *better spotting predators* hypotheses as the preferred explanations.

The current consensus is that the high-resolution, close-up vision of anthropoid primates was primarily selected because of the advantage it conferred in discriminating particular food items within dense foliage. The early anthropoid primates were almost certainly arboreal and lived on leaves, seeds, or fruits, as do many contemporary arboreal anthropoids. Good in-depth vision, favored by convergent vision of the two eyes, would be helpful for picking out all three food items but particularly seeds and fruits. To do so efficiently, good color vision would almost certainly have provided a selective advantage.[6]

The teeth of early fossil anthropoid primates support this idea. The dental characteristics of living primates, when compared with their main dietary items, provide clues to which dental features correlate with particular food items. From such hints, it appears that many of the early anthropoid primates were primarily fruit eating or **frugivores**; their molars and premolars possess fairly extensive flat surfaces, a characteristic that favors masticating fruit. Since both fruits and leaves are relatively poor in protein, animals that strongly rely on one or the other must consume fairly large amounts of such material each day to support themselves, as seen in these animals in the wild. For fruit, in particular, which has to be ripe and eaten in bulk to supply adequate nutrition, that requirement places a high premium on efficiently finding it among the foliage. That requirement, in turn, might help explain the special advantage of full-color vision of the catarrhine primates. In contrast to the monochromatic strepsirrhines and predominantly dichromatic platyrrhines, the catarrhines possess such vision using three pigments, which helps them to readily distinguish ripe from unripe fruits. This would have been a distinct advantage and is often cited as the selective explanation for their good trichromatic vision. Thus, the combination of good stereopsis and color vision would have been particularly effective in fruit gathering (see notes 5 and 6).

This is not the only possible explanation, however. An alternative is the predator-detection hypothesis, the idea that the excellent vision of the anthropoid primates was initially selected for avoiding a particular kind of predator. A type of predator that was quite near but especially quiet would be extremely dangerous. The principal class of predators of this type is deadly snakes, many of which are well camouflaged to fit in with foliage. Correspondingly, the ability to detect such animals with acute close-up vision would be an advantage. Good color vision, another property of cattharrhine vision, would be another useful property.

The evidence for the *snake-detection hypothesis* is circumstantial, as is that for the *foraging-for-fruit hypothesis*, but it is consistent with the known facts. First, both poisonous and constricting snakes had already evolved and may have been common in the late Paleocene and early Eocene forests when the protoanthropoid primates first appeared. Many anthropoid primates today are too large to be food for snakes, even if the latter are still potentially deadly threats, but the early anthropoid primates were much smaller and could have been direct prey items for snakes. In contrast, the major present-day predators of primates—namely, the larger raptors and the big cats—had not yet appeared.[7]

Of course, there need not have been only one selective factor involved in producing the excellent vision of anthropoid primates. The need for both improved foraging for particular food items such as fruit and spotting predators could have been dual selective pressures for improved binocular vision. Yet that initial improved capacity may have created both new pressures and opportunities for further changes, particularly in brain organization and specifically within the LGN and the regions of the neocortex involved with processing visual information. Such changes need only have been quantitative and hence relatively easy to amplify by natural selection. The net consequences, however, would have been enhanced capabilities for focusing on close-up objects. Initially, whether those items were food items or predators, additional subjects for visual perception and inspection might have surfaced. In particular, these could have included the faces of conspecifics and their expressions. In effect, enhanced face recognition could well have been a by-product of acute close-up vision that had initially been selected for something else. The potentiality for new uses as the consequence of an initial adaptive change is a common result of evolutionary change.

With the background on the brain and anthropoid primate vision previously sketched, we can now begin to look at the connections between the face and the brain and human sociality. We begin with face recognition, a fundamental yet complex attribute that is crucial to human social interactions.

The Human Capacity for Recognizing Faces

A first step in understanding the capacity for rapid and accurate face recognition involves knowing the particular brain areas that participate in and are essential for it. These include, of course, the two principal cortical areas devoted to vision, VI and V2, located in the posterior region of the occipital lobe, since without the initial processing of visual images by V1 and V2, one cannot see anything. Although necessary for facial recognition, these regions are neither sufficient for nor dedicated to this task.

Three additional principal areas of the cerebral cortex, however, have been identified as especially and specifically important for face recognition. Two are in the temporal lobe, and the third is in the occipital lobe. The best characterized lies in adjacent Brodmann areas 37 and 36 on the inside of the temporal lobe (Figure 7.4) and are part of a visible discrete structure, a fold of the temporal lobe, the *fusiform gyrus*. The particular region within the fusiform gyrus that is crucial for face recognition is the **fusiform facial area** (FFA), and it makes up a significant part of the fusiform gyrus. The second important area for face recognition is distinct from but relatively close to the FFA within the temporal lobe; it is a groove between gyres, a sulcus, the **superior temporal sulcus** (STS). The third is in the occipital lobe as part of the *visual association cortex* and termed the **occipital facial area** (OFA).

The first observations that demonstrated the crucial importance of the FFA for face recognition involved individuals who had suffered specific damage to this region. They lose the capacity for recognizing others by their faces, a condition termed **prosopagnosia**, although individuals with this affliction are not deficient in recognizing either faces as such or any other kind of object. From brain scans carried out by fMRI in people with normal face-recognition capacity, it appears that this area "lights up" in response to faces specifically, not to any other objects, and most strongly to human faces. The face-sensitive cells are grouped in a set of distinct *face patches* within the FFA, and electrophysiological studies show that they communicate with each other during face recognition. In the initial steps of face recognition, the strongest recognition signal as judged by the fMRI response is in the right hemisphere FFA, and studies of prosopagnosic individuals indicate that in most of them it is the right FFA that is the locus of damage. Furthermore, in normal individuals, it is the sight of highly familiar faces that elicits the strongest response, indicating that memory and learning are part of the mechanism. The basis for this ability appears to be that the more familiar a face is, the more neurons within the FFA that will be dedicated to it.[8]

The other two special regions for processing face images are the STS and OFA, as noted previously. These have also been studied by means of scans for brain activity in subjects who were asked to focus on face images and on observations of individuals with lesions in these areas. The cumulative results indicated some degree of specialization of role in face processing between the three key regions. The FFA seems particularly important for identifying faces as distinct from all other objects as well as some role in identifying individual faces. The OFA, however, seems to be more important for assigning both individual identities and gender identities from the configuration of the features, as well as processing the emotional content of facial expressions. The STS is also involved in gender determination and in assessing trustworthiness from facial images.

Despite such compartmentalization of function, it is clear that there is much neural communication between the various face-responsive regions in the brain. At one time the OFA was posited as the central clearinghouse for visual information about faces with the information being passed on to other regions successively. Today, the current pictures supports the idea that there is lots of cross-talk between the different regions. One consequence of these newer findings has been to render unlikely the earlier idea of separate processing of visual information about facial identity on the one hand and emotional state on the other. Rather, as described previously, such processing is more holistic and less compartmentalized, the net result of the whole process being some form of neural computation that assesses identity along with such aspects as (probable) gender and feelings associated with the particular face.[9]

From the noted neurobiological findings, it will be apparent that face recognition in humans is a highly intricate as well as—judging from the results—a fairly precise process. The great majority of people do it quickly, easily, automatically—and precisely. Nevertheless, like facial expressions, it must be a product of human evolutionary history. That thought, however, immediately prompts the question: is this capacity distinctively human, or is it found in, for example, our nearest cousins, the great apes, or perhaps even well beyond them in more distant mammalian relatives and even nonmammalian vertebrates?[10]

On the Evolutionary Roots of Facial Recognition in Humans

Given the highly social natures of the great apes, and indeed, the primates in general, we might expect similar face-recognition capacity and similar neural foundations in other primate species We have noted that the great apes also have distinctive, individual faces that, given the excellence of anthropoid primate vision, should be visible and distinguishable to them. If so, this should perhaps be an important feature of their social interactions. As we shall see, these expectations are fulfilled: the anthropoid primates have excellent individual face-recognition capabilities. In contrast, among our more distantly related nonprimate fellow mammals, there are fewer outward signs of such. Furthermore, nonprimate mammals occupy themselves with the faces of their conspecifics seemingly far less than humans or chimps or bonobos or primates in general.

We must, however, distinguish between what is casually observed and what may exist as underlying potential abilities. Recent work in several animal species shows that the capacity for recognizing the faces of conspecifics is far more widespread than the overt, high-frequency face checking in which catarrhine primates engage. In principle, individual face recognition would be a useful trait in any animal species that has three traits: (1) a distinct social structure in which recognizing individuals would be an advantage; (2) good vision, a prerequisite for accurate face recognition; and (3) variation in either facial markings or facial shape that would serve to differentiate individuals. While the members of most animal species spend little time looking each other in the face, such identification might take place in many species that associate in groups in a quiet and undemonstrative form. If so, its existence would only be found through experiments designed to detect it. In fact, such experiments have been carried out on a variety of animal species and indicate that the capacity for individual face recognition is far more widely shared than had previously been suspected. Most of the work deals with mammals, but several findings indicate that birds and even fishes have some ability to recognize faces of conspecifics.[11]

In fact, recognition of individual faces is not limited to vertebrates. There is one kind of invertebrate, the paper wasp of the genus *Polistes,* that has been shown to have some face-recognition capability. There are many species of the paper wasp, and most do not show this capacity, but one in particular, *Polistes fuscatus,* shows it in a marked way. These wasps have distinctive yellow facial patches and stripes on a dark green background, which give their faces individual identities. Altering those patterns visually by painting out particular patches or stripes was found to engender new hostile interactions among an individual's fellows, but with time there was full acceptance of those individuals again as their mates adjusted to their new facial identities. Control experiments showed that it was specifically the individual facial pattern and not any other variable that was the key factor in these social interactions. *P. fuscatus* has an elaborate social hierarchy in which workers know their places with respect to other workers and individual face recognition is evidently an important part of how these hierarchical relationships are maintained.[12]

Presumably, this specialized capacity for individual face recognition evolved from some more general pattern-discerning visual capabilities possessed by *Polistes.* The evidence shows that *Polistes* species that do not normally discriminate individual faces can be trained to preferentially recognize faces that belong to their own species. This indicates that all species in this genus probably have some latent capacity for recognizing faces. Yet that capacity is almost certainly a relatively recently evolved one specific to this particular group of insects since there is no evidence for it in all wasps or, more generally, social insects. What the paper wasp case illustrates is that similar selective pressures—in this case, for face recognition—can yield similar outcomes in widely different animals such as wasps and mammals even when the *substrates* for those evolutionary changes, their brains, differ strikingly from one another and the evolutionary trajectories must have been completely independent.[13]

Our focus here, however, is on human face recognition and its evolutionary roots. How widespread is the capacity to recognize individual faces among mammals? And for those mammals that show it, did it arise independently in different mammalian branches or is it an early, perhaps ancestral, general mammalian trait? To answer such questions, we must do as broad a survey as possible. The species that have been tested most thoroughly and for which individual face-recognition capacity has been found are humans, apes, monkeys, dogs, and sheep and cattle. There is also indirect but striking evidence of face-recognition capacity in two other kinds of mammal: dolphins and elephants. The dolphin and elephant findings involve the mirror self-recognition test in which the animals are shown themselves in a mirror and then taken through a series of tests in which each gets to demonstrate that they do or do not understand it is their own face staring back at them from the mirror. Any animal with the ability to recognize its own face should have, at minimum, some sense that its conspecifics have their own individual faces. Furthermore, dolphins and elephants are among the most socially interactive animals, a characteristic that makes conspecific face recognition a useful property. Elephants also exhibit facial differences readily observable by humans, making it likely that they themselves can recognize each other by their faces.[14]

Since the set of mammalian species demonstrated to have face-recognition capacity is still but a tiny fraction of the 5,400 or so in total, it might seem premature to conclude that this ability is likely to be a general mammalian condition. There is, however, a good reason for believing so, which stems from the distribution of the species known to have this ability within the mammalian family tree. Recall that there are four large groupings, supraorders, of the placental mammals (Figure 5.8) and that these diverged rapidly during the Paleocene and Eocene periods. The species that have been shown to or inferred to be able to recognize conspecifics by their faces are members of three of these four supraorders. Thus, primates (humans, apes, monkeys) are in the Euarchontoglira; dogs, dolphins, and cattle are in the Laurasiatheria; and elephants are in the Afrotheria. Among the placental mammals, only the sparsely populated Xenartha (armadillos, anteaters) are not yet represented by a species known to have face-recognition capacity, but none have been tested to date. Overall, the dispersion of face-recognition capacity among the placental mammals indicates that this capacity arose early, probably near the origins of the mammals themselves, and is a shared trait, at least for placental mammals.[15]

The best evidence that there was a single origin of the placental mammalian capacity comes from studies of face recognition in sheep. That sheep have the capacity to recognize their conspecifics by their faces is surprising, not only because of their poor reputation as thinkers but also because their social interactions seem of such a low grade—they appear to be simply milling around. Yet extensive testing of image discrimination among sheep reveals these animals not only recognize individual herd mates but also can learn to discriminate differences in faces of other sheep with whom they were not previously acquainted. Furthermore, sheep can recognize different human faces, though their ability to do so is poorer than that for discriminating faces of their own species. In this respect, too, face recognition in sheep seems similar to that of humans, for whom distinguishing faces within one's own "race" is initially easier than for other groups. Similarly, chimpanzees recognize individual chimp faces better than human ones, though they can be taught to individuate the latter. Thus, for sheep, as for humans and chimpanzees, training, accomplished with appropriate rewards, can improve discrimination of faces within groups that are normally regarded as the "other." Furthermore, some evidence indicates that sheep benefit from seeing each other's faces: sheep placed in isolation, which is stress-inducing, become calmer when exposed still in isolation to pictures of the faces of fellow sheep. This is what would be predicted for a social animal, whose individuals are in the constant company of their fellows, even when the degree of social interaction appears to be slight. Sheep, in short, are not nearly as dull as usually thought.

The crucial point, however, is that face recognition in sheep almost certainly has shared evolutionary roots with that of primates. From single cell recordings in the brains of sheep being tested for their reactions to different images of faces, it is apparent that they use comparable areas of the brain—within the right temporal lobe. Given the early separation of the four major mammalian supraorders, the fact that species in both the Euarchontoglira and Laurasiatheria have similar neural bases of face recognition indicates that the ability probably arose once near the origins of the placental mammals or of the mammals as a whole. The novelty of the mammalian neocortex makes certain that nonmammalian vertebrates with face-recognition abilities must use different neural circuitry.[16]

Given those common mammalian roots, we would expect all anthropoid primates to have this ability—and much evidence indicates they do. All tested species have a good capacity for recognizing individuals of their species and, in the few cases where this has been tested, a less efficient but real ability to recognize individual faces in other primate species. It should not be a surprise in light of the results with sheep that the neural basis of face recognition in macaque monkeys matches what is known about face recognition in humans.[17]

Yet this capacity does not mean that it is entirely hardwired in brain circuitry. Indeed, few if any abilities in mammals are wholly innate; experience is invariably important in the honing of each. In the case of face recognition, numerous experiments have shown that this ability, while present in young infants, improves from infancy to early childhood and depends on exposure to different faces. The well-established observation, referred to previously—that people can discriminate faces of people of their own ethnic group better than those of others and the fact that highly familiar faces give stronger brain signals in the FFA—indicates the role of experience in this capacity. This is confirmed from observations that children brought up among members of a different "racial" group lose the "other race effect" and learn to discriminate individual faces in the new group just as well as in their birth group.[18]

Nevertheless, it seems probable that the innate potential for discriminating faces is strongest in the hominoid primates than in other mammals. Furthermore, that capacity is far greater than would have been needed when first developed in anthropoid primate evolution. Apes tend to live in small groups, as was almost certainly the case for early hominin species as well as for human hunter-gatherer societies, these being the norm for 90 percent to 95 percent of our history as a species. For groups of these sizes, the capacity to recognize 200 or so different faces should have been ample; indeed, those are about the size of the largest known monkey troupes. Most people can discriminate many thousands of faces and probably readily remember at least 1,000 to 2,000. Assuming for the moment that face discrimination in hominoid primates, at least, was selected as an advantageous property—we will come back to this matter—it would seem that evolution can create surplus capacity. In fact, there are numerous examples of such. In the language of engineering, many traits, particularly mental ones, seem to be overdetermined, a consequence of natural selection's potential to deliver more than the minimum required.

Recognition of an individual's face, however, is not an end in itself but often the prelude to a social interaction. When two people meet, mutual recognition is usually followed by a short verbal exchange accompanied by appropriate facial expressions. Indeed, most facial expressions arise in social situations and, in particular, in conversations, although individuals on their own often exhibit expressions when surprised, frightened, or sad or when concentrating on something with emotional content. The link between conversations and facial expressions is so strong that when imagining a conversation yet to come, many people will show traces of the same facial expressions the actual meetings are likely to bring forth. Clearly, facial expressions are a major part of social intercourse, as we saw previously with respect to the origins of language. We have briefly noted the role of the mimetic muscles in facial expressions, but we should now take a closer look at both those muscles and the neural underpinnings that activate them to create facial expressions.

Making Facial Expressions: the Body Machinery Involved

The term *facial expressions* sounds like a simple unitary category. There is, however, one expression that is distinct from the rest. Unlike the primate expressions we will discuss, this one is shared widely among the amniotic vertebrates—namely, the reptiles, birds, and mammals—thus dating to at least 325 million years ago. Correspondingly, the capacity to interpret it instantly is also universal within this group. This is the threat display conveyed by a wide open mouth, usually accompanied by some form of vocalization as in the hissing of lizards and snakes, and it is triggered by an unexpected encounter with another animal. It involves action of the deep muscles that move the jaws and is triggered by the amygdala, the basal ganglion involved in fear responses. In contrast to the other facial expressions, it does not involve the mimetic muscles (which are not found in reptiles or birds) but just the jaw and oral musculature. Depending on the response of the animal to whom it is directed, the animal making it then either launches an aggressive attack or flees, the classic fight-or-flight response. The threat display certainly expresses an emotion, but it is an antisocial one whose function is to terminate an encounter.

The facial expressions we are concerned with here, however, serve to mediate close social interactions within a group. These are more subtle and more recent in evolutionary history and are made by the mimetic muscles. Exclusive to mammals, they must have originated with the true mammals or, more probably, in one of their ancestral synapsid precursor species. While their initial functions probably related to facial mobility and involved movements of the mouth and eyes principally to aid survival in different ways, they must at some point have been co-opted for expression making and thus for social communication.

Among nonhuman animals, facial expressions are most obvious and most varied in the great apes, as first observed and reported by Darwin.

Indeed, for the six major human expressions—which denote fear, sadness, happiness, anger, disgust, and surprise—there are similar expressions in chimpanzees, whose context indicates similar states of feeling. Furthermore, the comparisons can be extended using a method that detects movements of the underlying mimetic muscles. This technique was developed for humans initially by Paul Ekman and his colleagues, but it has been applied to chimpanzees, other primates, and more recently, in a major extension, to dogs. It is called the *facial action coding system* (FACS), and it confirms the use of similar mimetic muscle sets in humans and chimps for expressions that had been classified as signaling the same feelings. It is described in Box 7.2. Evidently, these major expressions are part of humanity's anthropoid primate heritage.[19]

To understand just how deep in evolutionary history the capacity for making these facial expressions is, however, we need to know something about the evolutionary history of their two physical components: the mimetic muscles and the nerves that innervate them. Let's examine the mimetic muscles first.

The Mimetic Muscle Sets in Different Mammals

Figure 7.7 offers a diagrammatic representation of the mimetic muscles in the human face. Given the attachment of these muscles to the underlying dermis of the skin, the function of each can be deduced directly from its location. Thus, specific muscles underlie and move the ears, the forehead, the eyes, the nose, and the mouth, respectively. For the great majority of expressions, those that move the eyes and mouth are the most important. Humans, as remarked previously, have twenty-one distinct kinds of mimetic muscles, most of which are present in bilaterally symmetrical pairs on the two sides of the face, the remainder being single and placed centrally.

In considering their evolutionary history, it is important to know the numbers of muscles and their properties in other mammals. Even if their initial function was not expression-making per se, one might predict that the extent of their development, their numbers, or both properties would roughly correlate with the degree of facial expressiveness in different mammalian groups. For all such comparative work, of course, the same (that is, **homologous**) mimetic muscles must be identified in the species being compared. Given their characteristic locations and structures, however,

BOX 7.2. Some Facts about FACS

The facial action coding system (FACS) is a taxonomy of expressions based on the underlying facial muscles involved. It was developed as a means of studying facial expressions with precision. First developed extensively to characterize the expressions of humans, it has subsequently been applied to several primate species and, in one study, to the dog. Since the details of facial expression are dependent on the morphology of the face, each application of FACS to a new species requires a recalibration and a check to determine which particular muscles are involved. Nevertheless, these comparisons work fairly well because of the gross similarities in facial muscle anatomy. (Of course, such similarities are far greater among primates than between primates and other kinds of mammals.) The basic system involves breaking down facial expressions into component action units (AUs), about forty-five altogether. Each AU is given a simple name and involves a particular movement, usually involving only one of the facial muscles (see Figure 7.7). Of these major AUs, thirty-five have been assigned to a particular facial muscle while others await that classification. In addition, there sare separate sets of AUs for head movements, eye movements, and certain basic behaviors—for example, sniffing or shrugging. Some expressions involve a single AU, hence a single muscle movement, but most—including the six major expressions of happiness, sadness, fear, surprise, disgust, and anger—involve several AUs. Thus, for example, happiness is expressed via the combined action of AU6 (*cheek raiser,* which involves the orbicularis oculis muscle) and AU12 (*lip corner puller,* which involves the zygomaticus major). In contrast, disgust is expressed with AU9 (*nose wrinkle*), A15 (*lip corner depressor*), and A16 (*lower lip depressor*), respectively involving the levator labii superioris, the depressor anguli oris, and the depressor labii inferioris.[1]

---

1   For the complete system, see Ekman, Friesen, and Hager (2002).

this is fairly easy. Hence, in comparing different species, it is usually possible to say whether a particular mimetic muscle is either shared or not.

When such comparisons between different kinds of mammals are carried out, the prediction that numbers of different kinds of mimetic muscles will correlate with the relative degrees of expressiveness is broadly borne out. Thus, the platypus, one of the two kinds of egg-laying mammals and hence one of the most primitive, possesses little evident facial expression and, correspondingly, has relatively few kinds of mimetic

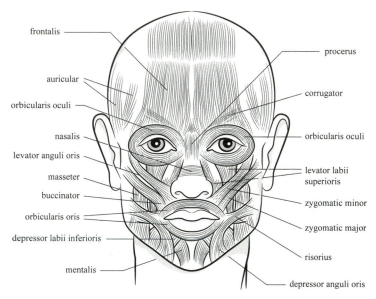

frontalis

procerus

auricular

corrugator

orbicularis oculi

orbicularis oculi

nasalis

levator anguli oris

levator labii
superioris

masseter

zygomatic minor

buccinator

orbicularis oris

zygomatic major

depressor labii inferioris

risorius

mentalis

depressor anguli oris

FIGURE 7.7 Diagram of the different facial (mimetic) muscles. Most of the mimetic muscles are in pairs, reflecting the bilateral symmetry of the head, but several are central and unique. Most human facial expression is concentrated around the mouth and the eyes, which will be apparent from the locations of these muscles. (Redrawn from McNeill 1998.)

muscles, specifically ten. This is distinctly fewer than three placental nonprimate mammals that have been examined for this muscle set, namely, the Norway rat, a tree shrew, and a flying lemur, which all have about fifteen or sixteen mimetic muscle types. Those animals, in turn, have slightly fewer than most of the primates that have been examined. The only prosimian species examined to date, two species of the genus *Otolemur,* have seventeen kinds of facial muscle, which is less than that of the visibly more facially expressive anthropoid primates, which have twenty-one to twenty-two mimetic muscles. Although this difference is not large, it agrees with the predicted correlation between number of mimetic muscles and extent of facial expressiveness.[20]

Within the anthropoid primates as a group, the extent of differences between platyrrhine and catarrhine primates is not yet clear. The numbers and kinds of mimetic muscles are essentially the same within both groups, yet the degrees of facial expressivity seem to differ. In general, the New World monkeys, the platyrrhines, have long been considered generally less facially expressive than the catarrhines. On the other hand, marmosets, small platyrrhines, have been found to exhibit a total of thirty-one different facial expressions, two of which—for pleasure and fear—have been shown to generate responses in conspecifics. The safest conclusion at this point is that strong facial expressivity is a basic anthropoid primate trait, although the cattarrhines show it to a larger extent.[21]

Yet the catarrhine primates also show different degrees of facial expressivity among themselves, with Old World monkeys and apes appearing to be more expressive than the gibbons and orangutans. The faces of bonobos and chimpanzees, in particular, seem to be in near constant motion during their social interactions. Yet these differences do not correlate with different numbers of mimetic muscles among these species. Perhaps they lie not in the numbers but in their structure, their physical robustness. Weaker muscles might make for less marked expressions. One study indicates that that may be true of the gibbons, the so-called lesser apes (**Plate 2**), which also have a visibly less extensive set of social interactions and hence a social structure in which facial expressions are probably less important. Differences in mimetic muscle robustness in general, however, are not likely to be the primary source of differences in expressivity. In general, the less expressive catarrhines do not seem to make feeble or ineffective expressions but simply make fewer over long periods.[22]

An intriguing correlation revealed by some comparative FACS studies is between degree of facial expressivity and body size; in general, the larger the animal, the more expressive it is. In particular, because platyrrhines are generally smaller than catarrhines, this relationship may be relevant to the difference in expressiveness between them. The correlation also holds generally within the catarrhines, although there is one obvious exception. (Lowland gorillas are larger than humans but less expressive.) Why should the size of an animal correlate with its facial expressivity? One possibility is that body size relates indirectly to the ability to perceive expressions since body size correlates with eye size; other things being equal, larger eyes have greater visual acuity than smaller ones. Better acute vision, in turn, should facilitate better perception of expressions, which may then have led to selection for greater expressive capacity in the larger primates.[23]

Such considerations lead into the neurobiology of reading facial expressions, but before we enter that arena we should examine the other way in which neurobiology may be relevant—namely, the activation of the mimetic muscles.

Differences in Neural Fine Control of the Mimetic Muscles

If the source of different levels of expressiveness within the catarrhine primates, however, is not in the numbers or (generally) the structure of the mimetic muscles, then it probably involves differences in their innervation. If there is less-extensive neural input with fewer nerve–muscle contact points, one would expect less fine control over muscle contractions. This possibility can be examined because neural control of the mimetic muscles originates in one place in the brain, a special pair of *nuclei* located in the pons, or within the hindbrain. This paired nerve, which has been mentioned previously, is cranial nerve VII, the *geniculate*. Because its neurons directly innervate muscles, they are termed motor neurons or **motoneurons** and the nerves themselves *motor nerves*. Altogether, there are twelve cranial nerves, of which the first two pairs, I (olfactory) and II (optic) develop from the cerebrum, while the others, including VII, originate in the hindbrain. Depending on the neural signals sent to the two halves of VII from the motor cortex in response to prior signals from the amygdala and other brain regions, VII sends appropriate signals to particular facial muscles, triggering them to contract and producing visible facial expressions.

Ascertaining whether differences in expressivity might be due to differences in cranial nerve VII requires comparative studies. Throughout the placental mammals, VII shows a highly similar general pattern of organization and cell types. It is divided into smaller physical units, *subnuclei*, each one innervating a specific set of the mimetic muscles, and there is a common pattern of innervation from these subnuclei to the muscles to which they connect. Yet there are differences in organization and structure of this nerve between species, and they provide clues to the basis of differential facial expressivity. In particular, there is a correlation between the numbers of estimated neurons in VII and the degree of expressivity between nonprimates and primates. Thus, tree shrews have fewer neurons in VII than prosimians, which in turn have fewer than the visibly more expressive anthropoid primates. Within the latter, however, there is no tight correlation between the number of neurons in cranial nerve VII or its subnuclei and degree of facial expressivity. The overall range in number of these neurons is only about threefold despite much larger differences in both body size and brain size, with much overlap in numbers between less and more facially expressive species, although apes and humans have somewhat more neurons in VII than expected from simple extrapolations of the size of the hindbrain. Hence, differences in neuron number are probably not the main explanation; differences in structure or degrees of connectivity to muscles are more likely to be.

One particular known structural difference is that larger-brained (and more facially expressive) primates have a more complex internal neural structure within cranial nerve VII than smaller-brained ones; that extra degree of complexity implies a higher degree of fine neural control of the muscles. In addition, the size of the motoneurons—in particular, subnuclei of VII—correlates with a certain kind of muscle fiber that is innervated by those neurons, the so-called fast-twitch fibers, which are activated rapidly as their name suggests. Mimetic muscles that are particularly active in certain species are controlled by larger subnuclei that have larger motor neurons associated with them.

The most significant neural difference between the generally less expressive platyrrhines and the more expressive catarrhines is that there are substantial *direct* connections between the cerebral cortex and cranial nerve VII only in the catarrhines. These connections link both the motor cortex and the *premotor cortex*, the part of the prefrontal cortex just anterior to the motor cortex, to the subnuclei of VII. Thus, in the catarrhine primates, there is the potential for direct, voluntary control of those expressions by the higher cognitive centers of facial expression. (In contrast, in all other mammals, the cortical connections are to a structure adjacent to VII, the *reticular complex*.) Such direct control in the catarrhines may well be related to the coupling of facial expressions with vocalization in these species. One reason to suspect the importance of direct cortical control of the muscles involved in vocalization is that there are similar direct links in songbirds between telencephalon-derived cortical regions and the motor neurons involved in the production of bird song.[24]

It is a relatively small extrapolation from direct links between cerebral cortical control of facial expressions and vocalization in apes, on the one hand, and the coupling of such expressions and speech in humans on the other. Our species, after all, is not only unique in possessing true language and speech but couples the spoken words with a whole range of subtle facial expressions that reinforce the meaning of the words and provide the emotional subtext of the verbal information. (The importance of this becomes obvious when one is talking with someone who is consistently affectless in his or her speech; the effect is eerie.) This connection between facial expressivity and complex verbal communication may well have its roots in some rewiring between the cerebral cortex and cranial nerve VII that occurred well before there were humans or human speech, in the ancient evolutionary origins of the catarrhine primates, perhaps 40 mya. This matter of how new neural connectivities arise and evolve is crucial to understanding brain–face coevolution, and we will return to it shortly.

Another further significant point about facial expressions is that the capacity to make them is universal among humans and must therefore in some sense be hardwired; in their fine-tuning, however, they often differ between human cultures. The universality of certain strong facial expressions in humans was first investigated and documented by Darwin as described in *The Expression of Emotions in Man and the Animals.* The basic human feelings—happiness, anger, sadness, fear, disgust, surprise—are expressed facially in basically similar ways in all cultures and, as we have seen, in chimpanzees, which testifies to their deep evolutionary roots. In humans, they are also seen in individuals blind from birth, so they are not learned by watching others. Yet these basic expressions often differ subtly between cultures, and the play of expressions on the face of someone who is speaking is affected by her or his cultural background. In European cultures, it is the expressions around the eyes that accentuate the feelings, while in Japan expressions around the mouth are more important. Thus, while the basic capacity for human facial expression is ultimately embedded in our genetic inheritance, the making of those basic expressions is modulated by cultural learning. The same thing pertains to the more fleeting *microexpressions* studied by Paul Ekman and his colleagues. Whatever the precise neural circuitry involved in conveying feelings via the mimetic muscles, those responses clearly can be modulated in strength and detail.[25]

The making of facial expressions, however, is only half the story of their role in social interactions. This ability would have little value unless those expressions can be instantly and automatically read by others. That interpretative ability must have coevolved with the capacity to make facial expressions. That capacity for instantaneous mental decoding of another's facial expressions is itself remarkable; we consider it next.

Reading Expressions: The Brain Circuitry Involved

Because the making of facial expressions via the mimetic muscles is a unique mammalian trait, their automatic visual interpretation or reading by other individuals must also involve special mammalian neural properties. Correspondingly, there must have been some specific evolution of neural capacities in the brain to make it possible.

A crucial point is that the reading of facial expressions is carried out by brain regions that are distinct from but physically and functionally connected to those involved in the recognition of individual faces. This is evident because certain brain lesions can eliminate or reduce one function without substantially diminishing the other. Thus, individuals who are afflicted with prosopagnosia and incapable of recognizing faces of particular individuals (including some they have known for decades) can still read and react to the emotional content of those persons' facial expressions. The converse case is that of individuals who have **Asberger's syndrome**, a form of high-functioning autism in which individuals often have highly developed intellectual capacities but poorly developed emotional ones; they can remember and recognize faces but cannot interpret the emotional content of facial expressions. Despite these separations, the brain regions involved in identifying faces also participate in reading the emotional content of images of faces, as noted previously.

Normally on meeting someone there is initially a complex act of distributed neural processing carried out by different centers in the brain that allow identification of the individual, tapping into the emotional associations with that individual, and reading her or his expressions as the encounter evokes recognition on his or her part. The initial neural responses are quickly followed by integration of the information within the prefrontal cortex, which often produces a new facial expression in response, mediated by the mimetic muscles, and an executive decision to do something such as walking toward or away from the individual or the situation that has evoked the emotional response. That executive decision is transmitted to the motor cortex, which initiates the signals for activation of the appropriate body muscles.

We have already looked at the first step, the neurobiology of individual face recognition, and will focus here on the second part, reading someone's expressions. A greatly simplified diagram of the neural pathways involved in the initial reading of expressions is shown in Figure 7.8. As with the basic visual and auditory (and language) pathways (Box 7.1), two initial pathways have been identified. One involves a relatively fast processing through both the **superior colliculus**, a midbrain structure derived from the mesencephalon, and the pulvinar nucleus of the thalamus; these send signals to the amygdala. Originally termed the *subcortical route* because its elements lie beneath the cerebral cortex, this pathway sorts relatively few of the details of the scene but serves to instantaneously detect whether there is an element of immediate danger; that information is then passed to cortical processing centers. The other pathway, relatively slower in processing the information, though only by small fractions of a second, takes place initially in cortical areas as indicated in the diagram. While the basic face-identification steps are carried out by the OFA and the FFA, the **superior temporal gyrus** located dorsally to the FFA, helps extract the information conveyed by the visible expressions. Both areas connect to memory centers in the prefrontal cortex, where that dual information is integrated.

These two routes are not wholly independent, of course, but show evidence of functional linkages to integrate the information, as also happens with the other dual-pathway systems. The fully integrated information (about identity and feelings about the person) leads to a rapid and immediate executive decision that determines the bodily response. That decision is first passed to the motor cortex, then to the thalamus and the basal forebrain nuclei, and from there to motor centers that control body and limb movements and to the orofacial motor nucleus (cranial nerve VII) in the brain stem, the latter resulting in appropriate facial expressions. When a potentially friendly encounter happens at a relatively short distance, the result is often movement toward the other, with some accompanying appropriate facial expression.[26]

That initial sequence is often just the prelude to what will become a more complex social interaction—namely, a conversation—in which there is not just exchange of verbal information but a stream of visual information provided by the flow of expressions that accompany the exchanged words. Since only humans have speech, that linkage of speech and appropriate facial expressions must be as distinctive a human characteristic as

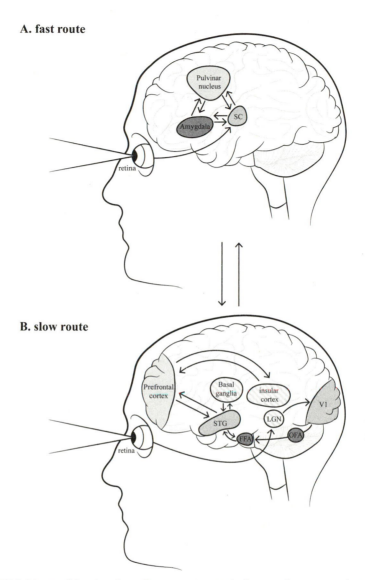

A. fast route

B. slow route

FIGURE 7.8 Diagram of flowchart for reading expressions. **A.** The fast route for processing the instantaneous emotional response; involves the superior colliculus (SC), the amygdala, and the pulvinar nucleus. This pathway is especially important for instantaneous appreciations of danger or possibly hostile encounters. **B.** The slow route in which the information is integrated with memory for a more thorough processing and understanding. It runs through the main visual processing pathways as also shown in Figure 7.6.

speech itself is, though undoubtedly it possesses deeper evolutionary roots in primate history.

A possible clue to those roots is given by work on the rhesus macaque, an Old World monkey, and the way it integrates *perception* of vocal information with faces and facial expressions in a distinct region of the cerebral cortex. Though macaques lack speech, they vocalize certain feelings, and those vocalizations are an important part of their social communication. The integration of aural and visual information about facial expressions takes place specifically within the lower, lateral regions of the cerebral cortex, the *ventrolateral prefrontal cortex* (VLPFC), and specifically in the area believed to be homologous to Broca's area in humans, one of the important regions involved in both the construction and processing of speech. The involvement of Broca's area in linking vocalization and facial expression had been suspected from brain scans in humans using fMRI via electrodes attached to the scalp. It has now been confirmed by physiological experiments involving direct tests of individual brain neurons in the VLPFC as carried out in macaques (under prescribed conditions that consider their welfare). The work shows that macaque Brodmann areas 45, 46, and part of the border region of the neighboring Brodmann areas 12 and 47 (Figure 7.4), which correspond essentially to Broca's area in the human brain while extending beyond it, participate in this integration. The electrode studies also show that some individual neurons are activated by and presumably integrate both the aural and visual information. Furthermore, behavioral studies of macaques show that they, like humans, look at the facial dynamics of vocalizing individuals and can link specific individual voices to particular individuals, even when the faces are not seen, for both familiar conspecifics and humans. These traits and features of macaque communication are seen in true speech in humans.[27]

Thus, rhesus macaques and presumably all or most catarrhine primates have the neural foundations for face-to-face vocal communication using combined aural and visual information and memories of the individuals involved in such communication. As mentioned previously, the basic rhythm of human speech may derive from the theta rhythms of such facial gestures as lip smacking as studied in macaques. Since the catarrhine primates diverged from the platyrrhines 30 million to 40 million years ago, these capacities must be at least that ancient. Might they, however, go back to the origins of the anthropoid primates? The answer is unknown, but it may be relevant that the platyrrhine primates have a smooth (*lissencephalic*) prefrontal cortex that lacks the visible Brodmann areas where these capacities reside in both macaques and humans. Of course, they may still possess equivalent functions, but from that difference in cortical architecture the simplest inference is that the human capacity to read the facial expressions that accompany speech evolved with the catarrhines and did not exist prior to their appearance.

Of course, speech is a human property, hence there must be some additional human-distinctive neurobiology—not possessed by nonhuman catarrhines—that joins speaking and facial expressions. In thinking about this, it would be helpful to know when speech itself arose in the hominin lineage. Although dating the origin of human speech from fossils is impossible, genetics might provide clues. If particular genes seem to have special roles in language or speech, then comparative studies on those genes might be informative if they compare humans and nonspeaking animals. One such gene is *FoxP2,* which encodes a transcription factor expressed in the brain. Although no longer regarded as *the* key to understanding language, as it was for several years, it clearly plays an important part in human language learning and speech. Its particular neurobiological roles are described in more detail in Box 7.3.

Since our nearest primate relatives, the chimpanzees and the bonobos, lack speech, the only certainty is that this capacity arose sometime after the split of the hominin lineage from that of the chimps about 6 million years ago and probably much closer to the present than to our australopithecine past. Indeed, the origins of speech probably occurred long after the chimp–human divergence and almost certainly sometime after the appearance of the genus *Homo,* whose first member species arose 2 million to 2.5 million years ago. Currently, the estimates for the onset of speech range from sometime in the late Pleistocene, perhaps 500,000 to 1 million years ago, in either *Homo erectus* or *Homo heidelbergensis,* to sometime well after our species, *Homo sapiens,* had arrived on the scene, perhaps only 70,000 to 75,000 years ago. Those latter dates bracket the point of origin of the oldest known art artifacts, whose manufacture almost certainly required the existence of complex, symbol-based language; it is hard to conceive how such things could have been made without the artists explaining to other members of the group what their purpose was. Of

BOX 7.3 *FOXP2* and Language

The clue that sparked interest in the *FOXP2* gene in humans as a possible key to under-standing the biological basis of language was the discovery of a large family (designated the *KE family*) many of whose members showed an inherited speech disorder that was trans-mitted over several generations; the disorder seemed specific to speech rather than a more general cognitive function. Genetic analysis showed that the source of the defect was a mu-tation in the gene *FOXP2*. This gene encodes a transcription factor, and the point mutation involved was in the DNA binding domain of the encoded protein; the gene is expressed abun-dantly in the brain, including regions known to be involved with language and speech. The implication was that the speech defects were a consequence of some partially dysfunctional gene regulation within the brain that affected language or speech centers. Subsequently, it was found that several other genes implicated in language disorders are part of a regulatory network, including *FOXP2*. Furthermore, *FOXP2* was implicated in both mouse vocalizations and bird song. Other work, however, indicates that neither *FOXP2* nor the other genes are required exclusively for language or the development of speech capacity. In particular, *FOXP2* appears to be involved more generally in certain kinds of motor circuitry, linking the cortical speech centers with a particular region of the basal ganglion, the striatum, which is involved in relaying messages from the motor cortex ultimately to the muscles, including the muscles of the larynx. In terms of its precise role in neural development, it seems to have two roles. First, it is essential for promoting the development of outgrowths from neural cells, or *neu-rites*. This is essential for establishing new connectivities in brain development to promote sufficient *plasticity* of brain development, not least during the period when language is being acquired by the child. *FOXP2* has also been implicated in promoting neurogenesis by in-creasing the numbers of secondary neuronal progenitor cells. Currently, it is not clear which role is more important or whether both are equally important in promoting the necessary neural connections for full speech capability. Finally, it was initially regarded as the key to understanding the evolution of language capability and thus was presumed to have unique properties in modern humans relative to Neanderthals, who are presumed on the basis of indirect and negative evidence to not have had speech. In fact, the gene sequence has been reported to be identical between *H. sapiens* and *H. neanderthalis*. Whether it is expressed more abundantly in modern human brains than it was in Neanderthals is unknown. *FOXP2* is no longer regarded as the so-called language gene, but clearly it and its associated genes contribute importantly to the capacity for language.[1]

---

1   See Enard (2011), Scharff and Petri (2011), and Graham and Fisher (2013).

course, language and speech may have developed long before *H. sapiens* took up artistic activity. Indeed, it is likely that simpler forms, **proto-languages** that involved only simple noun–verb relationships (such as in "he hit." "I eat"), existed earlier and were the precursors to true language.[28]

Nevertheless, while the time of origins of human speech are unknown, we can speculate about the evolutionary processes involved. The basic premise is that those origins lay in certain unusual properties of neural system development, followed by the conversion of those altered neural-developmental processes to hereditary ones by genetic changes. In the next section, we consider what those special neural-developmental processes are and how their conversion to hereditary properties may have taken place.

Connecting the Brain and the Face via Novel Neural Circuitry to Become an Expressive Speaking Creature

Framing the Problem

Let us begin with the various things that differentiate language from the vocalized communications of other primates. First is the existence of true words, which are the units of speech, and their great number. Most individuals whose native tongues are one of the major languages probably have working vocabularies on the order of 5,000 to 20,000 words, and some have estimated vocabularies in excess of 100,000. Even the lower numbers dwarf the number of known primate vocal signals, on the order of dozens, that convey some sort of specific meaning. The second distinctive feature of language is the variety of different kinds of words; these range from those that denote visible objects or entities (nouns) and specific kinds of actions (verbs) to those that serve as modifiers of such (adjectives, adverbs) to, finally, those that serve as linkers or connections of some kind (prepositions, conjunctions). Third, and most complex, are the intricate grammatical and syntactical structures that permit sentences to be made.

Furthermore, to recap a previous point, there is a set of laryngeal and mouth structures that together permit the conversion of mentally constructed language into speech. Most obviously is the capacity to form the wide range of different kinds of sounds that are needed in human speech; these involve the muscles of the tongue, larynx, lips, and lower jaw and the innervation patterns to move these muscles appropriately to generate sounds. This capacity is crucially important for speech: the range of sounds or phonemes that humans can make is much greater than that of any ape and certainly contributes to the ability to vocalize a large number of words. Only a few kinds of birds, such as parrots and mynah birds, and a still smaller number of mammalian species (perhaps just one or two seal species) can produce anything like the human range of phonemes.

Development of these capabilities would have required a great deal of evolutionary change from vocalizing but nonspeaking hominin ancestors. Most of this change would have been neural changes in the brain, while the physical capacity for speech would have required subtle developmental changes in the whole oral apparatus, in particular the repositioning of the larynx farther back in the throat and changes in the muscles of the mouth, including the tongue and lips, in addition to much neural change to permit the conversion of language thought into speech. If we step back to consider the amount of evolutionary innovation that speech required, it will be obvious that it is formidably difficult to envision how language and speech might have arisen through Darwinian step-by-step evolution, with each step creating slightly more function. The problem has long been appreciated. Darwin himself understood it well and saw it as a hard test case for his theory of evolution by natural selection, as did his severest critics. Its apparent intractability to simple explanation and the seemingly endless speculation the subject prompted even led two preeminent European learned societies in 1866 to ban their members from discussing it! Those prohibitions, of course, could not be enforced and were almost certainly ignored by some scientists, but they had a chilling effect on discussion of the subject. It would not be until the 1980s that it once again became a respectable subject drawing serious analysis.[29]

Today, because of progress on many fronts, the problem of the origins of speech no longer looks hopeless, even if there is still no agreed solution. Searching for the origins of speech today resembles more a purposeful swim toward a distant but visible shore rather than the situation 150 years ago when it might have seemed more a helpless and exhausting floundering about in deep ocean without any land in sight. We already noted, for example, some possible connections between primate lip smacking and other facial gestures that have the rhythmicity of human speech.

In particular, if we concentrate on the general aspects of neural evolution that must have been involved, we can pose three logically distinct possibilities for the neural evolutionary events necessary to integrate the different aspects of language construction and speech. Thus, it can be asked whether the changes involved (1) strengthening and deepening previously existing but minor neural circuits, (2) creating wholly new neural circuits, or (3) rewiring preexisting circuits to create new ones. In reality, possibilities 2 and 3 are not truly distinct because connecting previously unlinked circuits to generate a novel function is, by definition, to create a new circuit. Indeed, it is difficult to imagine how a new functional brain circuit could arise without some degree of rewiring. Hence, we may collapse the possibilities to two: (1) strengthening old circuits and (2) rewiring old circuits to create new ones. Yet evolution almost certainly employed both mechanisms in the creation of language and speech capabilities. The challenge is to work out the relative importance and specifics of their respective contributions.

The scope of the problem is fiendishly complex and more so than believed even until fairly recently. In particular, it was believed for many decades that the loci of language and speech hence of their evolution could be localized to a small number of key brain regions. These were Broca's area (in the inferior frontal cortex) and Wernicke's area (in the temporal lobe), which are connected by a looping neural cable, as it were, the arcuate fasciculus. It was believed primarily on the basis of lesion studies that Wernicke's was effectively the seat of language comprehension and Broca's the seat of language production and that the arcuate fasciculus provided the key communication between them. Broca's area has long been regarded as the center controlling vocalization per se while Wernicke's area was seen as the cortical center that was principally charged with word meanings and linguistic representations in general. In the past decade, however, this distinction has collapsed as evidence has accumulated that Broca's area participates in aspects of language comprehension and Wernicke's in language production. Beyond that, it is now apparent that the total set of neural connections required for language and for generating speech is vastly more complex, involving numerous areas and not only in the frontal and temporal lobes but also in the parietal lobes. The picture today of the neural bases of language and speech is similar (albeit more complex) to that for face processing: a highly complex, dispersed, multiply connected network involving many regions and parts of the brain and much neural computation.[30]

Without going into all the complexities of how the whole network might have evolved, we can illustrate the basic problem by looking at a principal feature of speech—the production of audible words—and asking how its neural circuitry might have evolved. The first consideration is that the production of words is carried out by the larynx and the apparatus of the mouth, including the tongue, lips, and jaws. This must involve neural signal directions from the parts of the cerebral cortex known to be directly connected with speech production such as Broca's and Wernicke's areas. Neural signaling emanates from these two cortical areas and travels to the basal ganglia, which sends projections to the thalamus and from there to the motor cortex. From the motor cortex, nerves go to the brain stem, from which motor nerves activate muscles of the larynx, tongue, jaws, and lips, which then produce the vocalized words.

How did such complex neural circuitry for making words arise? Although rhesus macaques and chimpanzees have regions recognizably homologous to Broca's area in humans (and to Wernicke's in chimpanzees), it was long believed that these regions had little or nothing to do with vocalizations by these animals. Rather, it was thought that monkey and ape vocalizations, in contrast to human speech, involved no initiating control from the cerebral cortex and were generated autonomously in response to provoking situations (danger, food, certain social situations) by a part of the brain located beneath the cerebral cortex named the **cingulate cortex** (see Figure 7.3) and specifically by its anterior region. If so, then it would be likely that the human neural circuitry arose de novo. More recent work, however, indicates that in macaques, *some* cortical connections to the muscular oral apparatus, including some from Broca's area, create vocalizations. There are also connections linking the anterior cingulate cortex and Broca's area and the so-called supplementary motor cortex in the macaque brain, which indicates the possibility of cooperative action between cortical areas and the cingulate cortex in controlling monkey vocalizations. Thus, the evolution of the human capacity to vocalize words may have employed the first-mentioned mechanism, the strengthening of the preexisting neural circuits that had previously only weakly linked regions of the cerebral cortex to the pharyngeal and oral musculature. Such a strengthening and expansion of the preexisting cortical circuit presumably occurred in late hominin evolution, during the past 1 million to 2 million years, the period of rapid enlargement of the cerebral cortex.

In contrast, the evolution of the capacity for large vocabularies and the creation of complex grammatical and syntactical structures that underlie spoken language probably involved some truly new circuitry. Consider the difference between chimpanzees and humans in the capacity for remembering new words. While chimpanzees simply cannot make the range of sounds humans can and therefore can never be taught to speak, they can be taught words using movable visual objects such as colored blocks of different shapes to act as their symbols. With substantial effort on the part of both the trainers and the animals themselves, a few chimpanzees have been taught 200 to 300 words in this way, mostly nouns and verbs. Those are, however, much smaller numbers of words than humans learn and who learn them with much greater ease and speed. Furthermore, the chimp-learned words lack the complexity of human word categories. That huge increase in human capacity, along with the potential to employ new mental structures not present in apes (grammar and syntax), seem to demand something qualitatively new in neural connectivity, as well as expansions of existing circuitry. (The chimp results show, however, that chimps have some capacity for symbolic thinking, which indicates that property itself was not original to the hominins.)

Furthermore, the sheer complexity of language structure would seem to demand some kind of neural circuitry that was not present in non-speaking hominoid or hominin ancestors. In particular, there are the demands of complex utterances in the form of sentences in which there is reference back to something contained early in the sentence. That embedded form of reference is termed **recursion**, and it has been argued that it is the central and perhaps defining property of human language in contrast to the simple sentences that would have featured in proto-languages. It requires greatly enhanced working (short-term) memory to keep track of the parts of the sentence, a mental ability that has been termed a **phonological loop**. Such increased short-term memory capacity is probably embedded in and distributed within the new circuitry itself rather than being localized in separate and specialized memory modules.[31]

Perhaps one of the most certain instances of rewiring to generate new circuitry for a new function involves the face, namely, the constant play of facial expressions that accompany speech, amplifying and reinforcing the meaning of the message. When two people are speaking face to face, they are also reacting to each other's facial expressions, carrying the emotional subtext, which is often just as critical as the overt meaning carried by the words. Yet that reading of the facial expressions is not wholly dependent on seeing those expressions. The same expressions often come into play in talking on the telephone, and relying on the subtle modulation of sounds created by the expressions around the mouth of the speaker, the listener can also, in effect, *hear the feelings*. Thus, the facial expressions subtly shape the production of the sounds.

In principle, this apparent coordination between spoken words and facial expressions denoting the feelings could be temporal coincidence: the speaker feels her or his feelings while speaking and since both feelings and words pertain to the same subject, the two aspects simply co-occur. Yet this seems unlikely because the coordination is too tight: if the speaker deliberately slows his or her speech for emphasis, the play of facial expressions are also produced with the new tempo even though the underlying feelings must be the same. Similarly, sped up, the tempo of expression change perfectly matches that of the words. There is almost certainly some neural coordination of the two processes.

Because speech itself is unique to our species while facial expressiveness is an ancient primate capacity, such active and continuous coupling of facial expressions to speech must have involved the evolution of some new circuitry. Specifically, this novel circuitry would have involved the creation of new connections from the key language-generating centers of the brain to the circuitry governing facial expressions. In effect, while the *phonological loop*, the neural circuitry that makes coherent speech possible, was evolving, new linkages from it to the neural circuitry controlling the mimetic muscles must also have formed. Previously we saw that catarrhine primates possess direct cerebral cortical connections to the vocalizing musculature, permitting some kind of higher level control of vocalizations. Similarly, there almost certainly must have evolved new neural connections connecting the vocalizing apparatus with the mimetic muscles that affect facial expressions.

The specific problem of how such new linkages came into being touches on a question that is central to the evolution of all animal behavior: how does the brain rewire itself, making new neural connections to generate new functional properties (behaviors)? As with all big evolutionary questions, this needs to be decomposed into more manageable ones, in this case, two: (1) How do new neural pathways form during the development of the individual, whether in embryonic or fetal stages or still later? (2) How do such changes become evolutionary ones? Conceivably, the evolutionary changes could take place in one step: a new mutation occurs and a new neural circuitry is formed and with it a new functional property during the development of any individual bearing that mutation. Yet for something as complicated as the phonological loop, it is difficult to imagine how it could have been created by one mutational change. It is more likely that multiple hereditary changes were involved. Furthermore, as argued shortly, it is more likely that the initial changes originated as developmental ones that were subsequently replaced and stabilized by genetic changes. Had these spread within populations to become fixed within them, they would have generated evolutionary change.

On the Evolution of New Neural Circuitry

Neural circuitry consists of chains of neurons that sequentially pass along electrical signals. All neural circuitry has this basis, hence all neural responses involving mental and emotional states are underlaid by it. Each neuron is connected with many others but tends to pass its message along preferred routes. The number and diversity of those routes is sufficient to generate the immensely diverse yet ordered complexity of transmission of neural information. The process involves small branches on the main body of the neuron termed **dendrites**, which receive a chemical signal released from the neuron immediately before it in the chain. This then triggers an electrical current that is transmitted—via directional surges of ions across the cell membrane—past the central part of the cell containing the nucleus, down a long and often branched process termed an **axon**, which makes contact with the dendrites of the next neuron(s). (These features of the neuronal cell are shown in diagrammatic form in Figure 2.3D.) When neurons first arise from cell divisions of neural progenitor cells, they lack both dendrites and axons. Dendrites form fairly quickly, however, while axons are slower to develop, growing out from the parent neuron, often only reaching their target cells (other neurons or muscle cells) after a long period of growth. It is axonal outgrowth, which is often accompanied by extensive branching of the axon to generate more connections with other neurons, that is central to generating long neural circuits that connect different parts of the developing brain.

The process by which axonal outgrowth connects different parts of the brain to form key functional circuits is simultaneously both highly specific and apparently wildly promiscuous. That oxymoronic description requires an explanation. First, the elements of specificity: axonal outgrowth is governed by a complex set of *guidance molecules,* of which there are four major classes, each encoded by a particular gene family: *netrins, slits, semaphorins,* and *ephrins.* Each set of these *canonical* guidance molecules (others have been discovered and still others may be found) has its own specific receptor molecules, and many of the guidance molecules can either act as attractants or repellents for axonal growth. Both responses involve the same receptor molecules, but the response depends on the properties of the responding cells. Furthermore, many act over long distances as diffusible molecules, while some act at short range as either attractants or repellents. When a particular match is determined by the molecular properties of the respective cell surfaces between the extending axons and the dendrites of the target cells they reach, neurons become connected to become a potential part of new circuitry.

In the wiring up of the brain, the initial contacts between two regions are often made by so-called pioneer axons. The physical pathway established by those vanguard cells then becomes a foundation for more axons growing out from the initial center to the ultimate target cells. The four main classes of axon guidance molecules are found throughout the animal kingdom from nematodes and fruit flies to mammals, so they probably originated early in the history of the animal kingdom—in the Cambrian period or perhaps earlier.[32]

These facts about guidance molecules and their context-dependent activities help explain the high chemical specificity of the process, even if they do not fully account for how pioneer axons from one region of the brain find their appropriate specific targets. This, however, is where the promiscuity in neural circuitry formation enters the picture. That feature is displayed in the initial outgrowth of axons, which is general and not always dependent on pioneer axons. As a result, many initial connections are made between different parts of the early developing cerebral cortex and also between parts of the cerebral cortex and other parts of the brain, including regions of the hindbrain. The extensive branching of axons that occurs during axonal outgrowth contributes to the profusion of new connections. Yet in contrast to some of the routes mapped out by the pioneer axons, many of these connections do not survive; there is a widespread dying back of such connections, *axonal pruning,* that leaves a relatively smaller number of final, stabilized connections. The processes by which the winners and losers in this axonal competition are established are not fully understood but they seem to involve, in part at least, competition between axons for so-called **neurotrophic substances** (literally, "neural feeding substances"). Apoptosis also eliminates a considerable number of neurons and also involves the neurotrophic substances; those that do not receive enough neurotrophic support are winnowed out, reducing the initial set of connections.[33]

At present, it is impossible to state precisely how specific new and important circuitries between different parts of the brain are formed, but there are two alternative general possibilities for how such circuits might arise. They are schematically shown in Figure 7.9. The first is that there might be subtle biochemical changes in one part of the brain that *initiate* or *extend* axonal growth to a new set of target cells (Figure 7.9A). Such new axonal growth could come about through changes in the *distributions* of one or more of the guidance molecules either via attraction of axonal growth cones to the new area or by some combination of redeployed attraction and repulsion. The second possibility is that the foundations of the new connections were already in place via the initial and highly unspecific efflorescence of axonal outgrowth, with more of the axons that would have been removed (or their parent neurons by cell death) now stabilized and remaining connected (Figure 7.9B). Such stabilization might indicate improved competitive ability by the axons, perhaps triggered by more neurotrophic factor production by the new target cells or increased ability to bind one or more of those substances by the axons that become stabilized.

A central question concerns the nature of the process that determines the apparent winners and losers among the neural connections. The whole process has a Darwinian feel to it by involving competition between neurons and their connections, with some winning out over others. Indeed, the whole process has been termed one of *neuronal selection*, though precisely what the basis of such selection is remains unclear. Much evidence now indicates that it is the individual animal's experience that is crucial: such experience can strengthen or weaken particular neuronal circuits. In effect, the process probably exemplifies the classic injunction of "use it or lose it," and pathways and circuits that are initially used become strengthened and are favored to survive.[34]

This accords with the facts of childhood development since much of the axonal pruning and neuronal apoptosis that winnows out many

A.

B.

A	B	C	D	E	F	G	H

✕ axonal pruning

✸ site where axonal pruning
does not take place

✳ apoptosis

FIGURE 7.9 Diagram of the two main mechanisms for establishing major routes of neural connectivity. **A.** Selective affinity for neuronal outgrowth in which the outgrowing axons of developing neurons are especially attracted to target cells B, D, and F. **B.** Selective pruning of neurons and axonal connections. This diagram indicates the pruning back of axons from neuronal targets A, B, C, , F and G, and apoptosis of neurons that extended their axons to E and H. (In such apoptotic events, the axons are subsequently lost.) In this diagram, the connection only survives with D.

connections and strengthens others begins in infancy and continues through childhood, the period when so much is being learned, including language. Two main centers where these processes seem especially predominant are the prefrontal cortex and the hippocampus, whose respective principal functions are integration (of different neural inputs) and memory; these processes are essential to the vital and rapid learning that takes place in childhood. A further distinct period of axonal pruning takes place in the teenage years. This takes place long after language has been acquired but corresponds to the period when much risky behavior is assayed and then, if all has gone well (the teenager has survived), more settled behaviors take their place. It is not hard to relate such facts to a winnowing of neuronal connections in favor of a select subset of such. This is also the period of the final maturation of the face. Perhaps, although this is only a tentative suggestion, this is the period when expressions that come to be typical of the individual become habitual; if so, this selective pruning of neuronal connections might also be involved in that.[35]

Whatever the developmental processes that shape and select certain neuronal pathways, those processes in themselves cannot account for new circuitries that become characteristic of a species. That requires hereditary foundations that, in turn, demand prior heritable change, hence *mutations* in the broad sense. Such permanent changes almost certainly require subtle and permanent changes in the underlying biochemistry of the neuronal circuits. There is nothing inherently improbable about the occurrence of such mutations. To the extent there is genetic regulatory control over the localized amounts of the guidance molecules or the neurotrophins—and much fine-scale regulatory control of these molecules takes place—the existence of mutations altering their amounts or distribution should occur. Such mutations would alter the spatial trajectories of axonal outgrowth, thereby producing new circuits.

Because the traditional view of evolutionary change depends on new mutations, such change has to wait on the occurrence of the right mutations or on factors that liberate the expression of preexisting mutations. There is, however, another broad possibility: that developmental change precedes and predisposes toward mutational changes that then become part of the animal's genetic inheritance. Imagine that in some developing embryos, some externally derived influence biases certain neuronal growth pathways to develop preferentially. If the animals that have these new neuronal circuits enjoy some advantage relative to their littermates and peers that do not have them, and if the external conditions are repeated over generations, there could be continued selection for those new circuits. Any new mutations that cause the same change, however, would probably do so more efficiently and would, therefore, have a selective advantage. The result would be the shift of a *plastic* developmental effect into a heritable one with evolutionary potential. This mechanism, which has had some experimental support, is known as the *Baldwin effect* after the man who first proposed it, though it also known by a later term, *genetic assimilation*. The consequences of a mutation that came via the Baldwin effect and helped create a new neural linkage—for example, one that helped create or stabilize neural circuitry linking facial expressions to appropriate moods and words—would be the same as for one that happened directly. The route, however, would be different, and if the initial developmental change were induced frequently, the chances of a favorable mutation might be relatively enhanced. In principle, the Baldwin effect speeds the rate of evolutionary change.

There is one final thought to be offered on this subject. By definition, mutations are alterations in DNA sequence. There is, however, another kind of stable heritable change, though its participation in evolutionary change is controversial. This consists of stable alterations in the binding of particular proteins to the DNA regulatory sites of particular genes. These are changes that affect the chromosomal neighborhood, the *chromatin*, surrounding those genes and thereby alter their regulation. These changes, termed **epimutations**, can often be induced by appropriate conditions with frequencies much higher than standard mutations. Any such changes that occur in and can be passed on by the sex cells to the next generation have evolutionary potential. If neural states, including neuronal circuit rewirings, can be affected by epimutations, such changes might be like the Baldwin effect and speed evolutionary processes. Both the Baldwin effect and epimutations will be discussed further in the final chapter.[36]

Conclusions

In Chapter 6, we briefly explored the possibility that social interactions and behaviors in the anthropoid primates were a major force in the evolution of their faces, and in particular that the evolution of the human face had been strongly influenced by social forces. Though not superseding the roles of eating and nutrition, which would have influenced jaw evolution in the early anthropoid primates, these may at least have rivaled them in importance. If so, then given that mental states and processes derive from brain activity, the brain itself must have become a major player in the evolution of the face. Those changes in the face, both in its morphology and expressive capabilities, could have influenced survival in such highly sociable animals; if so, they could have fed back into influencing the evolution of the brain. In effect, while the brain and face have been functionally connected since the first craniates, anthropoid primate evolution has featured a new impetus and new directions for brain–face coevolution in the form of sociality affecting survival of groups and their members.

In this chapter, we have looked at different facets of brain–face coevolution shaped in particular by sociality and the neural circuitries involved. In one respect, there probably was relatively little evolution from hominoid ancestors, namely, in the human capacity to recognize the faces of conspecifics. Instead, much of the neural innovation initially would have been involved in the making and reading of an increasing repertoire of facial expressions and the linkage of this expressive capacity with that for oral expression; this would lead eventually to the complex linkages that make language possible. In effect, new social requirements for behavior within a highly social species would have exerted its own special pressures for members to keep up with those demands—the process of *social selection* for new forms of cooperativity mediated by better communication. This process would have required selection for new neural linkages within the brain to facilitate those abilities.

Many treatments of human evolution end with the appearance of modern *Homo sapiens* no later than 200,000 years ago. Yet that point does not mark the end of human evolution. The past 70,000 years have witnessed many changes in our species, including changes in human faces. We turn to this history now, beginning with the remarkable migration of modern humans out of Africa between 72,000 and 60,000 years ago and the creation of a truly global species.

— eight —

"POSTSPECIATION": THE EVOLVING FACE IN MODERN HUMANS

Introduction: The Continuing Saga of Human—and Human Face—Evolution

In the typical history of an animal species, what it was in its beginnings is largely what it remains throughout its existence, in both physical attributes and behaviors. Although many species diverge into distinguishable subspecies, their differences are typically few and slight. In mammals, for example, such divergence often involves changes in coat color, particularly when an animal invades a new region or climatic zone. Such alterations may be accompanied by others—for example, in dietary items or certain activities—but most are relatively modest.

Not so for human beings. Some dramatic changes ensued well after our species' emergence. They can be termed *postspeciation changes*. The most important was the advent of complex language as discussed previously. Without language, our subsequent history might not have been especially different from that of all other hominin species. Modern humans arose in the latter part of what was one of the most climatically harsh geological periods in the past half-billion years, the Pleistocene. For a furless sociable mammal to survive in such conditions, coordination and planning would have been of great value and complex language an inestimable asset for facilitating such. Early *Homo sapiens* probably had some forms of *protolanguage,* as perhaps did our precursor species, or even our cousin species the Neanderthals, but their effectiveness for coordinating group action would have been far less than true language, which almost certainly arose later in our species' history. The latter probably not only ensured our species' survival during the late Pleistocene but also ultimately led to everything that distinguishes us from all other animals.

There were other notable postspeciation changes, as well. These include the partial genetic and phenotypic differentiation into visibly different ethnic groups characterized not just by skin and hair color differences but also by facial characteristics and overall body shape. Compare, for example, the faces and builds of Inuits, Congolese pygmies,

Australian aborigines, Polynesians, Somalis, and Nordic Europeans. These distinguishing features, although involving only superficial (outer body) traits, seem to be more marked than most mammalian intraspecies differences. As we shall see, that diversification was a by-product of something highly distinctive in our species' history: a rapid dispersal into Earth's major different habitable regions and splitting into small founder populations in different regions along the way.[1]

A third major kind of postspeciation change concerns the way humans now live: the development of highly complex, sophisticated, and large social arrangements that we term *civilizations*. These embrace nearly all humanity today, including people who live in remote rural settlements. Civilizations and their effects on human behavior are a major topic in anthropology and sociology, but they have largely been ignored by biologists because they seem a purely cultural phenomenon. Yet the *potential* to develop civilizations must lie in some aspects of human biology. If so, which ones specifically? To the extent this question has been considered, civilization building has been regarded as a quasi-inevitable outcome of our species' great cognitive abilities. Yet, it could not have been a product of superior cognition alone. Hands were equally important: the unparalleled ability of humans to use their hands to create tools and other artifacts must also have been vital. Yet even that combination of exceptional brains and versatile hands might not be the whole explanation. I will argue that a third property was essential—*self-domestication*—and I will discuss its possible biological bases and its possible telltale facial signatures.

In this chapter on postspeciation human history, there will be relatively little about language, which was discussed previously. The focus will be on the other two key features of our postspeciation history: the genetic differentiation of modern human populations and the phenomenon of self-domestication. As in the preceding two chapters, special attention will be paid to the roles that states of feeling and the behaviors that flow from them may have played in the evolution of the human face.

On the Matter of "Race"

It is impossible to discuss the evolution of modern humans and the way humans came to populate five continents without coming to the vexed matter of "race" and "races." It would be best therefore to acknowledge the ambiguities and difficulties of these terms at the start, before grappling with this phase of the evolutionary history.

The basic problem is that there is no clear definition of *race*, a term loaded with historical baggage and correspondingly diverse associations and meanings. As a result, it is almost impossible to discuss the subject without generating misunderstandings and antagonisms. This problem is not new, however, since there has never been a clear definition of "race." It was first used to denote what today we would designate as different nationalities, for example, the Italian or Irish "races." By the late eighteenth century, however, it was increasingly used by Europeans to denote the visibly different-looking physical types of humans initially associated with the different major geographical regions of the world (Europe, Asia, Africa, Australia, the Americas). During the nineteenth century, much of the ostensibly scientific literature devoted to this topic was based on the premise that the visible physical differences seen between different groups were indicative of more fundamental divergences involving personality and abilities. The research program associated with such observations, such as it was, was dedicated, in turn, to validating that premise, a circularity that instead undermined it from the start. *Scientific racism* is the oxymoron that denotes this branch of anthropology, and it underlay most of the nineteenth and early twentieth century writings on the subject. Undoubtedly, it was initially in part a response by Western academics to help justify the conquest of other peoples as the European powers extended their reach. Yet its conclusions were not safely quarantined within academia but, unfortunately, adopted by various xenophobic and racist movements—in particular, those in Europe and the United States—and contributed to those movements.

To the extent that one can pin a meaning to the word *races*, it denotes groups distinguished from each other by a uniform set of readily visible, qualitative physical characteristics. If one were to try to give it a more precise twenty-first century meaning, however, the demarcation of those groups would have to be based on genetic criteria and involve major population differences for alleles of different genes. But how many such gene differences are needed to distinguish putative "races"? That is never specified by advocates of "racial" distinctions, but presumably there would be more than one. Thus, for example, to take an example from another mammalian species, black leopards and typical African leopards look strikingly different but differ genetically by just one base pair in one gene, yet no one today would refer to this difference as signifying two leopard "races."

While distinguishing kinds of leopards without invoking "race" is easy, human beings have historically been a different matter. The general problem, though not its modern genetic dimension, was recognized by Darwin, who did the first systematic survey of the different racial classifications as done solely by appearance (phenotypes). As he remarks in *The Descent of Man,* there have been nearly as many different classifications of human "races" as there have been would-be classifiers. Any such taxonomy is bedeviled by subjective judgments as to where one draws the boundaries between the different groups, especially in border zones between the geographic areas where there have been contacts between peoples. This blurriness was the reason for Darwin's conclusion that all the "races" make up one human species, linked by much interbreeding between groups, which preserves the unity of the species. For him, the term *race* was useless, and the crucial fact was that all humans belong to one species.[2]

In principle, contemporary techniques for measuring extents of genetic difference today should be able to answer the question of how distinct different "races" are, provided one has designated groups to compare. Traditionally (and ignoring Darwin's cautionary words) demographers have favored a five-fold division into major "racial" groups associated originally with five different populated "continents" (one of which is actually a large part of the Pacific): Africans (sub-Saharan), Caucasians, East Asians, Native Americans, and Pacific Islanders. As in the nineteenth century, this classification is based on the phenotypic traits of these groups and their initial geographic localization. The physical differences that distinguish these groups involve four groups of traits: (1) skin color, (2) body size and shape, (3) head shapes and proportions, and (4) facial characteristics. For each trait, however, there is always some variability within a group and for many overlap with other groups. Consequently, it is the ensemble of characteristics and their perceptual *gestalt* that serves to visually place an individual in a particular group. Of course, individuals who have ancestry from more than one of these groups immediately fall out of the classification by definition and often on visual inspection.

Most significantly, when we compare the extent of genetic differences between these groups, we find the differences far less than what the visible phenotypic differences seem to suggest. The way such difference is measured involves a parameter termed **genetic variance**, and it is a quantitative measure of the extent of allelic difference between populations. Measured gene by gene or by DNA sequence region by region, the greatest fraction of human genetic variance—about 85 percent—is found *within* and not between individual "races." Only the remaining 15 percent of the variance is due to distinctive differences *between* "races." Furthermore, the primary set of those "racial" differences involves the *proportions* of the different alleles for those genes rather than qualitative all-or-none differences between allelic alternatives. Thus, a genotype containing a set of alleles that are normally found in high frequency in one ethnic group, A, but are low in a second, group B, is more likely to come from a member of the former group. Hence, the greater the number of genes or alleles tested when we know the relative allele frequencies between two putative "races," the more certain the identification will be. Usually, a set of alleles or DNA markers of thirty to one hundred, depending on those chosen, often have high diagnostic value for ascertaining ancestry from one (or sometimes more) of the five major "racial" groups demarcated by demographers. Those assessments usually accord with the self-reported "racial" ethnicity of the individuals involved. Apart from such shared polymorphic genes among the different groups, many alleles are found predominantly or solely within one particular group, but these are often in low frequency (1 percent or less), hence they cannot be responsible for any *general* differences between the ostensible major "racial" groups. Thus, at the genetic level, "race" is a statistical property, and the assignment of an individual on the basis of his or her genotype to one of the five traditional "racial" groups is both probabilistic and error-prone, especially when the subject has some mixed ancestry, which, in fact, is common.[3]

The genetic findings dissolve "race" as a scientific category. Accordingly, the fundamental tenet of scientific racism that there are distinct "races" is simply false. On the other hand, dismissing "race" as a purely cultural construct, as some do, is equally false. The visible differences, first perceived when different peoples from the different major geographic regions initially encountered each other, are real as is the fact that it is often possible today to judge at a glance whether someone has a major share of, for example, European or sub-Saharan African or East Asian ancestry. How can one square the scientific facts, which belie the idea of distinct "races," with the visual perceptions, which often do indicate something about ancient ethnic ("racial") origins? The resolution must be that there *are* a small number of strongly skewed genetic differences affecting the visible characteristics associated with the original major population groupings of humans as first described by seventeenth- and eighteenth-century geographers. These would fall into the 15 percent genetic variance differences that are associated with ethnic group origin as detected by DNA analysis. In contrast, there is no evidence for and, indeed, much against, the existence of genetic differences for mental or personality traits between the different groups recognized by demographers. Those groups often have some quantitative genetic differentiation related to certain disease susceptibilities, but supposed "racial" difference is primarily about the physical traits that gave rise to the concept of "races" in the first place.[4]

The questions raised by these considerations concern how human populations came to differ genetically from one another to the modest extent they do and whether there is *any* biological significance to those differences (and, if so, what they are). This question is bound up with the more fundamental one of human origins: when—and where—did our species first arise? We turn to that matter now.

Origin of the Human Species and Its "Races": The Historical Debates

Two general questions can be posed about any living thing: "What (or who) is it?" and "Where did it, he, or she come from?" The first concerns identity, namely, the essential properties that define something. Such questions are always interesting, hard to resolve, and often emotionally fraught as seen in contemporary debates about cultural or gender identities. In contrast, questions about provenance—where something comes from—are usually less contentious and easier to answer. Nevertheless, the answers to those latter questions frequently lead back to matters of identity. The query "Where did humans originate?" has just this character.

At first, the matter seems unproblematic. The fossil evidence overwhelmingly supports Africa as the place of origin of the hominins and specifically the genus *Homo*. Since fossils of no early members of our genus have been found outside Africa, a reasonable inference is that humans originated in Africa. The difficulties begin with what is meant by *human*.

Does the word embrace some of the later-arising member species of our genus or *Homo sapiens* solely and specifically? If the latter, the matter becomes complex. In principle, it is possible that an immediate precursor hominin species of *H. sapiens*—for instance, *H. erectus* or *H. heidelbergensis*—migrated from Africa into the Eurasian landmass and that our species first evolved there. Indeed, we know that some *Homo erectus* members first moved out of Africa about 1.8 million years ago and that their descendants spread through Asia from what is present-day Georgia to China and Indonesia. Conceivably, *Homo sapiens* might have evolved from one of those non-African *H. erectus* populations directly or via an intermediary species derived from *H. erectus*. Remember that for seventy to eighty years, until the 1930s, an Asian origin for our species was favored by many anthropologists. In evaluating the provenance of fossils of early modern humans, the approximate time of origin of our species becomes a relevant part of the needed information.

In turn, it is easy to see how the time and place of human origins connects with questions about when and where the different human "races" originated and how deep or shallow those differences run. The earlier that that divergence into physically distinguishable populations had occurred, the potentially greater their differences were likely to be and, conversely, the more recent that divergence was, the less significant, in number and depth, those differences are likely to be. The fact that the "inter-racial" genetic differences are comparatively small argues for a more recent separation into different groups. In contrast, and before the genetic data existed, there was a line of nineteenth century thought that the separation of humans into different "racial" groups possibly *predated* the origin of the species itself, which would have made them *de facto* species on their own. Darwin, as we saw, emphatically rejected this idea but not everyone did.[5]

By the early 1980s, the question of African versus Asian origins for *Homo sapiens* was still unresolved, and a novel hypothesis began to attract attention—namely, that our species had not a single origin but multiple independent origins. The initial version of this **multiregional model** of human evolution, the **MRE** hypothesis, was first put forward in the early 1960s by a prominent anthropologist of the time, Carlton Coon, but the recent version, which has garnered the most interest, dates to the early 1980s. The multiregionalists proposed that several *H. erectus* populations had evolved independently in the direction of and eventually became

H. sapiens. In effect, they suggested that *Homo erectus* had indepen-dently evolved in several places into and achieved the "grade" of *Homo sapiens.* Nothing like this had ever been proposed for any other species and, indeed, this would have been a remarkable instance of **parallel evolution** had it taken place.[6]

Despite the unconventionality of the MRE, two of its premises were widely accepted: first, that *Homo erectus* was ancestral to *Homo sapiens* (though not the immediate ancestor); second, because speciation events have local origins, a widely dispersed species might have given rise to new species independently in different regions. What seemed implausible to many evolutionary biologists was the idea that independent, geographi-cally dispersed speciation events could have given rise to the *same* spe-cies. Furthermore, the genetic studies, beginning in the early 1970s and brought to fruition in the late 1990s, rendered the initial version of the MRE simply untenable: the different "races" are too similar genetically to have had separate origins. Accordingly, some of the key multiregional-ists modified their initial position to argue that the high genetic similarity between the "races" could be explained by later interbreeding between these groups, including later arrivals from Africa, to achieve a degree of genetic homogenization.

By the mid-1980s, however, there was a clear alternative to the MRE, which was known originally as the *out-of-Africa hypothesis* and is now des-ignated the **recent African origin (RAO) hypothesis**. This contends that all living members of *Homo sapiens* are descended from a group that lived in Africa, some of whose members emigrated from that continent and whose descendants then proceeded to spread across the globe, evolving into visually distinguishable groups along the way. First proposed in 1976, it was given strength by fossil discoveries of early modern humans dating to between 90,000 and 120,000 years ago in Africa, the present-day Middle East, and the nearby Levant. No other equally old *H. sapiens* re-mains have been found farther east in Asia as predicted by the MRE.[7]

By 1985, these two alternative hypotheses of the origins of *H. sapiens* were deadlocked. It seemed that only the discovery of modern human skel-etons in Asia or Europe considerably antedating those in Africa or the nearby Middle East—thus, fossils more than 150,000 years old—could possibly resolve it. Such discoveries would have supported the MRE. On the other hand, failure to find such fossils would not disprove the MRE because it could be argued that they could yet be found. As it happened, the resolution came not from fossils but from an entirely different source: molecular genetics.

These findings involved DNA sequence comparisons from different ethnic groups from around the world. The analysis of these early data indicated that *all non-African groups contained a subset of the genetic variants of particular genes and DNA regions* found in African groups. (This finding has been largely upheld, but there is one set of major exceptions, which we shall come to.) Furthermore, there were many DNA sequence variants found only in native African populations. Such results only make sense if all non-Africans had African progenitors *and* that some, perhaps most, of the ancestors of present-day humankind had stayed in Africa. The first findings of this kind involved the autosomal beta-hemoglobin genes, but they were shortly followed by work on the DNA contained within the mitochondria, the small energy factories of the cell that contain their own small genomes. These genomes are inherited through the mother's eggs— sperm do not contribute mitochondria to the fertilized egg—hence they show pure **matrilineal inheritance**. One advantage of mitochondrial DNA for this analysis is that the genomes are small (16,569 base pairs long) and easy to purify. A second is that they do not undergo genetic recombination, hence they are altered only by mutation; the genotype of each mitochondrial genome is thus a block of genetic material, inherited as such, a **haplotype**. By comparing and analyzing different mitochondrial DNA haplotypes, each deriving from a specific known population native to a specific region, it is possible to reconstruct the probable temporal sequence in which successive mutations occurred. That information, in turn, opens a window on the historical past of these genomes. Although the original work did not involve complete mitochondrial DNA sequences, it identified hundreds of variable sites across the range of peoples sampled, enough to make the requisite comparisons. The results supported the earlier beta-hemoglobin gene work: the genomes of contemporary non-Africans are a subset of African genomes (though, of course, also showing genetic changes that took place following departure from the mother continent). Furthermore, the analysis made it possible to reconstruct a single ancestral mitochondrial DNA sequence for all humans. Since the results showed that humans originated in Africa, that ancestral genome would have had to have been African. Given that the mitochondria track matrilineal inheritance, the results were soon interpreted (too simply) to mean that there had been a single human female ancestor, soon christened, perhaps inevitably, the "mitochondrial Eve." The approach, results, and caveats are described more fully in Box 8.1.[8]

The result was exciting and controversial. Given that the analyzed mitochondrial genome region was such a small portion of the total human genome and that errors were found in the first analysis, further studies were required to see if other DNA sequences gave comparable results. Over the following fifteen years, later studies with improved methodologies included: (1) complete mitochondrial genome sequences; (2) the part of the male-specific sex chromosome, the Y, that has no homology to the other sex chromosome (the X), permitting tracking through the male lineage (**patrilineal inheritance**); (3) the X chromosome itself; and (4) more genes on the autosomes. The total evidence provides strong support for the African origins of humanity and a date of origin of about 200,000 years. By the late 1990s, the consensus was that the MRE, in both its original and modified forms, had been disproved. And the later discovery of a modern human cranium in Ethiopia that dated to 195,000 years, the oldest found to date, supported the conclusions from by the molecular analysis.[9]

The molecular genetic work not only constituted a crucial advance in understanding human origins but also helped to establish the field of *genetic paleoanthropology*, the analysis of genetic data of living humans (augmented by analysis of DNA from fossils) to reconstruct deep human evolutionary history. The approach makes it possible to sketch the likely routes of migration of the descendants of those first out-of-African emigrants as they spread around the world. In effect, by examining the DNA sequences of different contemporary peoples from their home regions, we can ascertain approximately when different areas were settled and even the approximate numbers of people involved in contributing to their respective initial populations. The genetic analyses used to construct the changing human demographics of the past need not be detailed here, but the general strategy is described in Box 8.2. What follows is a summary of that history as reconstructed in those studies.[10]

The Peopling of the World: A Species on the Move

The Beginnings of the Journey

Two critical questions posed by the RAO concern precisely where the journey began and when it took place. With respect to where, three possible

BOX 8.1 All About Eve: Estimating the Provenance and Age of
Homo sapiens from Mitochondrial DNA Sequences

The human mitochondrial genome recommends itself for tracking phylogenies because it is small (less than 17,000 base pairs) and readily isolated and purified, making its analysis comparatively simple. In addition, it contains many effectively neutral sites, those alterable by a base-pair substitution without creating a maladaptive effect. Furthermore, its DNA mutates at a sufficiently high rate to make it appropriate for studies of human evolution over relatively short time spans. Finally, because it is inherited matrilineally and does not undergo recombination, there is no scrambling of the DNA information, and each mitochondrial DNA sequence constitutes a heritable block of information, its haplotype, which carries much information.

The classic study by Allan Wilson and his colleagues (McCann et al. 1987) that established the out-of-Africa hypothesis as the most likely explanation of human origins used the fact that there is a range of enzymes known as *restriction enzymes* that cut DNA at different but highly specific sites (usually 6–8 base pairs long), one such site per enzyme. When a given piece of DNA is cut with such an enzyme, it creates an array of fragments of highly specific size. By analyzing the size and placement of the fragments, one can establish which specific sites are present in the molecule and where. Mutational events that change a single base pair can either render a restriction site resistant to cleavage—removing it from the sequence map and thereby creating a larger fragment—or create a new one in a new place. Thus, by comparing the restriction site maps of different mitochondrial DNA molecules from different individuals from different ethnic groups, one can effectively compare their sequences and then analyze them via the techniques of molecular phylogenetics (see Box 5.1) to obtain the most likely patterns of descent. When this was done with 147 different mitochondrial sequences from 145 individuals and two cell cultures (derived from individuals of known racial group), McCann and colleagues were able to construct a DNA phylogenetic tree that traced the evolutionary relationship of human mitochondrial genomes. They reconstructed one haplotype, L3, that was plausibly the ancestral one for all non-African peoples and, more dramatically, one that could be considered the mother of them all, which was soon dubbed the "mitochondrial Eve" by the press. This was interpreted by some as meaning that all humans were descended from one human female, but that inference is unjustified. The more probable explanation is that genetic drift led one mitochondrial haplotype to become the dominant one early in the history of the species, and that another haplotype (L3) likewise came to be the exclusive survivor in the descendants of the out-of-Africa emigrants. This would only be possible if the original population from which *Homo sapiens* is descended had been small.[1]

---

1   See Oppenheimer (2003,2012) and Richards et al (2000).

BOX 8.2 Deciphering Patterns of Migration and Their Timing from Genetic Data on Present-Day Populations

The scientific methods for deducing the history of populations from their present-day makeup and geographic distribution constitute the field of **phylogeography** First developed and applied to various animal populations, the methods can be readily applied to human populations, especially where there are long intact sequences of DNA that cannot be scrambled by recombination such as those in the mitochondrial genome and the nonrecombining part of the Y chromosome. One begins by doing a large DNA data-set collection from all the different regions whose ancient history we want to explore. This involves collecting DNA samples—preferably hundreds—within each region from people who are probably and who regard themselves as native to that particular region. In the case of mitochondrial DNA, this involves collecting blood samples, purifying the white blood cells, isolating the mitochondria, and then extracting the DNA from each sample. Then the DNA of each sample is sequenced. From the analysis and comparison of the haplotypes within each region, it is then possible to build a phylogenetic tree of those sequences to find its *root*, which is the probable ancestral sequence for that region. (Genotypes that look quite different from most of the local versions are probably from immigrants from other regions and are not used to build the local tree, though often their probable place of origin can later be deduced.) We then have a putative set of ancestral genotypes that we can designate, for example, as A, B, C, D, E, and F for the different regions. These are then compared for their probable evolutionary relationships in terms of their patterns of sequence similarity; those connections are then mapped to the geographical regions. For instance, if populations A and B are more closely related to each other than either is to any of the others and are in geographic propinquity, then one probably was founded by the other. Thus, if A and B are along the southern coast of Asia and A is west of B, then given that the out-of-Africa movement proceeded from the west to the east, A is likely to be the founder population of B. Or if one population, F, which is located well to the north of that route, has an ancestral genotype that is closest to C, even if populations D and E are closer to it, then C was probably the founder population of F.[1]

---

1   See Avise (2000) for a book-length treatment of phylogeography, Richards et al. (2000) for a specific treatment on human evolution, and Hickerson et al. (2010) for a recent review.

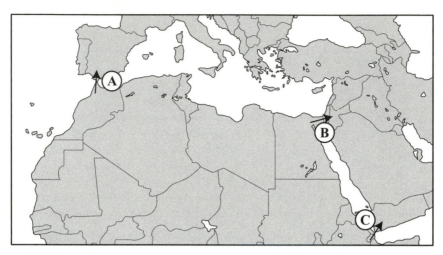

FIGURE 8.1 Three conceivable exit points for the out-of-Africa exodus. **A.** The straits of Gibraltar, which would have required a short sea crossing. **B.** An entirely land-based route from the northeastern Sinai. **C.** The "Gates of Grief," requiring a sea crossing from present-day Djibouti to Saudi Arabia.

geographic exit points from Africa have been identified and are shown in Figure 8.1. The first is the short, obvious route across the Strait of Gibraltar, which would require seafaring craft of some kind. The difficulty for this hypothesis is that there is no fossil or archaeological evidence of human presence either in northwestern Africa—the jumping-off point for Gibraltar—or present-day Spain until well after people had migrated into western Asia and then Europe. Genetic analysis, to be described shortly, indicates that *Homo sapiens* first entered Europe relatively late, sometime after 50,000 years ago and from the direction of Asia.

A second and more likely site of the "breakout" from Africa is across the northern Sinai into the modern-day Middle East and through Turkey. This would have been wholly a land route, hence easier and thus more probable than a sea crossing. As we have seen, modern human skeletons of 90,000 to 120,000 years have been found in present-day Israel and Turkey, suggesting that some emigration out of Africa via this route had taken place. Yet no other such human remains have been found north of there in Europe or western Asia until we come to those that have been dated to tens of thousands of years later. Furthermore, the genetic data already noted suggest that the peopling of Europe took place later and by groups that came into Europe from Asia. The simplest explanation of the older skeletal remains in the Middle East is that there was an early migration out of Africa via the Sinai route, but that it was abortive and left no descendant populations.

The third possible route is a more southerly one across the mouth of the Red Sea, a narrow strait between today's Djibouti in Africa and Yemen on the Arabian Peninsula. This passage is named the Bab-el-Mandeb, or the Gates of Grief, for the toll its reefs took on ships entering or leaving the Red Sea. Today it is a relatively constricted waterway, no more than fifteen miles at its narrowest point; 70,000 to 60,000 years ago, however, it was even narrower due to the extensive glaciation in northern altitudes that lowered sea levels, thereby potentially making traversal even easier. Another factor favoring the Bab-el-Mandeb as a crossing point out of Africa is the presence of a small chain of islands at its northern end. This could have facilitated the crossing by island hopping.

This route, however, would—like a passage to Gibraltar—have involved a passage over water and therefore some form of seagoing craft, presumably either rafts with sails or dugout canoes. Though comparatively simple, such technology could only have been possible if the migrants had possessed language and not merely a proto-language. Language would have been required to formulate the goals ("Let's build boats and go across this water!"), coordinate the logistics of the trip, maintain group cohesion, and, not least, carry out the construction of the rafts or canoes. Language and boat-building skills, however, were probably not the only requirements. The early out-of-Africa emigrants quite probably also wore some form of clothing made of animal skins for warmth. This is pure speculation, but it seems reasonable in light of the probable conditions of the late Pleistocene. It also accords with a rough molecular clock estimate of the length of time that human bodies have been accompanied by lice that specifically inhabit human clothes, approximately 70,000 years. Furthermore, that analysis indicates that body lice populations exhibit the same out-of-Africa pattern of genetic diversity—with more different genotypes in Africa than outside of it—as shown by humans.[11]

Two questions arise immediately. First, when did the exodus occur? Here there is controversy. One analysis places it at 72,000 years ago, others at 60,000 years. For the moment, it can only be placed in this range: 72,000 to 60,000 years ago. Second, how many people made the exodus? That number is also unknown, of course, but it is possible to estimate the size of the entire human population from which the emigrants sprung. These estimates employ genetic analysis and derive from the extent of shared ge-netic variability that exists today in the different populations of the world. From those measures, and making various assumptions about how it would have traced back in time, we can estimate not the absolute ancestral pop-ulation size of *Homo sapiens* but the **effective population size (Ne)**. This is the approximate number of individuals who left descendants in pres-ent-day populations. The estimates of the Ne of modern humans at the time of the out-of-Africa migration—both the migrants and those who stayed at home—are on the order of 10,000 individuals.[12]

Most probably, the great majority of the ancestors of modern humanity stayed in Africa and gradually spread throughout it, giving rise to most of the current native peoples of Africa from Bantus and pygmies to the tribes of bushmen and *some* of the ancestors of the Ethiopians and Somalis, the latter groups having some strong Caucasian admixture as indicated by their DNA sequences. (The ancestors of the present-day Berbers and Arabs of northern Africa were not in the original stay-at-home population but arrived in Africa much later.)

If there was a single out-of-Africa exodus, the emigrants would prob-ably have totaled no more than 1,000 to 2,000 and perhaps only a few hun-dred individuals. Furthermore, they need not and probably did not all go at once; the migration may have taken place in several small waves; if so, however, they probably came from the same ancestral population. The fact that single mitochondrial haplotypes and a small number of Y-chromosome haplotypes can be constructed for all the out-of-Africa emigrants indicates that the same source population was involved. (All non-African populations derive from the mitochondrial haplotype designated L3.) The key point is that the people who became the ancestors of all humanity today made up a relatively small group, probably not much more than 10,000, while the ancestors of all the populations that arose outside of Africa were a still smaller group. In effect, early in the history of our species, there was almost certainly a significant **population bottleneck** for the entire set of descen-dants of those who moved out of Africa. Furthermore, the fact that molec-ular phylogenetic trees based on either matrilineal mitochondrial DNA and patrilineal Y chromosomes can be constructed indicates that there was one relatively small group as the ancestor of the non-African populations.

What drove such an extraordinary and risky departure from Africa? The probable answer is food shortages. The interval from 120,000 to

10,000 years ago was the final phase of the Pleistocene period (2.5 million years ago to 10,000 years ago), the penultimate geological period in Earth's history preceding the current one, termed the "Recent." All periods in Earth's history have their distinctive features that leave characteristic signatures in Earth's strata; without such, they could not be delimited in the geological record. The Pleistocene, however, was characterized by something quite unusual: a series of ice ages in which much of the Northern Hemisphere was blanketed by glaciers, punctuated by warming periods of various lengths (*interstadials*) in which the glaciers retreated to varying extents. No preceding geological period for more than 600 million years had witnessed this particular kind of climate change. As a result, throughout most of the Pleistocene, the world as a whole was drier and colder, even in those regions that did not experience the glaciation directly and even in those periods when the glaciers that blanketed much of the northern landmasses were in retreat. As previously mentioned, one consequence was the lowering of sea levels during the periods of maximal glaciation because so much of the planet's water was locked up in the glaciers of the northern hemisphere. Another probable feature of this cooler world was reduced vegetation and, in consequence, smaller herds of game relative to warmer times. Modern humans were hunter-gatherers who had developed a heavier reliance on meat obtained from either hunting or scavenging.[13]

In addition to climatic factors, there were probably growing numbers of people competing for food resources. By several measures, including genetic estimates derived from studies of the amounts of genetic variance in contemporary African populations, it has been inferred that human populations had been growing in Africa for perhaps several tens of thousands of years. With a further cooling of climate, at the end of the last interglacial period, about 70,000 years ago, this would have created further pressure on food supplies.[14]

In all likelihood, therefore, game was in short supply relative to the needs of local populations in the region that was the jumping-off place for out-of-Africa migration. Of course, plant materials are an intrinsic part of the diet of hunter-gatherers, but supplies of those too might well also have been depleted by a population of moderate size during a period of stable occupation. Modern hunter-gatherers tend to circulate among a relatively fixed set of sites, but in periods of climate change such groups would almost certainly have been pushed to explore new regions. At the time of the African exodus, the invention of agriculture was still 50,000 years in the future, and it is not clear that the climatic conditions of the time would have permitted its development. For humans living in this region of Africa 70,000 to 60,000 years ago, the prospect of crossing the Red Sea at its narrowest point in primitive sailing craft might have appeared no more dire a prospect than moving north through the arid Sahara (as it was then, as it is today) or south through, quite probably, almost equally inhospitable regions.

The Onward Journey Around the Globe

If this explanation is correct, the same concatenation of circumstances could have provided a similar impetus for onward journeying after the Arabian Peninsula had been reached. What seems certain is that the arrival in Asia along the southern coast of what is today Yemen and Saudi Arabia was but the first stop on what became an extended journey along the edge of and into Asia by successive generations of descendants of those first out-of-Africa emigrants. The migration path initially stuck to the southern coast, first of Arabia, then farther eastward in Asia, eventually reaching India and later into southeast Asia. At many locations, local populations were founded and experienced different degrees of permanence while some subgroups moved on, either farther east along the coast or diverging away from the coast inland to the north.

A striking feature of both the initial and later eastward spread of humanity, however, is how much of it seems to have hugged the southern coasts. One hypothesized reason was that the migrants relied heavily on shellfish, a ready source of protein. This idea is known as the *beachcomber hypothesis,* and it was suggested by the discovery of large refuse dumps of sea shells, *middens,* along the presumed southern Asian migration route. (Other ancient middens, one as old as 125,000 years ago in present-day Eritrea, have been found.) Such piles of shells can only be produced by humans and suggest that shellfish were an important part of the diet of those who made them. Of course, supplies of shellfish might also have been depleted as local populations grew, prompting further onward migration.[15]

What has been described so far is the populating of Asia, especially along southern coastal routes into southeast Asia, but that was only one element in the spreading out of humanity following the exodus from Africa. The first major departure from the early southern coastal route toward more northerly regions of western Asia and eventually into Europe is estimated to have taken place approximately 50,000 to 45,000 years ago during a short period of relative global warming. That pathway was probably through the region that we now know as the Fertile Crescent in present-day Iraq. From paleoclimatic reconstruction, it appears certain that before that warming period it was cold and fairly barren of the kinds of living things needed to support humans in any number. Thus, at the time, it was anything but fertile. Presumably during that interglacial warming, however, there would have been sufficient plant material to gather and more game to hunt, permitting adequate subsistence during the migration. The descendants of these trekkers would eventually reach Europe as the first group of *Homo sapiens* to the continent sometime between 50,000 and 40,000 years ago. (They had been preceded there by the Neanderthals, descended from *Homo erectus,* who had reached Europe several hundred thousand years earlier, probably via the northern Sinai route out of Africa.) Other groups that had initially populated more western and central regions of Asia would later migrate from those places into Europe while others moved on, populating central Asia and later northern Asia. Still other groups penetrated southeast Asia and the islands to the south and west, in particular those of the Malay Archipelago and Indonesia and, beyond, to the island continent of Australia.[16]

The last big phase of this human migration was into the Americas from northern Asia, and it seems to have taken place via a northern land bridge between Asia and the Americas where the Bering Straits are today. Because of lower sea levels, it was then a substantial landmass now known as *Beringia*, and it embraced what is presently eastern Siberia down to the Kamchatka Peninsula, extending northerly hundreds of miles beyond the current Siberian shoreline over the present-day straits and connecting with Alaska. The existence of Beringia was made possible by the lowering of sea levels during the Pleistocene, just as much of its later submergence was a consequence of the rising sea levels toward the end of this period. The date of the initial human migration from Beringia into what is now Alaska is unknown but was probably about 15,000 years ago during the grip of the final ice age. The genetic data of present-day native populations in North America suggest several waves of immigration into that continent.[17]

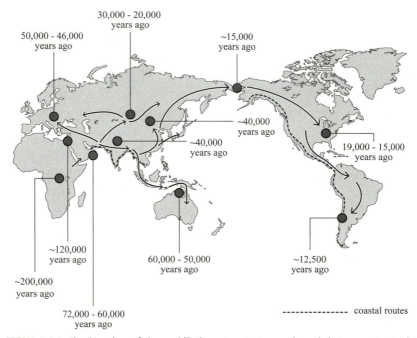

FIGURE 8.2 The "peopling of the world": the main migratory paths and their approximate dates. These routes are more fully described in the text. All dates are approximate, but where a range is given the uncertainty is implied; for the single dates, the approximation sign indicates it explicitly. (Modified from Oppenheimer 2012.)

Once what is now Alaska was reached, there was migration deeper into North America, perhaps down an ice-free corridor along the Pacific Coast, although some have argued for a more central route of migration farther inland. Quite possibly both routes were used by different emigrant groups. Later movements would take modern humans into what are now the United States and Mexico and then through Central America and into South America. The first migrants to reach the southern tip of South America did so about 12,000 years ago, perhaps no more than 3,000 to 4,000 years of the first arrival of humans in North America from Beringia. With that destination reached, all continents except Antarctica had acquired populations of *Homo sapiens*. The inferred pattern of the complete set of migrations, beginning with the initial out-of-Africa exodus, and their estimated dates as derived from both genetic and archaeological evidence is diagrammed in Figure 8.2.

In a synoptic view as shown in the figure, the spread of humanity around the globe looks like the manifestation of wanderlust, perhaps itself driven by risk-taking exploratory behavior. Certainly, a willingness to take risks must be part of the character of people setting out into the unknown, and both wanderlust and risk taking are well-documented human attributes. I have argued previously, however, that the more prosaic explanation was depletion of food supplies, followed by onward movement to obtain new supplies, particularly of animal meat. Having discovered how to hunt big game cooperatively, presumably on the African savannahs, humans were efficient at it, and the arrival of modern humans in new regions such as Madagascar, Australia, and the Americas was soon followed by fairly rapid mass extinctions of most of the large, native mammalian fauna in those regions. If human hunters were the direct cause of these extinctions—this is debated—it would testify both to the efficiency of such hunting and its inherent destabilizing effects.[18]

Nevertheless, judging from the genetic composition of present-day populations throughout Asia, it is striking that so many populations were founded along the way and achieved some kind of relative geographical stability long before agricultural practices had been invented. The whole pattern of migration and settlement formation along the way results in what is known as the *serial founder effect*, with each such founding event probably further reducing the original stock of genetic variation and contributing to a new stream of genetic differentiation in each place. Furthermore, the migratory paths shown in Figure 8.2 must have been generated by subgroups of each population. In particular, one speculation is that individuals in those traveling groups were preferentially endowed with alleles that favored risk taking. Those alleles, for which there is some contemporary evidence, may have "surfed" ahead in the onward moving populations. Thus, the peopling of the world may have been driven by a combination of two rather different human temperaments, the first characterizing those who were content to remain in their local regions and the second, the more adventurous ones, who pushed on to new territories.[19]

Whatever the forces that propelled this extraordinary expansion of humanity from its original African base, the net result was the spread of *Homo sapiens* around the globe within what is rapid in evolutionary terms—on the order of 50,000 years. In its geographical spread, it is probably unique in the history of land vertebrates; no other species is known to have covered so much ground so fast, apart from domesticated or commensal animals that were deliberately or accidentally transported by humans in later times.[20]

Archaic DNA in the Human Genome and
Its Implications for Human Evolution

During this time span in which humans came to populate Asia, Europe, Australia, and the Americas while the stay-at-home African populations began to spread within Africa, various populations began to diverge. Some large fraction of the genetic source material for much of that "racial" diversification was probably present in the initial collective **gene pool** of modern humans inhabiting the Horn of Africa prior to the exodus. It would have been parceled out during the onward migrations of people by the process of genetic drift—random samplings of alleles from the larger population—as small groups of onward migrants set up new local populations, revealing allelic differences that might have been hidden in the original group by the recessivity of certain alleles and the epistasis of others. That parceling out and reduction of genetic diversity is precisely what is expected with the serial founder effect, and this is mirrored in a general reduction of genetic diversity the farther populations are from Africa. Yet that initial genetic source material for humans, as they spread east and north, would almost certainly have been supplemented by newly arising mutations in the different groups, especially as local populations expanded, creating the unique alleles present in relatively small proportions in those populations.[21]

It is now apparent, however, that not all of the variations in the descendants of the out-of-Africa migrants came from either the original African source population or new mutations. There was a third source as modern *Homo sapiens* spread throughout the Eurasian landmass. Recent results show that in the genomes of modern Europeans and Asians, but not those of modern Africans, there are segments of **archaic DNA**, namely, DNA from other extinct hominins, specifically Neanderthals or in Asia a second late hominin species. The simplest explanation is that there were matings between migrating *Homo sapiens* and other native hominins encountered along the way, which left descendants who contributed to contemporary human populations in Europe and Asia. The investigations that made possible the deduction of the presence of such archaic DNA involved the extraction of DNA from fossil bones of the extinct hominins and the subsequent reconstruction of the genomes of the hominins that were their source, a tour de force of modern molecular genetics.[22]

Even before the full genomes had been reconstructed and compared, however, enough Neanderthal DNA had been sequenced to make comparisons with modern human genomes possible. These had established, after nearly 150 years of debate, the evolutionary relationship between Neanderthals and *Homo sapiens.* From the molecular clock analysis of Neanderthal and modern human mitochondrial DNAs, the initial divergence between these two lineages was calculated to have taken place between 500,000 and 800,000 years ago. Those dates would place the divergence of the two lineages as having taken place between populations of *Homo erectus.* That split almost certainly occurred in Africa: the subpopulation that would later give rise to Neanderthals migrated into Europe while the lineage that gave rise to *H. sapiens* had remained in Africa. (If so, it follows that *H. erectus* not only had initially migrated into Europe about 1.8 million years ago but also had seen at least one other but much later migration of *H. erectus* out of Africa.) These molecular clock dates are also consistent with more recent fossil evidence. The first European hominin skeleton with some diagnostic Neanderthal features dates to 400,000 years, while the oldest fully Neanderthal skeleton is 230,000 years old. The complete set of results show that Neanderthals were neither ancestors of *Homo sapiens* nor so-called degenerate modern humans (these two possibilities being the favored nineteenth-century views) nor a subspecies of *Homo sapiens* (the late twentieth-century hypothesis). Rather, the Neanderthals were our cousins, coming from a clade that branched off the *H. erectus* lineage 500,000 or more years ago.[23]

Comparisons of the complete DNA sequences of modern humans with those of Neanderthals and an Asian hominin established the presence of so-called archaic DNA sequences in the modern human genome. The physical appearance of the Asian hominin is unknown because the finger bone from which the DNA was extracted was a miniscule fraction of a whole skeleton. The DNA sequence, however, proves that it was a hominin species, which has been designated *Denisovian* for the cave in Siberia in which the fossils were found. One speculation is that Denisovian man was *H. heidelbergensis.* The crucial point is that the Denisovian genome sequence differs at various points from both Neanderthal and modern human genome sequences, permitting three-way genetic comparisons between *H. sapiens,* Neanderthals, and Denisovians.

The proportions of the contemporary human genome that come from archaic DNA are estimated to be 1 percent to 4 percent for European populations and as much as 7 percent for certain Asian populations. While both sets of figures are minor fractions of the whole genome, they are significant and suggest a more complex genetic heritage of our species than postulated by the RAO hypothesis. (They are not so large, however, as to eliminate the clear out-of-Africa signature in the descendent genomes of modern human populations.) Specifically, the results support one tenet of the revised MRE hypothesis: the occurrence of interbreeding between archaic humans and modern human emigrants from Africa. The actual result, however, is not that predicted by the MRE. That prediction was of incorporation, or **introgression**, of some DNA sequences from modern human genomes into separate *H. erectus*–derived populations, with the *H. sapiens* genetic contribution being the minority one. In contrast, the results show that introgression was in the opposite direction, that of archaic DNA sequences into the *H. sapiens* genome. Fatally for the MRE's claims that all modern humans originated from pre-*sapiens* stocks, modern African populations do not have this archaic DNA contribution; it is strictly a phenomenon of European and Asian populations.[24]

Intriguingly, different populations of modern non-African humans not only show different total percentages of archaic DNA sequence but also different pieces of these genomes. One study estimates that as much as 20 percent of archaic genomes (from both the Neanderthal and the Asian archaic hominin) is represented when the different DNA segments derived from them are summed across different modern human populations. If so, then encounters between *H. sapiens* and the other hominins must not have been rare. In addition, much in the other hominins' genomes must have been compatible with the *sapiens* genome.[25]

There are, however, two other possible interpretations. The first is that the archaic sequences in modern human genomes might have been inherited from the African ancestors of the Neanderthals and modern humans. If that were the case, then the out-of-Africa emigrants would have contained those sequences within their genomes from the start. This would, however, make it hard to explain why European and Asian populations possess different archaic DNA sequences. Furthermore, one would expect that some of the distinctive archaic DNA sequences in modern Europeans and Asians would be present in the genomes of modern

Africans, but no traces of these archaic sequences have been found in any modern African population. This possibility can therefore be eliminated. The other is that these sequences were ancestral yet preferentially lost from the African populations for some reason. A recent analysis indicates that this last explanation also fails: the size of the blocks of Neanderthal-type DNA in modern genomes, which are judged by their defining **single nucleotide polymorphisms** (SNPs) are too large to have been acquired well before the exodus because genetic recombination over many generations would have whittled them down. The analysis, in fact, permits a tentative assignment of the entry of these Neanderthal DNA sequences into the genetic pool of modern *Homo sapiens:* between 65,000 and 47,000 years ago, which is consistent with the time of arrival of modern *H. sapiens* in the Middle East and Europe.[26]

One final aspect of this phenomenon deserves mention: certain gene variants derived from the Neanderthals were apparently favored by selection within modern human populations while others were discriminated against. The former include certain pigmentation genes—including one favoring red hair—and certain keratin genes expressed in skin. (Keratin genes were noted previously in connection with skin and hair in the evolution of the early mammals.) These gene variants might have been part of the Neanderthal's adaptive success in extremely cold climates; if so, they would have been preserved and amplified in modern European human descendants once they had entered the human gene pool. In contrast, the Neanderthal genes that seem to have been disfavored and preferentially lost in modern human populations largely have to do with male fertility.[27]

To return, however, to our subject, the human face: did any of these archaic DNA contributions contribute to differences in facial characteristics between populations? The answer is unknown, but detailed analysis of those sequences may ultimately reveal the answer. Recent genetic analysis of the basis of facial features will be described in the next chapter. Intriguingly, one study has identified several genes that contribute to modern human facial diversity and that had previously been identified as differing in Neanderthal from *Homo sapiens*. We will return to this point in the next chapter.

Shaping the Face in Modern Humans:
The Roles of Natural and Sexual Selection

The findings about archaic DNA present in the genomes of modern humans reveal a hitherto unsuspected aspect of the evolution of modern humans. It should not, however, obscure the main point about the origins of different populations in different geographical regions: that the relatively slight genetic diversification of the descendants of the out-of-Africa migrants took place primarily or exclusively within the last third of our species' existence, hence during the past 60,000 or so years, and is thus a relatively recent phenomenon. (The contributions to the human genome from archaic human species also took place during this period.) Furthermore, these differences between populations did not arise simultaneously but were spread out over a span of about 50,000 years; hence, there must have been a temporal sequence in their origins.

In considering this matter, we inevitably return to pondering how these differences arose and what, if any, evolutionary significance they had. Were they purely a matter of chance and genetic drift? Or was something more involved? As we have seen, the visible "racial" differences, those involving skin color and facial characteristics, involves a relatively small set of genes whose predominant alleles must differ between the different main groups. What could explain the differences among populations for those genes while leaving largely unaffected the great majority of genes whose alleles are not associated with particular population groups?

Evolutionary thinking about the process of "racial" diversification began with Darwin in *The Descent of Man,* but he was attempting to solve the problem decades before there was a science of genetics and more than a century before today's detailed genetic information began to accumulate. Perforce, he concentrated on the heritable differences whose effects were the visible ones associated with different "races." His initial premise was that adaptive differences were the source of all such differences. Coming from the father of the idea of natural selection, that is not a surprising assumption. He had begun collecting information about the matter over many years, starting with his first encounter with the natives of Tierra del Fuego in 1833 during the voyage of *The Beagle.* In the end, however, he could not find plausible selective differences based on adaptive value for the different varieties of the major traits that distinguish the "races": skin color, eye shape, nose shape, and type and density of hair. (Today, for skin color differences, there is now much evidence for its adaptive value in both darker- and lighter-skinned people, as will be discussed, but Darwin did not have those facts.)

He was thus driven to an alternative hypothesis: that the supposed race-differentiating traits were initially selected not to enhance survival in different environments but for their power of attracting mates in particular groups. This was his hypothesis of *sexual selection*. In this view, "racial" divergence was an accidental by-product of sexual selection and involved different aesthetic standards in different groups. In his own words:

> Let us suppose the members of a tribe, in which some form of marriage was practiced, to [have] spread over an unoccupied continent. . . . The hordes would thus be exposed to slightly different conditions and habits of life, and would sooner or later come to differ in some small degree. As soon as this occurred, each isolated tribe would form for itself a slightly different standard of beauty; and then unconscious selection would come into action through the more powerful and leading savages preferring certain women to others. Thus, the differences between the tribes, at first very slight, would gradually and inevitably be increased to a greater and greater degree.[28]

The core idea is that individuals who are more successful in attracting mates by virtue of some distinct and attractive heritable feature or features leave more offspring; if those offspring inherit those same differentiating mate-attracting abilities, they too will tend to leave more progeny than their less fortunate peers. Just as natural selection for a survival-enhancing trait should slowly change the composition of a population and increase the proportion of individuals with specific adaptive advantages, sexual selection should similarly operate to alter population makeup to include proportionately more of those with stronger mate-attracting abilities. In the world of nonhuman animals, the most striking examples of sexual selection are those of the males of bird species with beautiful, elaborate plumages; the baroque but exquisite tail of the male peacock is the iconic example of sexual selection. Darwin thus quite explicitly attributed an aesthetic sense to animals, one in which members of one sex chose mates from the other whenever the latter had developed special visual lures to promote their selection.

The key to understanding the possible link between this process and "racial" diversification is that Darwin believed that human perceptions of beauty can be surprisingly local, with different populations developing their own aesthetic standards. If so, then sexual selection would inevitably shape visible features somewhat differently in different populations. Furthermore, in Darwin's view, the psychological element that drives such divergence is "that man admires and often tries to exaggerate whatever characters nature may have given him." Hence, just as animal breeders will breed selectively to exaggerate initial differences that appeal to them (the topic that Darwin begins *The Origin of Species* with to show the potential power of selection to amplify traits from small beginnings), humans will tend to favor sexual partners who show more extreme versions of characteristics deemed beautiful. An element of chance initially and inevitably enters into such judgments. If diverging populations arising in different parts of the world had acquired or begun to show even slight local differences, then different perceptions of what female beauty consists of would almost certainly have developed. Darwin cites Alexander von Humboldt, the great early nineteenth century scientist explorer of South America, as the originator of this insight about the aesthetic appeal of exaggerated traits, but it was Darwin who really ran with it.[29]

The hypothesis that sexual selection is potent enough to shape the evolution of species' forms was even more radical and original than that of natural selection. Darwin was almost certainly aware that the idea would provoke resistance, particularly in his own culture and its sensitivities about sexual matters. Hence he was careful to document its ubiquity in the animal world and did so in his book on human evolution, whose full title was *The Descent of Man and Selection in Relation to Sex.* The title itself must surely have puzzled his readers initially since the very idea of sexual selection originated with the book. Correspondingly, the coupling of the two subjects in the title could only have been mysterious at first sight. Furthermore, one suspects that few of those readers taking the book home with them would have predicted the amount of space given to the second topic; consisting of two volumes, there is far more in it on sexual selection, as Darwin documents its existence in one group of animals after another, than there is on human evolution. The book begins with matters concerning human evolutionary origins but is then, in effect if perhaps not as the original intention, taken over by the general theme of sexual selection.

Darwin only returns to human evolution in the final two chapters of the second volume, where he lays out his argument that the different races of human beings owe their differences primarily to the agency of sexual selection. A large part of that material is devoted to establishing the central premise that different human groups, whether tribes or major racial groups, can have profoundly different conceptions of beauty. Furthermore, different groups will often artificially enhance different physical attributes by various treatments—enlarged lips or earlobes in various African tribes, miniaturized feet in females (in China), artificially narrowed heads in certain South American tribes, the use of facial makeup in Western (and other) women. In this perspective, beauty is not so much in the eye of the (individual) beholder as in the collective eye of one's local culture. For Darwin, the roots of racial divergence lay in this phenomenon and the ways it would influence the patterns of who mated with whom, eventually shaping the appearance of populations.

If Darwin had worried that the idea of sexual selection as a force for evolutionary change in the animal kingdom might appear bizarre and unappealing to his readers, he was right to do so. In the first place, the basic premise that animals can have an aesthetic sense that can influence their behavior was just too much for many of them; in fact, it seemed laughable to some. The rejecters included Darwin's principal intellectual comrade in arms, Alfred Russell Wallace, as well as numerous other biologists. These critics, however, conspicuously failed to propose an alternative explanation for the fact that males and females in many animals differ strongly in certain traits, the phenomenon of **sexual dimorphism**, which is inexplicable in terms of conventional natural selection. Such differences in vertebrates are seen in body traits termed **secondary sexual characters**, ones not directly involving reproduction but that serve as fast visual identifiers of gender and are widespread in the animal kingdom. (Sexual characters directly involving the genitalia, gonads, and gametes of the two sexes are termed *primary*.) It would not be until the 1930s that the idea of sexual selection as a major driver of evolutionary change gained respectability through the writings of one of the founders of neo-Darwinism, Ronald A. Fisher, and it was only in the final decades of the twentieth century that it came to be widely accepted as a major force in evolutionary change.[30]

In a historical irony that is seldom noted, however, although Darwin was eventually vindicated on the widespread importance of sexual selection in evolution, he appears to have failed to make a convincing argument that sexual selection was a general driver of "racial" divergence in humans, even though the idea motivated so much of *The Descent of Man*. Today, his explanation is only rarely mentioned in the literature on the genetic basis of those differences.

Unlike the late Victorian rejection of his idea based on emotion and prejudice, there are two apparently good scientific reasons for its present-day neglect. The first is that sexual selection seems to predict only differences between the sexes: sexually dimorphic ones. Indeed, for such extravagant differences as the male peacock's tail, it would be difficult to find a convincing alternative explanation to sexual selection. In contrast, group characteristics of different major populations are shared by males and females of each "race." How could an idea designed to elucidate gender differences possibly explain, for example, either the shared dark skin color of both males and females of African (and Melanesian populations) or the characteristic eye shape of both males and females in east Asian populations?

Darwin, of course, was aware of this objection and addressed it. He began by noting that most traits are transmitted to both sexes. He then argued that while some initially sexually selected traits would retain their distinct sexual dimorphism, others by virtue of their equal transmission to both sexes—and, it was implied, some hereditary changes that allowed them to be expressed in the previously nonexpressing sex—would eventually appear in both sexes. Thus, a trait that began as one enhancing the attractiveness of one sex could end by becoming the property of both sexes—a species-characteristic trait rather than a sex-specific one. In so doing, it would lose some but not necessarily all of its value as a mating lure. For example, dark skin color in certain groups could be such a trait, one that began in males as a female attractor but which spread to both sexes without totally losing its power in males to attract females. In pursuing this idea, Darwin notes that in these "races," the female is often slightly lighter in color and that this comparative degree of relative lightness is often attractive to men. Thus, a degree of sexual dimorphism can be retained even if a trait came to be expressed in both sexes. In the prevailing absence of clear ideas of the basis of hereditary mechanisms, he could not take this idea further, but he had at least addressed the objection.

There is, indeed, nothing inherently implausible about the conversion of an initially sexually dimorphic trait to one possessed by both sexes. Features, especially facial features, that are attractive in one sex can be quite attractive to the other with only a slight degree of either masculinization or feminization that characterizes the gestalt of male and female faces, respectively. One can readily speak of a "handsome" woman—that is, a woman who has good strong features, the kind more frequently associated with men—or of a "beautiful" man whose regular facial features could be elements of a beauty in a woman but, slightly masculinized, are also attractive to women. In terms of sexually attractive facial features, what is sauce for the goose is (often) sauce for the gander, hence, a trait that initially arises as a lure for one sex can, if transferred to the other, become potent as an attractor for the sex in which it first appeared.

Darwin also hypothesized that human nakedness was just such a trait. He suggests that the anthropoid primates first lost fur on the face and on the buttocks of many species in males specifically because these bare areas were attractive to the females, a speculation made more plausible by the additional bright coloration of the skin in these two areas in males in certain primate species. Over time, the argument ran, the females lost fur in these two areas as well, promoting male attention and attraction. Continuing sexual selection could then have promoted further loss of hair or fur on the body, leading eventually to the naked human condition. Although there are other hypotheses based on natural selection for the comparatively furless state of humans, they all have difficulties. In general, there would seem to be a strong adaptive advantage in having a fur coat, as sported by the great majority of terrestrial mammals, not least in the chillier climates that characterized most of the Pleistocene. Hence, Darwin's sexual selectionist explanation for our comparative nakedness is at least as good as if not better than the adaptationist scenarios.[31]

More generally, the conversion of a sexually dimorphic trait to a general species characteristic, hence one shared by both sexes, is readily conceivable in terms of ideas about genetic networks. In vertebrates, the development of secondary sexual characters is triggered by the particular hormonal milieu found in one sex but not in the other. With appropriate and relatively simple genetic changes, a new non–sex-specific signal can come to the fore and, by alteration of some upstream element, take over expression of the trait from the original hormonal control. Such a rewiring at the top of the genetic network could convert a sex-specific trait to a species-characteristic trait. Furthermore, if it had either adaptive value or appeal in terms of sexual attraction in both sexes, it would spread within the population and carry long-term potential to become a characteristic of the species. Hence, to sum up the argument: "racial" traits could have arisen in this manner, and their lack of sexual dimorphism today is not a fatal objection to Darwin's idea. (Proving that this was the evolutionary trajectory for any particular trait is, of course, a different matter.[32])

The second reason for doubting the generality of Darwin's idea is perhaps more potent. There is one trait where a strong case can be made that adaptive value *is* associated with differences of that trait in the different races and hence natural selection would have been the major driver of change. Furthermore, this trait is the most obvious race-differentiating one: skin color. The closest Darwin got to a selective explanation for the fact that "races" found in the tropics or near-tropics have darker skin colors than people in temperate or near-arctic climes was that the former seemed more immune to tropical diseases. He thought it impossible that dark skin color in itself could have this protective value and proposed, therefore, that dark skin color was merely associated with the immunity-conferring properties. (There was, of course, no real theory of immunity at the time, although smallpox vaccination and later rabies vaccination confirmed the idea that immunity was a real property.) He also toyed with the idea that tropical climates might "induce" darker skin color but rejected this because the trait was clearly inherited; even after several generations in the tropics, descendants of Europeans were essentially still as "white" as their forebears (though perhaps a little more tanned).

Today, we know much about the selective reasons for differences in body pigmentation between the different races. It lies in the different levels of sunlight exposure experienced in equatorial versus more polar latitudes. Briefly, dark-colored skin has adaptive value in the tropics, regions that receive large amounts of sunlight all year round, because it helps screen out dangerous levels of one fraction of the incoming ultraviolet light, ultraviole radiation B or UVR-B, which reaches Earth's surface with high efficiency. UVR-B can damage skin, either in the short term in terms of

BOX 8.3 Genetics of Human Skin Pigmentation

The genetics of skin and hair pigmentation are complex and still incompletely understood. The basic facts of pigmentation are that there are two major kinds of pigment, eumelanin (black) and phaeomelanin (yellows to reds). These are produced by special pigment-producing cells, *melanocytes*, and passed to the skin cells in the form of granules called *melanosomes*. Extremely dark skin has both more and larger melanosomes dominated by eumelanin, while lighter skin has less eumelanin and more phaeomelanin. The complex genetic foundations of human pigmentation differences affect the respective amounts of the two pigments and the numbers and size of the melanosomes, while different genes can be involved in setting hair, eye, and skin pigmentation separately. One key gene (TYR) encodes the enzyme tyrosinase, which is essential for synthesizing melanin (in both forms); mutations in this gene cause albinism. Three other genes—OCA2, TYRP1, and SLC45A2—also can, when mutated to complete loss-of-function mutations, produce albinism but less severe mutations, with partial function, and can contribute to pigmentation diversity in different populations. One gene of major importance, the melanocortin receptor 1 gene (MC1R), is probably essential for the especially dark skin of sub-Saharan Africans. Different alleles of MC1R in European populations have been found, but these seem to affect hair color more than skin color. Intriguingly, one variant MC1R allele found in Neanderthal DNA is associated with red hair, strengthening earlier ideas that Neanderthals had lighter skin than their African forebears. A skin color gene discovered in zebrafish and named *golden* (because its mutant form, when homozygous, gives lighter stripes) is also found in humans, and genetic analysis shows that golden is of major importance in producing the light skin color of Europeans. Other genes, however, are involved in the lighter skin pigmentation of Asians. The full elucidation of the genetics of human skin pigmentation is still a work in progress.[1]

---

1 A good review of the molecular genetics of skin color is that of Sturm (2009), while the population genetics of pigmentation genes is discussed by Rees and Harding (2012).

---

burns and blistering or long-term because of its promotion of skin cancer. There is also a third way in which UVR-B is destructive of health: it destroys folic acid in the skin, and folic acid is important for health, especially for the full and normal development of fetuses in pregnant women. Thus, the net effect of darkly pigmented skin is to protect against the various kinds of damage inflicted by high levels of UVR-B in sunlight. The genetics of pigmentation are complex, so it is not surprising that different ethnic groups showing different skin colors differ in the genetic basis of their skin, hair, and eye pigmentation (Box 8.3).

Yet UVR-B also does something of positive value: it can promote the synthesis of vitamin D in the skin. This additional source of vitamin D is a valuable supplement for health in northern latitudes, where dietary supplies are often not sufficient. Thus, where reduced levels of sunlight are experienced most of the year, UVR-B stimulates the synthesis of vitamin D in the skin and does so most efficiently in "white" skin, which has less melanin to absorb it. Darkly pigmented skin screens out most UVR-B, but dietary supplies of vitamin D are usually sufficient in the tropics.[33]

Hence, in this particular trait, which is the crucial one in rapid perception of racial difference, Darwin was wrong: selective (adaptive) reasons do exist for differences in skin color. Still, it is legitimate to ask: does the explanation of physical "racial" traits have to be *entirely* in terms of one selective process—natural versus sexual selection? In principle, both factors could have operated to either darken or lighten skin. It is human nature to yearn for simple single- cause explanations, but the sheer complicatedness of life guarantees that causation often involves multiple contributing factors, with the role of each in generating a particular property or outcome influenced heavily by the occurrence and magnitude of others.

In the case of facial shape characteristics, however, many of the features altered in racial diversification—in particular, nose and eye shape—continue to elude convincing adaptive and hence natural selectionist, forms of explanation. One can, for example, make equally good cases for both the small noses of Eskimos and the relatively large noses of Europeans as providing selective advantage when inhaling cold air. Furthermore, it is possible that both are correct but that each neglects a salient difference at play in the two groups that conduced to the two different outcomes. Nor would some involvement of sexual selection in affecting nose shape be excluded by them.

Altogether, Darwin's hypothesis of sexual selection as a factor in the genesis of many "racial" traits remains at least as convincing as pure adaptationist explanations, although it is clearly not the entire explanation for differences in all those traits. To the extent that he was right about this, his idea gives new force to the old adage that "racial differences are skin deep." Traits that evolved under the influence of sexual selection and involving the choosing of mates need have no deeper or more general significance. And that, in turn, accords with the failure to find significant overall genetic differences between the main "racial" groups.

Are We a Self-Domesticated Ape?
The Domestication Syndrome and Facial Evolution

We now come to another set of possible relationships between mental states and the evolution of the human face: the idea of humans as a *self-domesticated* species, as briefly introduced at the start of this chapter. We begin with the fact that human faces resemble those of juvenile chimpanzees more closely than those of adult chimps (Fig. 8.3). The similarities involve the virtual absence of a muzzle and the presence of a forehead, the two characteristics that jointly produce a more vertical face and a rounder head. It is only in growing up that chimpanzees acquire their apelike facial features, a fact first noted in the nineteenth century by several observers. In this light, humans may be seen as ape-like animals that retained juvenile physical features while becoming sexually mature (as well as, of course, developing other distinctive attributes).

Such evolutionary retention of juvenile physical features into the adult stage is the widespread phenomenon among animals known as **neoteny**. The classic case of neoteny is the axolotl, a species of salamander that remains a water-dwelling larval form complete with gills even while becoming sexually mature. In humans, neoteny takes a less dramatic form as seen in the comparison with chimps, and its significance has correspondingly been subject to more debate. The central question is whether a true slowing down of body maturation in the course of human evolution, with prolonged retention of juvenile characteristics (including behaviors),

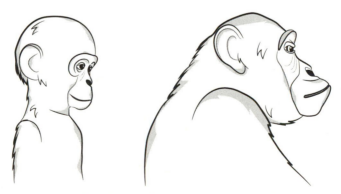

FIGURE 8.3 Comparison of juvenile and adult chimpanzees, illustrating the more humanlike face of the young chimpanzee. (Redrawn from Gould 1977.)

actually took place, and if that occurred whether it was accompanied by prolonged brain growth and maturation relative to body growth to produce the large human brain. The consensus answer today to both parts of the question is "yes." The closely related question is whether such a prolonged period of brain maturation might have been important in the development of our distinctive mental capacities, especially our cognitive abilities since a larger brain, produced by prolonged growth, could in principle facilitate the acquisition of new abilities. This, too, seems likely, but it has long been unclear how these ideas could be taken further.[34]

Observations of neotenous development in other mammals might, however, provide a clue. Specifically, there is a large group of mammals that show juvenilized features into adulthood, along with various other shared features. These animals are united not by close phylogenetic relationships but by their shared historic fate: they had all been domesticated by humans. In general, domesticated breeds of wild animals show the retention of juvenile features in ways and extents not seen in their wild progenitors. These features are exhibited by domesticated animals as diverse as dogs, cats, camels and llamas, elephants, mice and rats, guinea pigs, (real) pigs, horses, rabbits, and goats and sheep, thus including representatives of three of the four supraorders of mammals. Apparently, something inherent to domestication favors neoteny. The tame breeds in general not only show a degree of retention of juvenile physical features such as flatter faces and rounder heads but also exhibit prolonged periods of juvenile behaviors, in particular exploratory behavior and the high sociability and playfulness of the young. The extended juvenile phase includes a delay in the onset of the adult fear-startle response to potential threats, a response that takes place at a characteristic point in the maturation of every wild mammalian species and curtails the high sociability of young animals.

In addition to those mental and behavioral traits, domesticated breeds often show a mix of odd and unexpected features. These include floppy ears, shorter or curly tails, areas of lighter pigmentation—usually white spots, sometimes light brown on a darker background—smaller teeth, reduced adrenal glands and blood glucocorticoid levels, and expanded sexually active seasons due to more frequent estrus cycles. Of these physical traits, only floppy ears are readily linked to the juvenile state since many species whose adults bear relatively long erect ears begin life as infants possessing floppy ears. Thus, the neotenous features that accompany domestication are a subset of the ensemble of domestication-related ones. A final and particularly surprising trait is that domesticated species typically have reduced forebrains relative to their wild progenitors.[35]

The collective set of those traits is now termed the **domestication syndrome**, but it was first described, as with so much else in biology, long ago by Darwin. The description and discussion are in his two-volume study of the mechanisms of heredity, *The Variation of Animals and Plants under Domestication,* first published in 1868. There he ascribed it to the gentler "conditions of living" experienced by domesticated animals relative to their wild progenitors. He was unable, however, to distinguish which traits were induced anew in every generation due to the continuing conditions involved in domestication from those that were hereditary and would appear regardless of environmental constraints. Perhaps that distinction was perhaps not even always clear to him, since he believed that acquired traits could sometimes become hereditary. He reports that for a few domesticated breeds accidentally released into the wild, there was some reversion of traits toward the ancestral state, thus indicating a plastic developmental condition for some of them, while for other domesticated animals so released, there was apparently retention of the domesticated characteristics, indicating a true hereditary basis. Today, it is clear from experimental work in several animal species that both the physical and the behavioral traits of the domestication syndrome are genetically fixed, not simply induced every generation by the conditions of "easier living" provided by domestication.[36]

An indication of the differences between domesticated and nondomesticated silver foxes, the species with the most extensive experimental history on domestication, can be seen in **Plates 16** and **17.**

The possible relationship between these special characteristics of domesticated animals and humans is that several of the special traits that differentiate our species from most other mammals are those associated with domestication: juvenile facial and head features, a prolonged period of juvenile development and brain growth, frequent female sexual cycles, and a generally high and continuing sociability throughout life or at least tolerance for the close presence of many conspecifics. Other domestication traits such as floppy ears, spotty pigmentation, and relatively smaller brains are not seen, but the ones that are exhibited fall into the domestication syndrome. In addition, humans show reduced sexual dimorphism for both overall body size and size of teeth (particularly the canines), which constitute another two signature traits of the domesticated state.[37]

The idea that humans are, in some way, a domesticated species has a long and fractious history. French philosopher Jean-Jacques Rousseau (1712–1778), who appears to have been the first to propose it, felt that it was obviously true but a lamentable fact. For him, the domesticated state connoted loss of alertness, vitality, and resilience compared to untamed, wild forebears. Indeed, he saw modern civilized humans as degenerate descendants of a primeval ancestral stock. In contrast, the contemporary *noble savage* (as envisaged by Rousseau) was not just nobler in mien and behavior but stronger and tougher than his civilized cousins. Rousseau's pessimistic view of humankind was later taken up by various biologists and political thinkers in the eighteenth and nineteenth centuries and regrettably, though perhaps inevitably, fed into Nazi doctrines about declining human "racial health."[38]

Rousseau's interpretation was only one line of thought about the postulated state of humans as domesticated. The other view was positive—that domestication permitted the high level of social intercourse and cooperation that must have been prerequisites for the building of human civilizations. Two men who thought so, although with different emphases and approximately seventy years apart, were Walter Bagehot (1826–1877), the influential English political theorist and founder of *The Economist,* and the eminent German-American anthropologist Franz Boas (1858–1942), the father of modern social anthropology. For these writers, the domesticated state of humans, with its inherent qualities of sociability and cooperativity, made it possible for humans to construct complex human societies.[39]

The difference of opinion on the nature of the human domesticated state between Rousseau on the one hand and Bagehot and Boaz on the other hand, however, highlights a crucial question: what precisely does the word *domesticated* mean? Clearly, no other species could have domesticated humans as we have done to various animals. While Darwin recognized the parallels between domesticated animals and humans, he fretted over this question extensively. In *The Descent of Man,* he described the traits of humans that reflect domestication, thus arguing for the proposition but simultaneously dismissing the idea because no external agent could have carried it out. His discussion of this matter exudes frustration.[40]

If the human state bears some genuine relationship to the state of domesticated animals and that condition could not have been imposed externally by some other species, it could only have involved an element of self-taming. From that perspective, the modern human species can be seen as a "self-domesticated ape," one whose early social evolution put a premium on tameness and lack of aggression, thereby promoting cooperation and sociability. With the later development of the capacity for complex language, these characteristics could have provided the psychological and behavioral foundations for building the highly complex social constructs that we term *civilizations*. This was the essential point of Bagehot and Boas. Such pacific characteristics would, of course, neither have ensured perpetual sweetness and light in human life nor precluded a major role of coercion in the construction of civilizations exemplified by the building of the Egyptian pyramids by slave laborers. Yet there must be some element of truth to the Bagehot and Boas thesis: it is inconceivable that civilizations could have developed without large measures of social cooperation over long periods of time.

The concept of self-domestication of a species might seem fanciful, but there is at least one other living animal species that seems to exhibit it, indeed one whose characteristics seem almost inexplicable without invoking it: the bonobo, one of our two closest living animal relatives. In both its physical traits and high sociability, it seems a model of how sociable apes would be if they had been successfully domesticated by humans. Compared to chimpanzees, the bonobo retains a more juvenilized face and head, reduced brain size, prolonged juvenile behaviors, a much greater level of sociality in adults, increased sexual activity, reduced competitive fighting, specific brain changes signifying reduced inhibition of aggressive responses, reduced sexual dimorphism, and thinner (more gracile) bones. That last characteristic is seen in many though not all domesticates and is an interesting difference seen between the skeletons of archaic and modern *Homo sapiens* (see below). For the bonobos, the probable selective cause of these changes was increased food supplies as their branch split off from the chimpanzee lineage and occupied new territory about 1 million to 2 million years ago (as judged from molecular clock estimates). Less competition for food—ergo more benign conditions of living—would have favored the development of a less competitive and male-dominated primate society and a more egalitarian one between the sexes, which these smaller chimpanzees possess. To the evident embarrassment of some commentators, they also exhibit more frequent sexual activity than chimps, both same-sex and heterosexual, at least in captivity where food is provided by the human caretakers, eliminating the work and time of foraging for it and thus perhaps providing more leisure time for such things as sex.[41]

The case of the (possibly) self-domesticated bonobos returns us to the question of what the basis of the domestication syndrome might be, although bonobos (like humans) do not exhibit all its facets. Lacking a theory of heredity, Darwin could not explain it, yet even after the modern science of genetics developed in the twentieth century, the mystery persisted. A major difficulty was that all the domesticated animals had initially been domesticated millennia earlier. Whatever genetic changes had been essential to the early stages of domestication could well have been obscured by subsequent genetic changes in their long subsequent history. For most domesticated animals furthermore, there could have been many such changes, both accidental and selected, the latter in conjunction with breeding for particular desired qualities such as the wool of sheep, egg production in chickens, and speed or strength in horses. In addition, the selected changes might have brought others in their wake via the process of Darwin's *unconscious selection*.

One possible explanation is that there is a large genetic regulatory network involved in the normal specification of all the affected traits and that the domestication syndrome involves mutations in this network, presumably in some of the most upstream elements and affecting all the downstream branches. This, however, would demand an improbably large genetic network. Furthermore, mutations in upstream genes of gene networks, which would seem to be required in this case given the multiple effects, are usually lethal. The domestication syndrome, however, is not a lethal condition, and its evocation with taming and breeding in many species indicates a relative mildness of the condition.

An alternative conceptual track starts with cells, not genetics. Is there a common cellular type involved in the traits affected by domestication? Initially, it would seem not, given the diversity of cell types involved: pigmentation cells in the head and the trunk, connective tissue cells that produce bony elements (in the face) and cartilaginous elements (in the face and the external ear, the pinna), various neural cells in the trunk, sensory cells in the head, neuroendocrine cells, and brain cells (in the forebrain).

This is quite a range of cell types and locations and they lack obvious functional links. There is, however, a developmental fact that connects them: all either derive from or depend in their development on neural crest cells, whose importance for the development of the face, in particular, we have thoroughly examined. With one major exception, neural crest cells are direct precursors for all the affected cell types. Thus, the pigmentation defects arise in both cranial and trunk melanocytes, which are derivatives of the neural crest; the pinna derives from neural crest cartilaginous cell precursors; and the jaws develop, as we have seen, from neural crest cells in the lower facial primordia. The one exception is the brain: neural crest cells make no direct contribution to the formation of the brain, yet as we have seen, the reduction of the forebrain is one of the most consistent if not universal characteristics of domesticated animals. Yet as it happens, neural crest cells play an indirect but essential role in the development of the forebrain: if most cranial neural crest cells are excised in early developing chick embryos, there is extensive cell death of the precursors of the forebrain, both of the telencephalon and the diencephalon. In principle, *some* natural reduction of neural crest cell input in the embryonic development of domesticates could produce smaller forebrains, though this has yet to be shown.[42]

These connections lead to the hypothesis that what distinguishes domesticated animals from their wild forebear strains is a mild decrease of neural crest cell contributions to the various sites to which these cells normally contribute. Strong reductions of neural crest cells are lethal, while moderate reductions involving mutant genes known to be required in neural crest cell development produce various clinical syndromes known as *neurocristopathies*. In principle, however, still smaller deficits of these cells at the key sites could produce detectable changes but not lethality or serious defectiveness. Such slight reductions could, in principle, be produced by the collective effects of slight loss-of-function mutations in some of the many genes that are known to be crucial for neural crest cell development or migration. A hallmark of such genes is that many should show dosage-sensitive effects of their wild-type gene products on development. Such effects are frequently revealed by the property of **haplo-insufficiency**, the fact that one active dose of the particular wild-type gene (for an autosomal gene) is insufficient to produce normal development. (In diploid cells, most genes are present in two doses, reflecting the presence of two chromosome sets, but for the great majority, one dose is sufficient to produce the normal phenotype.) In addition, if the syndrome is a complex function of reduced activities of several such genes, then we would expect additive or synergistic effects by mutations in different neural crest cell genes. Indeed, many of the neural crest cell genes examined show these properties.[43]

The aspects of the domestication syndrome that may seem unexplained by this idea are the behavioral ones: docility, the prolongation of juvenile behaviors, and the alteration of female sexual cycles. These, however, might be explicable by indirect effects of neural crest cell deficiencies on the adrenal glands, the forebrain, or both. In particular, the docility might reflect diminished stress reactions mediated by the adrenal glands since the adrenal glands are smaller than those in wild strains in the domesticated animals that have been examined for this trait (rats and foxes), who also show reduced levels of the adrenocorticotropic stress hormones. If smaller or slower-developing adrenal glands are accompanied by a delayed onset of the fear-startle response as seen in domesticated animals, then this might account for the neoteny of juvenile behaviors. Alternatively, these behaviors might reflect subtle alterations in the forebrain, which regulates the fear-startle response through the amygdala. Similarly, the altered sexual cycles might reflect minor alterations in the hypothalamus, which regulates sexual cycles and which derives from the diencephalon.[44]

The natural extension of the neural crest cell-hypothesis to humans is that the domesticated state of people might derive from a similar genetic basis affecting neural crest cells. There need not, however, be one exclusive road to domestication or self-domestication. Another possible explanation, for example, is that the human condition originated in induced hormonal changes brought on by social pressures to cooperate in large groups as human population densities increased. A recent study has made just this case. Involving detailed measurements of human skulls from before 80,000 years ago in Africa (hence prior to the out-of-Africa emigration) and those from 30,000 years onward, the authors found that the latter skulls were less robust and indeed more feminized. Such changes could result from lower rates of androgen induction and perhaps also higher rates of serotonin production, conditions known to be associated with more sociable—indeed, feminized—behavior. A schematic

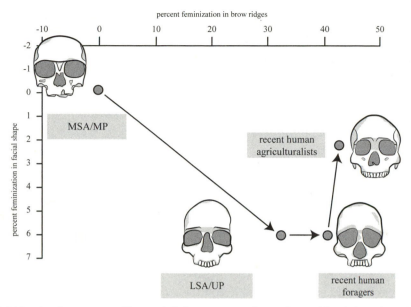

FIGURE 8.4 Feminization of human skulls by two criteria—brow ridges and facial shape—between 80,000 and 30,000 years ago. As discussed in the text, such changes might be signs of the domestication process. MSA/MP, Middle Stone Age/Middle Pleistocene; LSA/UP, Late Stone Age/Upper Pleistocene. (Modified from Cieri et al., 2014.)

diagram of these trends in feminization of the human skull is shown in Figure 8.4. Such changes might have initially been developmentally plastic ones but might later have been fixed by genetic assimilation. The premise of the authors of the study is that higher population densities would have prompted more competition but also created pressures for better socialization with concomitant physiological changes. This idea fits with an earlier genetic analysis of present-day African populations that suggested there had been a human population explosion in Africa between 80,000 and 60,000 years ago, which we have already noted might also have propelled the out-of-Africa migration about 60,000 years ago.[45]

A Summing Up: Mental Processes and the Evolution of the Human Face

In Chapter 7, we looked at the ways that evolving mental, social, and neurobiological processes might have affected the evolution of the face from the early anthropoids through the hominins. I argued that selection pressures for social integration in species whose members are heavily dependent on their fellows would have influenced the expressivity of faces and the ability to read expressions. These abilities would also have contributed to the capacity for expressing thoughts via language. The accompanying physical changes in the face such as furless faces and re-duced muzzles, initiated at the start of anthropoid evolution but taken further during hominin evolution, would have both fostered these devel-opments and been influenced by them.

In this chapter, I have further proposed that mental predispositions ex-pressed through specific behaviors that can themselves be promoted by selection have continued to shape the human face since our emergence as a species. In part, "racial" features of the face might have been selected for via sexual selection of different preferences for those facial features— as suggested by Darwin—while the juvenilized ape face of humans might have reflected a much earlier self-domestication. The latter, it was argued, may have some similarities with the domestication process that humans have imposed on a large number of animal species. It would, presumably, have influenced our species behavior from its early days but may have only come into its full development with the rise of complex societies and their demands for high levels of cooperative interaction.

By taking the evolution of the human face up to the present, this chapter would seem to complete the history of the face. The story is not over, how-ever. There is a history yet to come: the future of the human face. Though any thoughts on that history still to unfold are necessarily speculative and provisional, it is possible to extrapolate from the past and present and sug-gest possible elements of that future history. We turn to those matters in the next, and penultimate, chapter.

— nine —

ON FACE CONSCIOUSNESS AND
THE FUTURE OF THE FACE

Introduction: On the Relatively Late Emergence of Face Consciousness

In contemporary modern life, human faces and their images are ubiquitous: we see them on the streets and in shops, at school and work, and from
numerous advertisements. The advertising industry deserves special note
because of its significant cultural impact. It cannot resist faces: beautiful
faces, happy faces, serious faces, interesting faces, homely and humorous
faces, not to mention animal faces in all those modes. In particular, if one
spends any time with electronic media, the plethora of face images guarantees you are never alone. In general, most faces flit by unnoticed, but
familiar ones usually register. Furthermore, one's own face matters, with
most people quite aware of theirs, either with contentment or dissatisfac-
tion. While individuals have been unconsciously reacting to each other's
faces since time immemorial, most people today are fully cognizant that
the human face is an important signifier of individuality and personality,
a state of knowledge I will call *face consciousness*.

Such consciousness is so widespread today that it may be surprising to
realize just how relatively new in history it must be. In modern Western
culture, it undoubtedly grew with the Renaissance and its stress on the
importance of the individual personality and expression. Yet those ideas
were foreshadowed, as implied in its name, in the worlds of classical Greece
and Rome. Subsequently, the manufacture of mirrors and other reflecting
surfaces during the Industrial Revolution permitted many more individuals to monitor their own faces. It also fostered the growth of publications
with pictures, thus increasing exposure to the faces of the great and the
famous. These developments must have increased face consciousness from
the late eighteenth century onward.

Of course, awareness of the faces of others has deep evolutionary roots.
As we have seen, even cows and sheep register each other's faces, while
anthropoid primates have undoubtedly been visibly reacting to each

other's faces and facial expressions for the past 50 million years or so. I have argued that their social interactions mediated by face-to-face inspection were a strong influence in shaping the physical properties of their faces, in both social selection for more expressive faces and sexual selection for more attractive ones. If anything, these influences would have been accentuated in hominin evolution. Yet these factors would have been *unconscious* forces in influencing the evolution of the face. Might face *consciousness* provide a further impetus in shaping human faces in the future?

Addressing that question involves speculating about things to come. In this book so far, we have concentrated on the past, but here we will turn toward the future and examine two distinct aspects of the future of the face. The first concerns the matter just raised: the cultural sources of face consciousness and their potential for shaping the face in the future. Necessarily, it begins with a look at face consciousness in the recent human past but then will proceed to the effects of globalization and evolving cultural attitudes as shapers of the physical features of the face. The second aspect concerns the future of scientific investigations of the face, specifically at attempts to identify the genes whose alleles affect facial diversity. This new genetics of facial difference has arisen in recent years, and it aims to answer the question of how many and which genes generate the huge diversity of human faces, a topic that was introduced in Chapter 4. Intriguingly, in part this work is morphing into studies designed to uncover links—if they exist—between facial features and personality. We may call this endeavor *physiognomic genetics*. What sort of future might it have, and what are its implications?

First, however, we will look at the history of face consciousness as a prelude to thinking about its possible future.

A Brief History of Face Consciousness: The Evidence from Absence and Presence in Art Artifacts

When did humans start becoming consciously aware of faces as badges of physical individuality and personality? Alas, we cannot interview the dead, so it is impossible to know. In principle, ancient written records could provide an answer, but the history of writing is only about 5,000 years old, hence its reach into the human past is limited. Furthermore, even for this period, there is a sampling problem: for most of that history, writing was restricted to relatively few individuals, and the writings that survive surely do not reflect all of the concerns and interests they had, let alone the vast majority of those individuals who left no written record. Certainly, a few eminent writers were keenly interested in the face, in particular Aristotle and Hippocrates, but there is no indication that such interest extended beyond a relatively small number of thinkers.

Fortunately, even without the testimony of ghosts or writers of antiquity, the quest is not hopeless because a different form of record exists, one that extends several tens of thousands of years into the past: artistic artifacts. Much of this record has undoubtedly been lost: depictions on perishable materials such as wood or animal skins or on exposed stone surfaces that would have been subject to loss through weathering. Yet the ancient artistic record, incomplete as it is, provides clues. After all, any form of figurative representation must reflect *some* degree of interest in the subjects depicted on the part of the artisans, and their interest presumably reflected a wider concern within their cultures. The main inference we can draw from this evidence, to be discussed soon, is that the human face had relatively little interest for artisans until approximately 5,000 years ago with the appearance of the first great civilizations. Prior to that, the depiction of human faces was rare indeed.

Of course, there is a perfectly possible trivial explanation for that scarcity of face pictures from earlier times: the artists lacked the skills to draw them. Indeed, what we call figurative art seems in general to have been a comparatively late developing skill in human artistic endeavor. This is indicated in Figure 9.1, which shows a summary time line of prehistoric artistic endeavor. If we set aside the making of stone tools with utilitarian functions (which does not exclude the possibility of aesthetic pleasure for those who made them), the earliest known artisanal endeavors involved the making of colored beads and the use of ocher pigment for body coloration. The oldest of these date to roughly 165,000 years ago and are suspected to have been made by members of *Homo heidelbergensis* or *Homo erectus,* not by modern *Homo sapiens.* Most of the traces of artisanlike activity for the next 130,000 years or so are similarly decorative beads or painted shells or, in a few instances, abstract designs, the earliest from about 77,000 years ago in southern Africa. During this long stretch of tine,

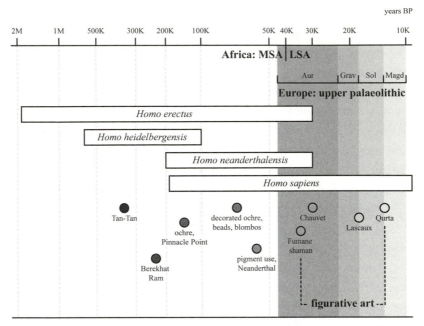

FIGURE 9.1 Time line of artistic endeavors within the archaeological record. Artistic endeavors preceded the appearance of modern *Homo sapiens*, though they only flowered relatively late in our history. The existence of artistic artifacts implies some level of communication via language in the community in which they were made, though possibly proto-languages sufficed in the earliest stages. Art thus did not arise with the European Upper Paleolithic cave paintings, but figurative art may first have appeared at that time. As indicated in the figure, it was a later development. The abbreviations Aur, Grav, Sol, and Magd denote successive stages of stone tool development; MSA = middle stone age; LSA = late stone age. (Adapted from Morriss-Kay 2010.)

the surviving artistic artifacts do not include figurative portrayals of humans, animals, plants, or landscapes, whatever the reasons for the deficiency might be.[1]

When figurative art finally appears in the archaeological record, it is stunningly evocative and beautiful. Furthermore, with no surviving precursor work, it appears to have developed rather suddenly and sui generis, although again the possibility that earlier creations of similar nature were lost cannot be ruled out. These manifestations of sophisticated artistic ability are the celebrated **Upper Paleolithic** cave paintings and drawings that were found predominantly in present-day France and Spain. These have been dated to between 33,000 to 12,000 years ago and were painted on the surfaces of the walls of limestone caves, many deep within the caves rather than the entrances where the artists might have worked by visible daylight. They show a profusion of animal images, many quite large and highly detailed, and all of them clear representations of big animals of the late Pleistocene. Among the animals shown are lions, rhinoceroses, bison, horses, wooly mammoths, cave bears, reindeer, red deer, and ibex; some but not all of these would have been game animals for Paleolithic hunters. Indeed, large nongame animals predominate while larger animals known from bone deposits to have figured prominently in diet are underrepresented; small mammals and birds, which might well have also been part of the diet of these people, are not represented at all.

The images were made with red ocher and black pigments and frequently incorporate features of the surface of the rock face, thus adding three dimensionality and affecting the play of light on them. The images are even more impressive when we consider the conditions in which they were made, painted by torchlight and almost certainly by small groups of artisans, who must have used some form of scaffolding to reach the upper regions of the walls. That this is figurative art of a high order was attested to by Pablo Picasso, who saw the images in the cave of Lascaux, created 19,500 years earlier, and remarked, "Nous n'avons rien appris." The literal translation is, "We have learned nothing," but its sense is closer to "There is nothing that we modern artists know that was not known to these cave artists."[2]

What is conspicuous to the modern eye in these great cave paintings is the lack of representation of human figures and the complete absence of human faces relative to the much later figurative art of the first great civilizations of the world (Egyptian, Assyrian) and, of course, the omnipresence of this motif in art over the past six centuries of human history. Given the exquisite depiction of animal faces, which would have required no less skill, this absence of human faces cannot represent an inability to make such images. The faces of many of the animals are not only immediately identifiable but also show individual detail, as in one of the representations of groups of lions, and even expressions in the famous panel of horses (**Plates 18** and **19**) in Chauvet Cave. Evidently, some Paleolithic cave artists were fully aware that faces can express emotional states, and they found this interesting. The work shows they had the skills to draw lifelike human faces had they so wished. Furthermore, there are actually a small number of depictions of full humanlike figures that have faces, but intriguingly in each case the face is that of an animal. Such figures are called **therioanthropes** and probably represent either something from the spirit world or shamans wearing masks of animal faces. Altogether, it is difficult to escape the impression that these gifted Paleolithic artists were more interested in animal faces than human faces, whatever the reasons for that preference were.

Even that modest conclusion, however, raises a big question—indeed, the central question—about these Paleolithic cave paintings: what purpose or purposes did they serve? We refer to these pictures as *art*—indeed, the separate branches of the major caves where these paintings were made are termed *galleries*—but it is almost certainly a mistake to see their manufacture as motivated primarily by the desire to give aesthetic pleasure, a goal of many Western artists since the Renaissance. In Paleolithic times, there were probably several functions, as there is in the art of recent human history, just as there are many different kinds of images on the cave walls. These range from the highly naturalistic to the fantastic to that of seemingly abstract design. Yet the idea that these were created simply for visual delight ("Art for art's sake") is unlikely. Given their relative inaccessibility, it is far more probable that they played some role in the religious life of the communities that created them. One possible function is that they were part of shamanistic practice; some of the abstract depictions correspond to visions in mental fugue states associated with shamanism. Although these make up at most 10 percent of the painted images in the major sites, public ceremonies and rites may have been conducted near the more naturalistic images, presumably under torchlight, the same lighting conditions needed for their creation.[3]

It was once believed that the major function of the pictures involved some form of *sympathetic magic* to help hunters kill these kinds of animals, several of which were hunted for food, as judged from the remains of kitchen middens. There are, however, two arguments against this. First, the proportions of the painted images of different animals do not match the inferred respective abundances of the animals that were hunted for food, a major disconnect if the function of the paintings was to aid hunting. Second, there is a dearth of hunter figures in the cave paintings, unlike, for example, many of the much later African rock-art paintings in sub-Saharan Africa. Even for that African art, it is not clear whether sympathetic magic to increase the success of the hunt was a major motivation, but the high frequency of hunters in these paintings makes this one of its probable functions.[4]

The key point with respect to this discussion, however, is the frequent representation of animal faces in contrast to the absence of human ones. Of course, that absence does not mean the artisans had no awareness of or interest in human faces. They almost certainly paid attention to the faces and facial expressions of their fellows, just as present-day hunter-gatherers do, and perhaps with some degree of awareness of doing so. Yet such consciousness in the tribal societies to which the European Paleolithic cave artists belonged was, as argued earlier, almost certainly at a lower level than it is in people generally today.[5]

One kind of representation of human figures was common in the Paleolithic period, however, and first appears in the archaeological record about 40,000 years ago at many European and Asian sites. These were not paintings but carved stone figurines of female figures, specifically and predominantly of pregnant women. These have come to be labeled generically "Venus" figures, though they hardly convey the sense of classical Greek female beauty associated with that name. The scholarly consensus is that these figurines were used by women as worn or carried amulets, good luck charms for fertility and successful childbearing. Perhaps their most striking feature, however, is not the heavy reproductive bodies but the absence of faces. Indeed, in many cases, the head is reduced to a nubbin. The entire emphasis is on the body, hence on generic femaleness. There are exceptions to this general rule, of course. These include several figurines that show human faces carved in various stones and in ivory from mammoth tusks, principally from a few Asian sites. These, however, tend to be highly schematic—eyes, nose, and mouth—without individuation, and they are a minority in the overall assembly of Paleolithic statuary. Again, it is difficult to escape the impression that the human face did not count as a subject of major interest in the worlds of Paleolithic Europe and Asia.[6]

What about other regions of the inhabited world—Africa, Australia, and the Americas? As judged by their surviving art, were the peoples of these regions more interested in human faces than the European cave artists? In these regions, the Paleolithic lasted until much later than in Europe and the Middle East, where it began to end around 10,000 years ago with the invention of agriculture. There, the Paleolithic yielded to the Neolithic, which was succeeded in Europe and Asia by cultures that employed metals. Yet despite the longer run of the Paleolithic in Africa, Australia, and the Americas, there was nothing like the European cave paintings in terms of artistic detail created in any of these regions—at least any that has survived. Nevertheless, in Africa and Australia, there is an extensive archaeological record of rock paintings. Unlike the cave paintings of Europe, these were made outside of caves or other fully enclosed spaces and thus were exposed to the elements. As a result, many undoubtedly have been lost, particularly some of the older ones. Those that survive, however, show animals and often representations of hunters. In the cave paintings that survive in present-day Zimbabwe, for example, human figures are the dominant form depicted. For these, however, the humans also tend to be represented fairly schematically, with male and female figures distinguished by their sex-specific structures (penises, breasts) but with no attention to the details of heads or faces. There is also a richness of depiction of different human activities and situations that is distinctive. Similarly, the animals are drawn with skill but often with less three-dimensionality than the European cave art and with no distinguishing facial detail. Though dating these rock paintings is difficult, it has been possible to date a few of the African rock paintings where the pigments had charcoal admixture (allowing radioactive carbon dating), and the estimated times of creation for some of the older ones obtained range from 5,000 to 13,000 years old. This form of art is clearly ancient (see note 4).

Art was also apparently fairly stable as a cultural phenomenon in Africa for thousands of years and carried on without the human face becoming part of the repertoire. That element only began to emerge in African artifacts in the past 1,000 to 2,000 years and then primarily in carved wood and stone, not as paintings. Though comparatively recent, falling within the period of written history in the Middle East and Europe, it was well before Europeans reached those regions, hence it was fully indigenous African peoples who made them. Nevertheless, it was apparently a late development in the (unwritten) history of these peoples.

One might well ask, when did the first stirrings of face consciousness as shown in art take place? There is certainly no unique time at which it suddenly appeared, instead varying from region to region, but one generalization seems safe. With the beginnings of each major civilization—whether in the Middle East, Europe, Asia, or the Americas—there was a great deal of accompanying artistic work that accompanied its creation; correspondingly, depictions of the human face start becoming far more frequent, with animal faces (often representing deities) also becoming common. As far as I am aware, none of this civilizational early art included portraits in our contemporary sense in which the created object is concerned exclusively with the face. Such portraiture would come later and then not in all of these early major civilizations. In the Aztec and Mayan civilizations of Central America, furthermore, the faces were not those of ordinary mortals but of godlike entities who nevertheless bore distinct human resemblances.

The first of these depictions of faces, which were parts of statues showing the whole body, were those of early Egyptian civilization, specifically the Old Kingdom, which existed roughly 4,500 to 3,500 years before the present (BP). The earliest Egyptian art depicting faces focused on the pharaohs, their wives, their servants, and their gods—the latter almost invariably therianthropic figures—and, indeed, that emphasis continued throughout classical Egyptian times until the end of the New Kingdom, approximately 2,600 years BP. These faces are stylized, rendered in profile, and only part of whole figures. Therefore, they do not constitute portraits per se. Nevertheless, they often show signs of individuality. Indeed, a few artifacts from the earliest period depict ordinary people whose faces have clear individuality such as the famous statue of the "sitting scribe," which is now in the Louvre (**Plate 20**).

In the period that followed the Alexandrian and later Roman conquests of Egypt, both occurring within 300 years of the start of the Christian era, the interest in faces became fully conscious and explicit. The portraits of well-to-do Alexandrians around the start of the Christian era, approximately 2,000 years ago, affixed to their coffins were probably the world's first extended exercise in individual portraiture by a culture. These were done by artisans, not special artists attached to the royal court, and many of these paintings show a high level of detail in the individual physical features and often a hint of expression (**Plates 21 and 22**). Thus, by the start of the Christian era, there had been roughly 2,000 years of artistic depictions of individual and their sometimes expressive faces, first in statuary and later in paintings.

Of course, it is possible that the connection between depicting faces and the rise of civilizations is simply an ascertainment artifact: each of these civilizations left substantial ruins and thus extensive archeological evidence. In consequence, more of all kinds of objects, including those made of more perishable materials, were preserved than would have been from earlier settlements. Yet that is probably not the entire explanation. I submit that these more enduring human settlements, whose cultural maturation was progressive, fostered more social contacts and more different kinds of such contact, one of whose consequences would have been a growing general awareness of the face as a medium of exchange of social information. That awareness, in turn, would have translated into greater interest on the part of the artists and artisans in depicting faces. These early civilizations, with their dense social architecture, no matter how hierarchical and repressive, would have inevitably fostered the recognition that people are not wholly defined by their social roles but have individual personalities. Awareness of the faces of others as signs of those personal differences could have been both a by-product and a contributor to that sense of people as individuals.[7]

One can further imagine a technical reason for the connection between the rise of civilizations and a growing interest in human faces. The early civilizations, with their new technologies for working metal, would have witnessed the creation of more reflective surfaces, particularly those of smooth metal. We can be fully aware of the faces of others simply by looking, but to know one's own face takes a reflecting surface. Still water provides such a surface, but there is little evidence, for example, that contemporary hunter-gatherers spend much time in Narcissus-like inspection of their own images in puddles or ponds or lakes. Modern societies, with their plethora of manufactured reflecting surfaces, make facial self-knowledge easy.

Impossible to quantify, face consciousness has undoubtedly varied from place to place and time to time, and it has almost certainly been more common among the prosperous than the poor, the latter making up the greater part of humanity, both in the past and today. Correspondingly, as prosperity spreads and more people rise out of poverty, consciousness of faces, both one's own and those of others, will undoubtedly grow and spread.

The Future of the Face: I. Toward a (More) Globalized Human Face

As the late Yogi Berra, the homespun American philosopher, once noted, "It's tough to make predictions, especially about the future." Furthermore, the more distant the imagined future, the greater the odds it will have a different character. For would-be prophets, however, there is the consolation that they will not live long enough to be embarrassed by failed long-term predictions, which will almost certainly have long been forgotten anyway.

With such reassuring thoughts in mind, let us try to predict what the future holds for the human face. An implicit assumption in the exercise is that humanity *has* a future—namely, that neither natural events (a devastating meteor impact, for example) nor multiple interlocked global calamities of humankind's own making will prematurely end our species' existence. To focus our thinking, let us concentrate on the specific question raised at the start: given the earlier role of mental states and behaviors in shaping the evolution of the human face in the past, might face consciousness, which is now effectively a global phenomenon, affect human faces in the future?

It certainly is affecting faces now in terms of what people do to, for, and with their faces. This is not merely in terms of the use of conventional makeup or the addition of facial tattoos and metallic studs to noses and lips but even more dramatically through plastic surgery for cosmetic purposes. Millions of people in North America and Europe, and increasing numbers in Asia and South America, have taken advantage of this technology to alter their facial appearance. In the mid-twentieth century, cosmetic surgery was focused on changes to nose shape, primarily in young women—"nose jobs"—but is now also used to alter eye and jaw shapes in both men and women of all ages, while other techniques (e.g., *liposuction*) are employed to make lips fuller. The effects of all of these methods on the look of a face can be subtle or pronounced but always transformational. Even more dramatically, in the past decade, there have been several face transplants for individuals who had previously suffered serious disfigurement. This technology, however, is still too new for its impact to be clear, including its psychological effects on those living with a new face.

Altogether, in more than half a billion years of animal life on Earth, we are the first species to be able to dramatically alter our faces physically to enhance their attractiveness. Nevertheless, none of this fiddling with faces will have direct evolutionary consequences because such changes cannot be transmitted to offspring. In evolutionary terms, all forms of face decoration, plastic surgery, and face transplants are dead ends. What about gene therapy, however? Apart from the ethical questions about alterations of the *germline* to alter future generations, this technology is far more difficult to implement than was initially expected. Consequently, its applications have been far more limited than first projected. Furthermore, deliberate genetic alteration of faces would require not just a highly reliable technology but also a much greater knowledge of the genetics of the face than currently exists (to be discussed shortly). Finally, even if these prerequisites were to materialize over the coming decades, the technique would surely only be employed by a tiny fraction of people, far too few to affect the evolutionary future. Gene therapy is therefore highly unlikely to contribute to evolutionary changes in the human face.

One major phenomenon, however, will certainly affect the evolutionary future of the face: globalization. Though usually discussed in terms of its economic importance, it involves not only the movement of goods and financial assets but also people. And when people move to new locations, they often meet and marry or, at least, mate with locals who may and often do have different kinds of faces from those of potential mates back home. And such interactions have important psychological and cultural impacts, as well as demographic consequences. A dramatic instance of this was the large number of offspring of joint Vietnamese and American parentage that followed the huge influx of U.S. soldiers into Vietnam between 1963 and 1972. Yet this general process is taking place all over the world as globalization disperses people from their traditional homelands to new ones. The result is that many countries attracting large numbers of foreigners are becoming less ethnically homogenous and far more varied in their "racial" composition and in the look of offspring who result eventually from some fraction of the inevitable encounters between individuals of different ethnic origin.

In effect, this global melting pot phenomenon reverses the process of genetic diversification that took place over 50,000 years (approximately 2,000 generations) as the original out-of-Africa pioneers and their descendants spread out and populated different regions of the planet. Perhaps it can be called "racial de-diversification." If there were to be complete random mating of all peoples and if the different genetics of all present-day "racial" groups were essentially subdivisions of the collective genetic inheritance of the original out-of-Africa migrants, then that remixing would lead to a sea of faces similar to the original out-of-Africa immigrants. In reality, of course, homogenization to that extent is completely improbable. Furthermore, current genetic human global genetic inheritance is not simply a partitioned ancestral inheritance. In the more than 60,000 years that have elapsed since the original out-of-Africa exodus, we can be certain that new alleles affecting the shape of different facial features will have arisen. Finally, and perhaps most significantly, there have been different amplifications of different portions of the original out-of-Africa genetic inheritance among the current different "racial" groups, in particular those of different Asian groups whose facial types were not at all represented in the pioneering humans who left Africa 72,000 to 60,000 years ago. Thus, a complete global homogenization, were it to take place, would not re-create the look of our African ancestors but show a distinctly more Asiatic cast. While such a global homogenization is not in the cards for anytime in the near to medium term, humanity is almost certainly moving toward a more general and generic human face. The collective genetic inheritance for individual facial differences will not be diminished thereby, but the phenotypic component of those differences associated with different ethnic groups will diminish.

In turn, this may be expected to drive an interesting change in ideas of human beauty. Recall that Darwin viewed "racial" diversification as driven by sexual selection and involving different local ideas of beauty, especially as manifested in the face. If so, it follows that the obverse side of such preferences as members of different racial groups came into contact from the sixteenth century onward would have been a degree of aesthetic recoil when confronted with the "other"—and frequently on the part of both groups in such encounters. That element of recoil, in turn, probably contributed to the genesis of racist beliefs, particularly in Europe and East Asia. Yet increasingly in the modern world, especially in major urban centers, people of different "racial" groups now encounter each other all the time. The result is that the "other" no longer seems so different; she or he is often simply "one of us." Thus, being of "mixed race" has essentially lost its stigma and can even be a source of pride, as reported by young people in the United States and the United Kingdom.[8]

There is even a linguistic clue to these changing attitudes: the obsolescence of the word *miscegenation,* which denotes the offspring of "mixed" marriages, a term whose emotional color is given by its prefix. This word was used without reflection by many writers in the Western world for more than a century and through the 1950s. Today, however, its use is almost unthinkable, not simply because of its embedded racism but also because of its implied exceptionalism, a rare deviation from a norm. As more and more individuals of mixed race parentage appear on the scene, mixed race marriages and their resulting offspring are simply no longer exceptional. Furthermore, more and more of those offspring occupy positions of high visibility and influence. The most obvious example is U.S. President Barack Obama, but the presence of countless mixed race film and sport stars and professionals of all types make the same point. These men and women are quite obviously perfectly fit and fine human physical specimens and as intelligent and talented as any other group. Nor would any other result have been predicted by current genetic knowledge, which, as we have seen, shows the close overall genetic relatedness of all major "racial" groups.

These considerations bring us back to Darwin's ideas about beauty and "race." Just as it often used to be said, "All politics is local," it was also true for tens of millennia that the sense of beauty was also local, with different standards for it in different geographical regions, as Darwin pointed out. Such differences in local standards, however, are less and less marked today; individuals of mixed race parentage are often highly attractive physically. Long recognized in Europe and North America for Eurasians, this is increasingly perceived to be true for all "interracial" combinations. In effect, ideas of beauty have gone global, driven by the forces of globalization and popularization in the mass media. As the collective human face increasingly reflects those changes, the prospects are for a better world, in this respect at least, one in which "racial" distinctions—and racism—have faded greatly.

The Future of the Face: II. The New Genetics of the Face

The future of the face also includes its further scientific exploration, through genetics in particular. In Chapter 4, the genetic basis of facial differences was discussed, and I pointed out that the enormous variety of human faces could, in principle, be specified by a relatively small number of genes if most had even a comparatively modest number of alleles of distinct effect. Yet that is only a numerical argument; the genetic basis of human facial diversity could involve, at the other extreme, many thousands of genes. After all, something on the order of 18,000 genes are expressed in the developing facial primordia of mice, and the number is probably similar in human facial development. In principle, all or many of these could have allelic variants that affected facial properties, including details of shape.

To decide the question of how many genes contribute to facial diversity, we need actual data. Until recently, however, those data did not exist. Today they can be obtained, if not easily or cheaply. New methods have been developed that make it possible to identify such genes and, if the work is sufficiently extended, to answer the question. Though this whole endeavor is still in its early stages, its potential has been demonstrated. The approach involves the combination of methods from two distinctly different fields: morphology and genetics.

Before this work is discussed, however, a question presented earlier should be addressed: can we truly speak of *different facial traits,* each underlain by specific genetic differences, in the way, for example, that Gregor Mendel's garden pea traits (e.g., yellow versus green, wrinkled versus round peas, white versus purple flowers) or blood group types (e.g., the ABO and MN systems) can be distinguished cleanly? For such classical traits, there are distinct alleles of particular genes that *qualitatively* specify visibly different versions of those traits. In contrast, facial features seem to be *quantitative* ones. Thus, for traits, such as nose width or shape, the distance between the midpoints of the eyes, or fullness of lips, the differences involve relative and absolute distances between the midpoints of facial features, hence, their relative proportions. The fact that computer imaging allows us to morph one face into another indicates that facial traits can be treated as quantitative ones. In general for quantitative traits, where there is a clear genetic basis, the particular values of such traits as human height or the weight of chicken eggs in a given individual is set by the cumulative result of numerous essentially interchangeable small gene effects, where the identity of the particular genes is unimportant. Hence, before attempting any analysis of facial genetics, it would be helpful to have an indication whether facial traits are underlain by Mendelian-type inheritance—that is, genes whose alleles give clear, alternative phenotypes—or, in contrast, by quantitative and additive inheritance of numerous genes of small effect.

For decades, only one kind of finding spoke to this question. It argued in favor of the existence of genes with clear, visible allelic effects on facial characteristics—namely, Mendelian genes—though the evidence was anecdotal and impressionistic. This was the general observation, noted previously, that many children resemble one parent or the other in specific facial features. That result would not be expected if most individual features are underlain by dozens to hundreds of genes with alleles of individually minute effect. Furthermore, in a handful of cases, there seemed to be highly specific facial features that could only be explained by the inheritance of discrete, single alleles such as the previously noted distinctive Habsburg lip, evidently a dominant Mendelian trait. The new approach has confirmed and made general such observations: the face is built up of facial traits in which individual gene effects participate, each altering the trait in a slight but distinctive way.

The first of the two methods that makes possible a true genetics of facial difference is termed **geometric morphometrics**. In the past thirty years, it has revolutionized the formerly largely qualitative and descriptive discipline of morphology, converting it into a quantitative science. It consists of a range of somewhat different techniques, each sophisticated and demanding, but whose general underlying principle is simple: shapes of related objects can be depicted as subtle spatial transformations of one another, and those degrees of transformation can be accurately measured on a coordinate system onto which the images have been projected. Thus, the method allows complex shapes to be treated quantitatively and hence subject to objective comparisons. The basic principle was first laid out nearly a century ago by the illustrious British polymath, Sir D'Arcy Thompson, whose areas of expertise included linguistics, mathematics, physics, and biology. In his classic text on how organisms acquire their characteristic shapes, *On Growth and Form,* Thompson depicted how shape change could be approached quantitatively. In his illustrations, Thompson first placed a typical fish drawn on a conventional Cartesian grid of two axes and then showed a version of the same fish in which part of that underlying grid was deformed by being stretched in one or two new directions. The result is the characteristic shape of another kind of fish (Figure 9.2). Thus, related shapes can be seen as spatially transformed co-

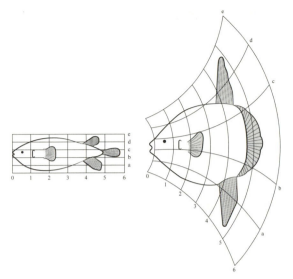

FIGURE 9.2 Changing the shape of fishes by transformation of Cartesian coordinates, à la D'Arcy Thompson. His insight and graphic depictions of shape change were an essential early step in quantitating such changes in organisms, the discipline now known as *geometric morphometrics*, which is an essential tool in measuring facial features as affected by different genes. (Adapted from Thompson 1917.)

ordinate grid systems of one another in which the same characteristic points of their anatomy are placed at relatively different distances from one another. Thompson, however, did not convert this insight into a method for actual measurement of shape differences in animals. That development had to wait another seven decades until the development of geometric morphometrics in the late 1980s. Nevertheless, Thompson's basic insight was crucial—that qualitatively different appearing shapes can be studied in terms of quantitative difference.[9]

To perform a quantitative geometric morphometric representation of facial differences, four things are required: (1) a readily identifiable set of physical points—*landmarks*—on the faces to be measured; (2) a way to standardize the images of the faces to be compared so that all are essentially the same size or fill the equivalent space; (3) accurate measurement of distances between landmarks to within about half a millimeter; and (4) proper statistical analysis.

Figure 9.3 depicts an idealized human face with a set of commonly employed facial landmarks, each a readily identifiable point on the face. With a set of images of different faces and the same landmarks indicated

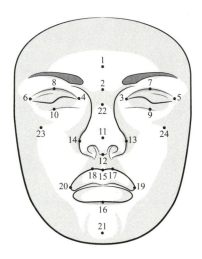

no.	abbr.	name
1	g	glabella
2	n	nasion
3	enl	left endocanthion
4	enR	right endocanthion
5	exL	left exocanthion
6	exR	right exocanthion
7	psL	left palpebrale superius
8	psR	right palpebrale superius
9	piL	left palpebrale inferius
10	PiR	right palpebrale inferius
11	pm	pronasale
12	sn	subnasale
13	alL	left alare
14	alR	right alare
15	ls	labiale superius
16	li	labiale inferius
17	cphL	left crista philtri
18	cphR	right crista philtri
19	chL	left cheilion
20	chR	right cheilion
21	pg	pogonion
22	men	mid-endocanthion point
23	zygR	right zygoma
24	zygL	left zygoma

FIGURE 9.3 Some facial landmarks used in geometric morphometrics of the face to assess distinct genetically specified facial traits. The precise distances between different landmarks can be used to determine whether specific genetic changes can alter human facial traits. (Redrawn and modified from Paternoster et al. 2012.)

precisely, and with the images standardized to the same frame and thus registered spatially with respect to one another, distances between the landmarks, especially those that are relatively close, are measured. (The curvature of the face does not become a serious problem for relatively short distances, while some of the more recent three-dimensional imaging techniques yield still more accurate measurements, even for close landmarks.) From the fact of visible difference in facial morphology between individuals, it must be that many of those interlandmark differences will differ between individuals, apart from identical twins, many of whose slight differences in facial features are below the level of detection of these methods. Different individuals, of course, differ in head and face size, but those differences will drop out in the standardized images; measured differences in distance between landmarks on different faces, whose images have been so standardized, thus should be able to detect significant and real underlying differences.[10]

Although the recently developed techniques are the most accurate, more than seventy years of analysis of craniofacial differences has amply confirmed the possibility of detecting and measuring such differences. Furthermore, where degrees of relationship between individuals are taken into account, statistical analysis of the measurements shows a strong hereditary (genetic) basis for those differences as would be expected from the similarity of the faces of identical twins.[11]

In terms of developing a reliable genetics of facial difference, the crucial need is establishing whether such discriminable differences in interlandmark distances can be associated with differences in specific genes. Thus, for example, if a characteristic and heritable difference in an interlandmark distance was found within a family, one whose inheritance pattern matched that of a single gene allelic difference, how could the gene involved be identified? With between 21,000 and 23,000 traditional (protein-coding) genes and (perhaps) thousands or even tens of thousands of nontraditional (that is, noncoding RNA product-specifying) genes in the human genome, that might seem an impossible task. And so it was until the development of a highly sophisticated technique for identifying genetic differences throughout the genome. This method is known as **genome-wide association study** (GWAS) and is the second technique that makes possible the genetic analysis of facial difference. It is based on the ability to detect **single-nucleotide polymorphisms** (SNPs) within DNA sequences throughout the genome. Each SNP constitutes a single base-pair difference at its site relative to the typical DNA sequence found in the population. Those SNPs are a major source of the genetic differences that physically and visually distinguish human beings from one another.[12]

Thus, for any gene whose variants contribute to differences in human faces, one or more SNPs would be expected to be the source of those allelic differences in most cases. If we can match a particular SNP that is near or in a particular gene to a phenotypic facial trait difference (as detected by geometric morphometrics), then one facial diversity gene has been provisionally identified. With more than 2 million SNPs identified in the human genome—about one for every 1,500 base pairs on average—the challenge is not small. Yet GWAS makes it possible to connect specific genetic differences with high probability to particular genetically specified facial features. Its limitation is that it does not in itself prove that a particular SNP statistically tightly associated with a gene is actually in that gene; that demonstration requires detailed sequencing of the DNA surrounding the SNP, and only when the SNP is shown to be within the gene or nearby regulatory sequence is that identity established. Even without such proofs, however, a highly significant association between an SNP and a genetically specified trait indicates some functional connection between the two. SNPs found to be associated by GWAS with a particular phenotypic difference often turn out to be in noncoding genetic regulatory regions.

To date, four studies of this kind have produced informative results. In the first, the major result was the identification of *PAX3* as a gene, some of whose variants could affect nose shape. This gene, which is downstream from the sonic hedgehog (SHH) pathway, is expressed in the facial prominences and encodes a transcription factor belonging to the *paired box gene* family (named for a gene in the fruit fly whose loss-of-function mutants are affected in adjacent pairs of body segments). It had been previously shown in developing mouse embryos to be expressed first in the neural tube, then in neural crest cells, and finally in the facial primordia of those embryos. *Pax3* had also been shown to affect facial morphology in one strain of mutant mice. Furthermore, in humans, certain *PAX3* mutants had been implicated in some cases of cleft palate. In normal development (in both mice and humans), its expression is turned off at a certain specific point in early face development, but if that shutoff does not take place, the result is a cleft palate. The gene's role as a contributor to normal facial morphology was indicated from a set of detailed measurements of facial images of 2,185 white British fifteen-year-olds. The results identified two *PAX3*-associated SNPs that were associated with, respectively, slight increases of nose height (as measured by the distance between landmarks 2 and 12—see Figure 9.3) and prominence (the distances between landmarks 12 and 11).

The second study looked specifically for SNPs in seemingly unaffected relatives of individuals who had suffered cleft palates or cleft lips as specific defects (as distinct from broader syndromic effects in which multiple structures and traits are affected). The reason for concentrating on such relatives is straightforward: whether or not a specific genetic defect is expressed to give a phenotypic defect and how severe that affect is depend on its interactions with other genes and their alleles, the *genetic background* for that allele. Hence, in nonaffected relatives, the disease allele might be present but its effect diminished by the actions of other genes in the genetic background and yet still potent enough to give some (often slight) effect on facial shape. Because cleft palate can be produced by mutant alleles in many different genes, in principle the approach can identify a number of those genes. In this study, alleles of two genes were provisionally identified as affecting facial morphology. One was a gene in the small *Gremlin* gene family, *GREM1,* which was found to be associated with an SNP that affects nose width. These genes, it will be recalled, inhibit and thus modulate the activity of bone morphogenetic proteins, a few of which are key molecular players in the genesis of the face. The second allele found in this study is in a gene named *CCDC26,* which was previously identified from earlier discovered stronger mutants as a cancer risk factor for glial cells in the brain. This allele was found to be associated with increased facial width—namely, the interzygomatic distance—the distance between the central points of the cheekbones (that between landmarks 23 and 24; see Figure 9.3).

The third and largest study to date involved eight distinct cohorts of people from five different European-derived populations embracing a total of nearly 10,000 people, whose facial images were measured and their DNA sequences analyzed by GWAS. The study identified five different candidate genes; all showed some effects on elements of nose shape or size. One of the genes found was *PAX3,* thus confirming it as a gene affecting facial morphology. In addition, alleles of two of the other genes previously implicated in building the face were found: *PRDM16* had previously been shown to be expressed in the developing palate, while mutations of the other, *TP63,* had been implicated in certain orofacial dysmorphic syndromes. The other two genes, involved in basic cellular functions, had not been previously identified as playing any roles in setting craniofacial morphology, and their activities need further attention.[13]

The fourth and most recent study on genes affecting facial diversity involving this approach and applied to over 6,000 Latin Americans may be the most informative to date. It also implicated five genes, but a different set, as having effects on facial diversity. Four showed variants that affected aspects of nose morphology; these genes are *DCHS2, RUNX2, GLI3,* and

PAX1. The fifth, *EDAR*, was associated with differences in chin protrusion. All these genes are expressed in neural crest or neural crest cell derivatives in the developing face, as judged from mouse studies. Of special note is *GLI3*, which is the major downstream gene of the SHH pathway, encoding a transcription factor. These are all genes, however, whose products play major roles in the development of the face, which accords with the earlier suggestion that such genes might be particularly important as sources of facial diversity in humans. Furthermore, three of these genes—*GLI3*, *RUNX2*, and *DCHS2*—have been found in Neanderthal and Denisovian genomes to have variants characteristic of those genomes and exhibiting molecular signs of selection in those groups. The implication is that these three genes might have been important in facial diversity in those hominins, as well as in our species, and that there might actually be selection for human facial difference, a point to which we will return shortly.[14]

These studies fully validate the use of combining geometric morphometrics and GWAS to identify genes involved in facial diversity. Several of the genes picked out by the analyses had been previously identified as either being expressed in the developing facial primordia or, when severely defective, creating visible craniofacial defects. That last connection supports the idea, mentioned previously, that the Online Mendelian Inheritance in Man data base may provide a guide to finding *some* of the genes involved in the generation of differences in normal facial morphology. Just as significantly, the results also confirm that some of the major genes involved in building the vertebrate face are also sites of genetic variation that contribute to normal human facial diversity. Including the cichlid jaw results discussed in chapter 4, this includes genes of the SHH pathway, which plays such a key role in facial development. Of course, there may well be many other genes involved. The technique has succeeded so far in identifying numerous other SNPs that are associated with high probability with aspects of facial difference, though these have not yet been affiliated with particular genes. The first of the studies discussed above identified a further fifty-three such SNPs. A large fraction of these may be in *cis*-regulatory sequences that act at large distances to regulate particular genes, but connecting those SNPs with the specific genes whose regulation they affect remains a large challenge. Some may be in nontraditional genes—that is, genes for long, noncoding RNAs. Ultimately, if their actions can be precisely identified, they should make a real contribution to the study of the genetic diversity of faces.

Before we leave the subject of genes and faces, however, two further aspects merit attention. The first concerns *physiognomic genetics*, and the question of whether genetically based facial differences are related to personality differences. The second concerns the evolutionary significance of facial difference. We consider these two issues next.

Does the Face Predict the Personality?

At several points, we have seen that the relationships between the brain and face are intimate. One set of these links involves early development and how the early developing forebrain induces development of the facial primordia. The initial clues to this connection were obtained in the late 1950s and early 1960s from the study of newborn babies with abnormal faces. In one landmark study, the facial dysmorphologies were deduced to result from defects in brain development, the condition being named *holoprosencephaly*. This work gave rise to the adage in clinical medicine, "The face predicts the brain."[15]

Today another question has arisen, thanks in part to the new work on facial genetics: "does the face predict the personality?" That question, of course, is an old one, being at the heart of the ancient "science" of physiognomy, whose practitioners answered it strongly in the affirmative after having set out to do so. In the long and checkered history of physiognomy, however, there were always two different strands of thought, though they were not always clearly distinguished. The first emphasized facial expressions as a key to understanding personality, and the second emphasized physical features of the face as personality predictors. It is hard to disagree with the first: in general, individuals who smile a lot are happier than those who go around with perpetual frowns. This is simply a commonplace observation. Furthermore, as noted previously, characteristic expressions tend to become imprinted on the flesh of the face as individuals age, hence the personality traits that correspond to those expressions can become reflected in the face.

The second and far more contentious proposition is that there are intrinsic, distinct, and readable connections between fundamental facial structural features—such as nose shape, mouth width, and forehead height—and particular character traits. This belief was the core proposition of several distinct schools of physiognomy, a belief that lives on in so-called folk wisdom. For example, fleshy round faces are often take as a sign of a warm and friendly personality while narrower faces often suggest a "lean and hungry" look that denotes a lack of trustworthiness.[16]

Such beliefs are not the sole provenance of conventional thinking or ordinary people. Francis Galton (1822–1911), a half cousin of Charles Darwin and a distinguished scientist of his time and a truly eminent Victorian, believed that there was a distinct criminal face. He did something, however, that no physiognomist had done before: he tested the idea. He did so by superimposing pictures of known criminals on one another to make composite face photographs. If there were a fixed set of presumptive criminal features, then they should have been apparent in the composite portrait, revealing the characteristic face of a born criminal. Instead, to Galton's surprise, the composites looked like rather average faces. To his credit, he admitted that his prediction had failed. The result clearly showed that there was no one facial type associated with criminals. Though undoubtedly discouraging to his fellow physiognomists, Galton's result was not the end of the matter, however. For example, in his *Our Face from Fish to Man* in 1929, William Gregory devoted much of the last chapter to discussing facial types as indicators of personality. The chapter included a delightful tongue-in-cheek account of how he traced his own various head and facial characteristics to his various animal ancestors. Even though the account was done with a light touch, the reader senses that Gregory was half-serious about the basic proposition.[17]

In our time, the question is being aired again: are certain facial features connected to certain personality traits? Its revival is probably due to a combination of the growing general interest in faces and facial expressivity and in the new ascendancy of genetics in biology. Indeed, the new quest has a genetic dimension that is far beyond anything that Galton, a student of heredity, or Gregory, who was fully aware of genetics, could bring to the matter. Given what we know about the high degree of genetic specification of faces and the vastly greater knowledge of genes and their roles of development, the modern version of the question is: are gene variants that help shape the face also involved in the development of personality traits and, if so, can they be detected? If that should prove to be the case, then indeed the face might predict the personality, at least to a degree.

To contemporary scientific ears, the supposition sounds fantastic; indeed, the first impulse is to dismiss it because there are too many exceptions to all of the conventional beliefs about such correlations. Most people, for example, have met overweight individuals who were mean (not jolly), thin-faced and lean individuals who were kind (not mean), or redheads who do not have choleric tempers. Hence, any such correlations must be about statistical likelihoods at best, hence *propensities,* not hard and fast genetic determinations of personality traits. As such, if taken as nothing more than a set of predicted and probably weak correlations, it is neither completely ridiculous nor harmful when the high context dependence of such associations is remembered.

There is, in fact, a genetic reason that lends the idea of facial feature character-trait association a wisp of possibility. That fact is the phenomenon of *pleiotropy,* or multiple usage of the same gene for different processes in development. We have seen several important examples of this when comparing face and limb development that involve SHH, the fibroblast growth factors, the Wnt's, and more. In fact, such multiple use for slightly different purposes in development is the rule for genes, not the exception. Its possible extension to linking faces and personality traits is the idea that specific alleles of particular genes that are expressed in both facial development and neuronal development of the brain might affect those processes in correlated ways. Thus, any tweaking of neuronal circuits by an allele of a gene that was expressed in areas of the brain that affect emotional responses, such as the amygdala or particular association centers of the cerebral cortex, might reveal itself by a correlated effect in facial development, showing up in a facial morphological feature. The reasons why any such correlations are bound to be somewhat weak are the two that come into play for all genetic effects: effects of environment, which includes experience in influencing brain development, and of genetic background, since the latter so often modifies the effect of a variant allele.

That, of course, is just an a priori argument, and it is reasonable to ask whether there are known alleles of any specific genes that give closely correlated alterations of aspects of both the personality and the face. Currently, there are none. Two major genetic alterations, however, are associated with characteristic and readily recognizable facial and personality types: Down syndrome and Williams syndrome. Individuals with Down syndrome possess an extra chromosome copy of chromosome 21 (hence three instead of the normal two). In contrast, Williams syndrome is caused by a smaller but still nontrivial genetic alteration—in this case, the loss of a segment on human chromosome 7—of about 2 million base pairs containing some twenty-five genes, which shows up as a haplo-insufficient effect (that is, it occurs when paired with a normal chromosome 7). Both conditions are associated with distinctive and different cognitive impairments and equally characteristic facial and personality traits, which are diagnostic of the two conditions.[18]

One basic caveat, however, must be considered in any attempt to infer the normal functions of the normal (wild-type) form of a gene from the phenotype of the mutant condition—especially if this involves loss of function. The relationship is highly uncertain and often leads to spurious conclusions. Such uncertainty is especially strong when, as in both Down and Williams syndromes, multiple genes and altered dosages of those genes are involved. (Since Down syndrome involves an extra chromosome, it might seem odd to classify it as a loss-of-function disorder, but it seems probable that the extra dosage of genes on chromosome 21 leads to the repression of some normal functions and hence their reduced activity.) Nevertheless, as we saw with the cleft palate and cleft lip syndrome, a more extreme condition associated with severe loss-of-function mutations can signal the possibility that less extreme deficiencies for certain gene activities can affect the same processes, albeit in less dramatic ways.

As yet, there is no comparable information on correlations between particular facial trait differences, particular genetic differences, and specific personality traits. Several research groups are interested in the question, however, and have produced preliminary studies that take the first step—trying to correlate particular facial features with particular personality traits. The particular trait that has been looked at is the facial width to height ratio (fWHR). In terms of the landmarks shown in Figure 9.3, this is the ratio of the interzygomatic distance, the distance between landmarks 23 and 24 to the distance between landmarks 1 and 15. Briefly, men tend to have higher fWHR values; to that extent, the value can be taken as one measure of masculinity, although it must be stressed that this is an average difference, with much overlap between men and women in the measured values.

Earlier work attempted to correlate the fWHR values with testosterone levels in the blood in accord with the supposition that they are a surrogate value for maleness, and it found some degree of correlation. More recent studies, however, take the matter into the realm of physiognomy. Several have now reported correlations, often just within standard measures of statistical significance, between higher fWHR values on the one hand and increased aggressiveness, decreased trustworthiness, a tendency to express overt prejudice, and a propensity for engaging in deceptive behavior on the one other hand. Some women might be tempted to exclaim, "I told you so!" but these results should be treated cautiously. The correlations are partial and weak and hence allow for many exceptions in the claimed connections, and all of the reported results require confirmation. Furthermore, none of the personality traits has been linked to a particular gene or genes. Yet if reliable measures of the psychological traits can be measured, their connections to possible candidate genes should be measurable. As we have seen, facial width, the interzygomatic distance, has already been connected to one gene: *CCDC26*.

With time and the accumulation of both genetic data and facial morphometric data, we can expect more "physiognomic genetics." It will probably never be a major field of investigation and does not deserve to be so because the connections will always be statistical and provisional. Yet it might prove to be part of the future scientific analysis of what makes faces—and individual personalities—different.

The Evolutionary Significance of Human Facial Differences: A New Study

As briefly discussed at several earlier points, one intriguing feature of the human face is its range of variability and the extent of human facial differences. The fact that such facial diversity is a trait shared specifically with our nearest anthropoid primate cousins, the great apes, suggests that it has some function and hence some evolutionary significance. The obvious function of such differences in a species that has excellent close-up and intermediate-distance vision is that it allows ready, often instantaneous identification of individuals. Could selection for rapid individual identification have been the basis for facial diversity in the hominoid primates and in humans specifically? The seeming difficulty in that hypothesis is that selection for most traits leads to a narrowing of values clustered around an optimal value. How could there be selection for differences?

In fact, one mode of selection promotes differences. It bears, indeed suffers from, a rather awkward name: **negative frequency-dependent selection** (NFDS). In plain English, it denotes situations in which there is selection for rare variants, where the property of unusualness gives the individuals showing those unusual variations some advantage simply by virtue of being different. Examples are known that involve such properties as reduced susceptibility to predation or increased appeal to members of the opposite sex. (We have already noted the appeal of exaggerated characteristics in sexual selection; NFDS for such novelty would be part of the process.) Thus, NFDS automatically promotes the perpetuation of differences in populations. Until recently, however, there was no evidence that NFDS operated with respect to human facial differences.

A study published in late 2014 now provides some support for this idea. The authors analyzed a data set obtained by the U.S. military and measured the extent of human facial differences relative to comparable intertrait differences for other body traits such as hands. The analysis showed that human faces are more variable than the body traits, thus confirming the subjective general impression that faces are indeed more individuating than other body regions or traits. The authors then studied the amount of genetic diversity in relatively narrow DNA windows surrounding SNPs associated with facial diversity, comparing them with SNPs associated with height and others associated with neutral DNA regions. Altogether, fifty-nine face-associated SNPs from the previously mentioned studies were compared with 365 height-associated SNPs and 5,000 putatively neutral SNPs. If NFDS holds for face-associated genes and not for height genes or neutral DNA sequences, then their neighboring sequences should have more DNA-sequence diversity associated with them. By several measures, this was confirmed: DNA sequences associated with faces are islands of relatively elevated genetic diversity.[19]

The work supports the intuition that facial diversity is not a neutral trait—an accidental feature of humans without significance—but that it has been selected, presumably to facilitate ready identification of individuals in incipient social interactions. In principle, such ready differentiation would help in many kinds of social interactions, but these studies have not illuminated which ones might be especially important. The results, however, lend further support to the main thesis of this book that the human face has been particularly shaped for aiding social interactions. One further result from this study provides a further evolutionary dimension worth mentioning. When the neighboring DNA sequences of two face-associated genes were examined in a Neanderthal and a Denisovian genome, those sequences were found to possess subsets of the modern human DNA sequence variants and did not match the equivalent chimpanzee sequences. These results, though limited to two genes, are suggestive that some human facial diversity genetic differences go far back in the hominin lineage but not into that of the great apes lineage. Thus, while selection for facial diversity is probably an ancient anthropoid primate phenomenon, the particular variants are not themselves ancient but arose separately in different lineages.

Two Modest Conclusions about the Likely Future of the Face

While the preceding chapters focused on the past history of the human face and how it came to be the way it is, in this chapter we have ventured a tentative look toward its future in both its physical attributes and its exploration by science. As mentioned earlier in the chapter, whether the human face has a (long) future depends, of course, on whether we as a species do. Despite all the problems and dangers posed by the twenty-first century, most of them of our own species' making, I think it likely that we do. The reason for modest optimism is that, unlike all previous species that have dropped into the abyss of extinction, we have some sense of the kinds of things that might lead us there and can take the corresponding actions to avoid that fate. It is, of course, precisely our collective intelligence that has gotten us into the multiple fixes that we and Earth's other complex life forms are now experiencing. Yet that same collective intelligence has the potential to help us muster the needed wisdom to avoid self-termination. In the worst contingency, there should at least be lucky survivors for virtually any imaginable catastrophe; they probably would use their intelligence to restart the journey of the human race.

Assuming that humanity has a future, there are two reasonable projections for the future of the human face. First, human faces in the future will increasingly be more homogenized—in effect, globalized. Just as humans spread out of Africa 60,000 to 50,000 years ago and diverged into physically recognizable types, as a species we are in the process of merging the different groups and lessening, though not eliminating, those differences earlier associated with geographic provenance. At the same time, we are adding and mixing in new allelic differences that were that not there when the out-of-Africa migration took place 72,000–60,000 years ago.

Second, if the genetics of the face continue to develop—which will depend on the interests of investigators, the vagaries of fashion in science, and appropriate levels of funding—it is bound to be informative; this is a reasonably safe prediction. Given the importance and intrinsic interest of the face, its genetic foundations are intriguing. In particular, we should be able to obtain estimates of the number of genes and the number of their alleles that affect the formation of the face. More distantly, the genetic foundations of the connections between human brains and human faces will probably be further illuminated; as a consequence, so will the connections between faces and personalities. If so, that understanding will be far more complicated and provisional than anything dreamed of by the Greek physiognomists although sounder; as such, it will give us further insight into what we are as a species and as individuals. Those insights, in turn, will surely deepen our sense of wonder at the extraordinary facts and basis of human existence. The recent work on the probable evolutionary significance of facial differences is an example of the kind of insights that might be expected.

One final point about the future of the face deserves a brief mention, although the matter is cultural, not biological. Technologies are now being developed that allow highly accurate identification of individuals from their faces. These technologies are in increasing demand by governments and businesses. It is not beyond the realm of possibility that within a few decades the movements of most people on the planet will be readily traceable by these systems, which are made possible by the individuality of the human face. It would be ironic if the face, our badge of individuality, were to come to be used in surveillance systems designed to reduce expressions of that individuality, no matter how the purpose is rationalized. Perhaps it is worth thinking about this particular aspect of the future of the face before it arrives.[20]

PLATE 1 An Old World monkey, the vervet, *Chlorocebus pygerythrus*. (Photograph courtesy of Dieter and Netzin Steklis.)

PLATE 2 A "lesser ape," the eastern hoolock gibbon, *Hoolock leuconedys*. (Photograph courtesy of Thomas Geissmann.)

PLATE 3 A male Bornean orangutan, *Pongo pygmaeus wurmbii*. (Photograph by Anna Marzec.)

PLATE 4 A mountain gorilla, *Gorilla beringei beringei*. (Photograph courtesy of Dieter and Netzin Steklis.)

PLATE 5 Two cichlids with differing jaw morphologies. **A:** *Labeotropheus trewavasae.* (Photograph by Justin Marshall.) **B:** *Metriaclima mbenjii.* (Photograph by Reed Roberts.)

PLATE 6 The reconstructed face of an early synapsid from the early Permian, the pelycosaur *Dimetrodon*. (Illustration courtesy of Dmitry Bogdanov.)

PLATE 7 Picture of the reconstructed face of a Triassic cynodont, *Cynognathus*. (Illustration courtesy of Dmitry Bogdanov.)

PLATE 8 Reconstructed putative ancestral placental mammal. (Picture by Carl Buell; from O'Leary et al., 2013.)

PLATE 9 A picture of the ancestral primate as derived from reconstruction of fossil evidence and comparative studies of living primates. (Courtesy of Dr Robert Martin.)

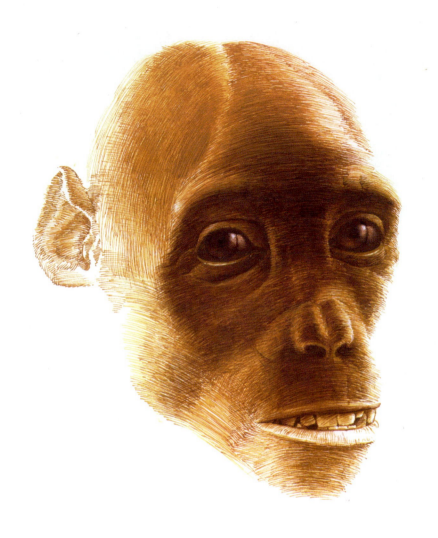

PLATE 10 The face of *Proconsul* as reconstructed from fossil evidence. (Courtesy of John A. Gurche.)

PLATE 11 The face of *Australopithecus africanus* as reconstructed by fossil evidence. (Courtesy of John A. Gurche.)

PLATE 12 The face of *Homo habilis* as reconstructed from fossil evidence. (Courtesy of John A. Gurche.)

PLATE 13 The face of *Homo erectus* as reconstructed from fossil skulls. (Courtesy of John A. Gurche.)

PLATE 14 The reconstructed face of a Neanderthal, *Homo neanderthalis*. (Courtesy of John A. Gurche.)

PLATE 15 Chacma baboons, mother and child, *Papio ursinus*. (Photograph by Anne Engh.)

PLATE 16 An aggressive, nondomesticated, farm-bred silver fox. (Photograph courtesy of Lyudmila Trut.)

PLATE 17 A domesticated silver fox. (Photograph courtesy of Lyudmila Trut.)

PLATE 18 Multiple images of horses from the famous horse panel in Chauvet Cave. (Photograph courtesy of Jean Clottes.)

PLATE 19 Cave painting of a hunting group of lions from Chauvet Cave. (Photograph courtesy of Jean Clottes.)

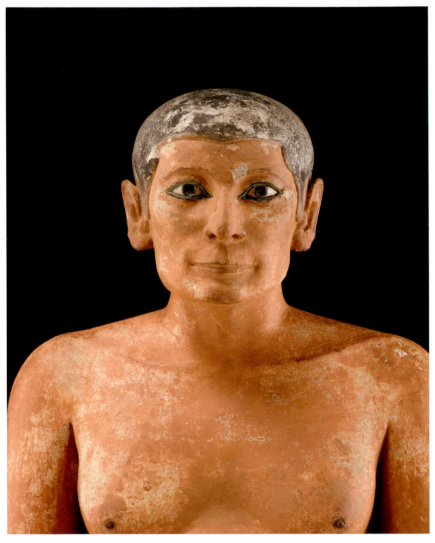

PLATE 20 Photograph of upper portion of the statue of a seated scribe dating from the Old Kingdom of Egypt, circa 2,500 years ago, in the collection of the Musée du Louvre, Paris. (© bpk / RMN-Grand Palais Musée du Louvre, Paris / Franck Raux.)

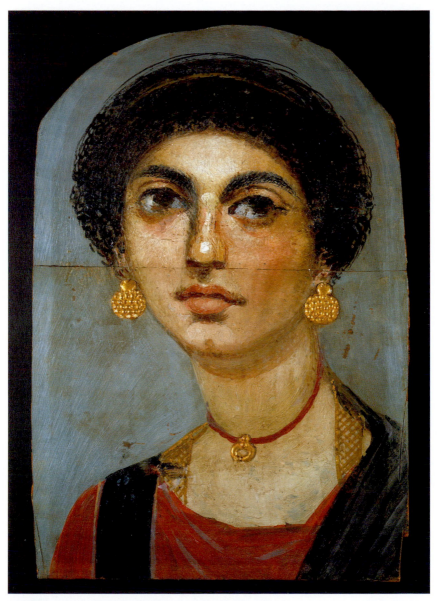

PLATE 21 Fayoum mummy portrait of a young woman, early Christian era. (© bpk/ Aegyptisches Museum und Papyrussammlung, SMB.)

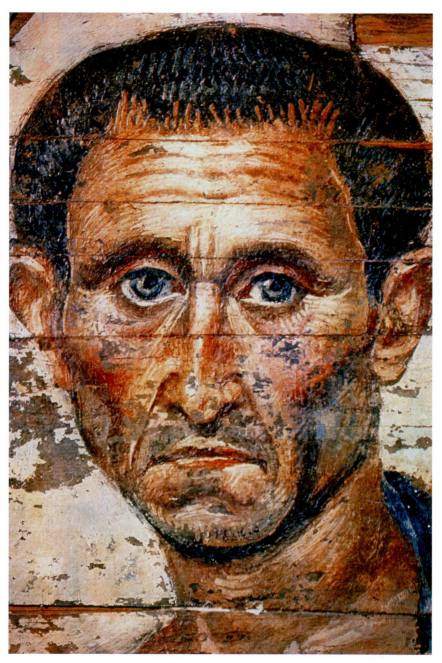

PLATE 22 Fayoum mummy portrait of a man, early Christian era. (© bpk / Alinari Archives Staatliches Puschkin-Museum der Bildenden Künste/Alinari.)

SOCIAL SELECTION IN THE
SHAPING OF THE HUMAN FACE

Introduction: On the Evolutionary Genesis of
the Human Face, an "Organ of Perfection"

The first book devoted to the evolution of the face was William K. Gregory's *Our Face from Fish to Man,* published in 1929. This book may be seen as its tardy sequel. There are, of course, many differences between them, not only in style and emphasis but also in detailed content, the latter reflecting more than eight decades of research and discovery in the biological sciences. One result of that huge expansion of information is to make possible the exploration of many areas and ideas that were unavailable to Gregory. A second but regrettable consequence is that the reader may feel that the end result is a rather sprawling treatment that extends well beyond the subject's anticipated boundaries. In this final chapter, therefore, I will attempt to bring things together and emphasize and elaborate on the central conclusion.

In considering the evolution of the human face, where should we start? I posed this question at the beginning of the book, and there I argued that it should begin with the first vertebrates and their faces. For this summing up, however, I will modify that conclusion: I think the best place is slightly *before* the beginnings of the first animal faces. At that time, more than half a billion years ago, there existed animals with mouths—that is, apertures for the ingestion of food. A mouth, however, does not constitute a face: it takes two eyes above a mouth to have a face, as all children know and as contemporary emoticons show. It was therefore the evolution of paired eyes above and in close proximity to the mouth that created the first animal faces in both the earliest arthropods and the craniates. Indeed, a pair of eyes alone is sufficient to connote a face. For example, when certain butterflies perched on bushes spread their wings to show two wing eyespots, the visual suggestion to birds that prey on those species is that a larger animal is hidden in the foliage; the consequence is that the birds refrain from attack.

Eyes themselves, of course, have long been a source of rich interest for evolutionary biologists. As Darwin famously remarked in a letter to one of his scientific correspondents, the human eye gave him "nightmares." Of course, it was not the eye itself he found terrifying but rather the possibility that the origins of this "organ of perfection" (his phrase) might not be explicable by his theory. Could something as complicated and marvelous—indeed "perfect," as the vertebrate eye, he wondered—have actually come into being via natural selection of small, incremental adaptive changes? He posed this question explicitly in *The Origin of Species*. If the answer should prove to be "no," Darwin's theory of evolution by natural selection would have been in mortal jeopardy. Yet, he argued, after raising the question, that indeed such an origin for the eye was fully conceivable; in a paragraph, he explains how small successive improvements could have generated something as complex as the eye.[1]

If Darwin had chosen the human face as the demanding test case of his theory, would his sleep have been less troubled? I doubt it. The face, after all, contains the eye and a great deal more; in its complexity and function, therefore, it might also be seen as an "organ of perfection." Yet, as I hope this book has shown, we can account for much in the half-billion year history of the face in terms of step-by-step changes, many of which can be explained in terms of some initial selective advantage they conferred, in accord with Darwin's basic theory. The discoveries that have made this expanded view possible have not come primarily from new fossil finds—though these have contributed—but from comparative studies of living animals as made possible by advances in molecular genetics and neurobiology in the late twentieth and early twenty-first centuries.

This concluding chapter will attempt to knit the disparate threads of the story of the evolution of the face into a unified picture. It has two parts. The first is descriptive, summarizing the main events in the evolution of the morphology of the face from the earliest fishes in the Cambrian period to the origin of *Homo sapiens* and the modern human face. It will give the basic outline of the story by focusing on landmark events and stages. Although this short overview will largely omit the genetic details, it is knowledge of those details that has substantially deepened our understanding of the morphological history of the vertebrate face from the first fishes to modern humans.

In contrast, the second and longer section of this chapter is primarily interpretative and focuses on the period of primate evolution in which the distinctive characteristics of the human face arose and developed. Its main emphasis is on the various selective forces that propelled this sequence of events, and its goal is to show how they fit within the larger picture of human evolution as a whole. In particular, I will return to the argument sketched in Chapters 6 through 8 that, with the advent of the anthropoid primates, social and psychological factors increasingly shaped the physical evolution of the face and continued to do so even after our species' appearance approximately 200,000 years ago. Finally, I will describe how the evolution of the face exemplifies some basic if all-too-often neglected general aspects of evolution.

The Evolutionary Origins of the Face:
A Capsule History of the Morphological Events

From Faceless Chordate to the First Faces

In the beginning was the face of a fish—a tiny, jawless fish, an animal probably an inch or less in size. Its existence was made possible by the evolutionary invention of neural crest cells, which produced the first cartilaginous-based or cartilage-like facial structure. The proto-neural crest cells had evolved from nonmotile cells at or near the midline of what was probably an amphioxus-like creature, though it was probably a larval form of a urochordate, not a cephalochordate. Much of the molecular machinery for the neural crest cells would already have been in place, but a relatively small number of genetic changes would have empowered the crest precursor cells to acquire the ability to move via an epithelial-mesenchymal transition to new locations and then to differentiate into new kinds of cells in those sites. The evolution of neural placodes, which permitted the development of paired sensory structures in the head, in particular visual and olfactory ones, accompanied and completed the evolutionary origin of the archetypal vertebrate face. Much of the genetic machinery that creates the properties of neural crest cells and neural placodes is known, and it is possible to begin to see how they evolved from simpler ancestral systems.

Those first steps in craniate evolution were also accompanied by a major change in the anterior neural circuitry, the primitive brain of the ancestral chordate—namely, the addition of a telencephalon-like structure to that brain. It is not clear when that step occurred, but given the dependence of craniate face development on the telencephalon through well-characterized molecular signals, these evolutionary changes were almost certainly closely linked in time.

Those first craniates would have been a comparatively late addition to the world of animals. The earliest animals were probably the stem Porifera and Cnidaria, but they were soon followed by some of the early bilaterian invertebrates, the latter arising somewhere between 610 million and 543 million years ago and before the Cambrian. Animal life, however, only really bloomed with the rapid diversification and multiplication of later bilaterally symmetrical animal forms during the first 10 million to 20 million years of the Cambrian "explosion," which commenced about 543 mya. In this period, stem forms of most of the present-day animal phyla originated, including the earliest chordates, which possessed primitive neural, muscular, and digestive systems as well as mouths and some anterior sensory apparatus but which lacked true heads and faces.

The first craniates, small jawless fish (agnathans), probably appeared between 510 million and 490 million years ago during the final phase of the Cambrian. They would have had two eyes and a mouth and the three sensory capabilities—visual, gustatory, and olfactory—associated with the face that all craniates possess. Yet in the structure of their mouths, which were suited only for suction and filter feeding, these animals were severely limited. Had that construction persisted, even as the agnathans increased in size and diversity, the vertebrates would probably have continued as simply one more phylum of the animal kingdom and been numerically dominated by the main invertebrate classes—namely, the arthropods, mollusks, and nematodes. Had vertebrates not developed beyond the agnathan construction—in retrospect, one might term it the *agnathan constriction*—their most complex members would probably have been comparable in complexity and mode of life to that of present-day large squids and octopuses. The latter are interesting, complex, and intelligent animals, but their body structures and physiology limit them to life in the sea with all the constraints that entails.

The Gnathostome Revolution

The change that threw open the gates to the far more extensive vertebrate history that actually ensued was the evolution of jaws, which took place sometime between 440 million and 410 million years ago during the Ordovician period and probably in just one lineage of the agnathans since the molecular data indicate that the gnathostomes are a monophyletic group derived from a common ancestor. The underlying genetic and developmental changes involved and required in the development of movable jaws were probably comparatively slight—relatively minor changes in the domains of expression of one or a few genes in the developing lower facial primordia and perhaps involving the recruitment of a new gene activity to generate the joint of the first pharyngeal arch. The net effect of these shifts in gene activity was to permit the folding and bending forward of the two resulting halves of the first brachial arch to form the lower (mandibulary) and upper (maxillary) regions, which would become the jaws. Despite the relative modesty of the underlying genetic-molecular changes, the consequences of this transformation to the jawed state were large. They included many new feeding possibilities at first, with the prey consisting of both larger marine invertebrates and fellow fishes. This, in turn, led to new and undoubtedly faster rounds of competition among an ever-diversifying group of fish species, with selection for increased size, stronger jaws, and teeth for grabbing and ingesting prey; and better defenses against predators in the form of hard dermal plates to form a kind of dermal armour; and undoubtedly new behaviors. Many of the predators, of course, were themselves the prey species of other larger predators. Some of these armored gnathostomes, collectively termed the *placoderms,* grew to more than thirty feet in length and thus perhaps fifteen times or more longer than the largest primitive agnathans. Of course, the correlation between having jaws and attaining large sizes is highly suggestive. While some gnathostomes such as whale sharks have re-evolved filter feeding and attained large sizes, the paleontological data support the idea that the initial breakthrough to large sizes in vertebrates was connected to the evolution of jaws and presumably more efficient nutrient intake as a result.

From Fish to Amphibian Faces and on to Synapsid and Mammalian Faces

The evolution of jaws and their increasingly improved function and variety—an accretion of capabilities as a result of new selective opportunities and selective pressures—was thus probably a critical feature that helped make possible the eventual opening up of terrestrial environments to vertebrates. Lungs for air breathing and legs were also essential, of course. Judging from currently known fossils, the transition period from sarcoptyrigian fish, the immediate ancestors of **tetrapods**, the land-living vertebrates, to indisputable amphibians is judged to have taken place within a period of 20 million years, or between 385 million and 365 million years ago in the late Devonian period (410–355 mya).[2]

Considering the enormous changes in the way of life that the first amphibians exhibited and experienced relative to their solely water-dwelling piscine ancestors—in respiratory physiology, food sources, and locomotion—the initial architectural changes in the face were comparatively slight. The basic structural facial features remained essentially the same, though selection-propelled remodeling of the sensory apparatus for olfaction, taste, and vision were required and took place. There were also modifications in the properties of the outer dermis (from fish scales to amphibian-type skin capable of absorbing oxygen) that helped enable the transition to life on land and in the air. In the general structure of their faces, if not in their skin, the first true amphibians were still quite fishlike, although their bodies had been transformed for movement on land and their heads were flatter and redesigned internally for air breathing.

The next big event was the evolution of the amniotic egg and the consequent expansion and diversification of the amniotes, creatures with more leathery skins suited to independent living on the land. Again, however, the initial changes in face structure from precursor amphibians were relatively modest. Several amniotic lineages arose, but the two that thrived and diversified highly were the Diapsida, which gave rise to the dinosaurs in the Triassic, and the Synapsida, which gave rise to the first mammals during the Jurassic. The long sequence of evolutionary transformations of the face that began with the appearance of the first synapsids probably around 325 million years ago and which culminated with the first mammals 130 million to 200 million years later, however, were more dramatic in their outcomes than earlier ones. Initially, most of those changes concerned the internal features of the mouth, with secondary but important consequences for the shape of the face. The crucial initial changes in the synapsids involved the development of different kinds of teeth—a mammalian characteristic. As milk and nursing evolved in the more mammal-like synapsids, these eventually led to the development of two distinct sets of teeth—milk teeth and adult teeth—which appeared sequentially. This form of tooth development supplanted the continual tooth-replacement mode possessed by both the early amniotes and the sauropsids. Another change was the development of a secondary palate that permitted the physical separation of breathing from eating. A third was the simplification of the jawbones in which two of the bones (the malleus and the stapes) that made up part of the jaw joint became separated from the jaw, migrated inward, and became part of the hearing apparatus; one of the lower jaw bones, the dentary, became the principle structure of the lower jaw. The resulting simplification of the jaws provided both increased strength and efficiency such that smaller jaws could accomplish the jobs that previously required larger mouths.

As jaws became both simpler and stronger in the synapsids, the basic mammalian muzzle evolved. Though showing a great range of sizes and shapes, it is a feature of most mammals, including most primates, and is generally smaller as a fraction of the head than the jaws in most reptiles. It also exhibits an important developmental change whose equivalent is not seen in crocodiles, lizards or snakes: for most mammals, the muzzle is relatively smaller at birth—sometimes much reduced—and grows during the juvenile stages to reach adult proportions. (For crocodiles, lizards, and snakes, the proportions of the head remain essentially the same in development.) Shorter mammalian muzzles permit newborns and infant mammals of many species to nurse more efficiently; as nursing declines, the jaws grow and full muzzles develop. The rates at which these events happen are characteristic of the individual species and a function, in part, of the time of transition from nursing to an adult diet.

A further distinguishing facial feature of the mammals is the mimetic muscles, which allow an animal to move its facial skin independently of the underlying bone. That flexibility would have probably been most important for moving the mouth in more subtle ways to permit better handling of food items with the lips and the beginnings of the production of more diverse sounds using the lips, tongue, and velum. (It also permits nursing in a way that rigid lips would not.) Much later in the anthropoid primates, these muscles would become particularly important in the creation of facial expressions, which are used as signaling devices for mood and intention.

Other changes that helped create the generic mammalian face included the development of external ears (pinnae) for better directional sensing of sounds, the elaboration and improvement of the olfactory apparatus and the eyes, and, not least, the acquisition of hairy coats. The latter, of course, included fur-covered faces. By at least 125 million years ago, in the early Cretaceous period, there were indisputable mammals (if not yet modern or crown mammals) exhibiting, as we now know, a large variety of shapes, forms, and species and a correspondingly wide range of habitats and modes of life. Despite that variety, all their faces were distinctly mammalian, unlike their initial Permian ancestors, the first synapsids.

From First Mammals to First Primates

The next evolutionary changes along the long hominocentric lineage path (LHLP) that eventually led to humans—namely, the segment that ran from the first mammals to the first primates—were slight. The dates, places, and specific evolutionary changes that led to the primates are still obscure, an obscurity connected with the events of the end-Cretaceous mass extinction and the immediate aftermath of that global catastrophe. The event itself is generally believed to have been triggered by the impact of a giant meteor, approximately 6 miles in diameter, on what is now the Yucatan Peninsula, which sent huge plumes of dust and smoke into the atmosphere, creating the equivalent of a nuclear winter by blocking sunlight and creating a darkened Earth for months. A preceding or accompanying set of events also appear to have been major volcanic activity in India, the site of the Deccan traps, with further reduction of sunlight around the world. Although the relative contributions of these two sources to the end-Cretaceous mass extinction are debated, the net result was a major crash in the biota of the world, especially much of the complex animal life on land and in the seas.

That mass extinction took with it not only all the nonavian dinosaurs but also much of the plant life, amphibians, sea life, and almost certainly a large part of the diverse forms of mammalian life that had existed in the shadows of the dinosaurs. Since mammals are high in the food chain, they are by definition dependent on many other species lower in that chain. The putative nuclear winter-like conditions of Earth would have extinguished much of the plant life and the animals that were directly dependent on that plant life, thus undermining the livelihoods of dependent mammalian species. Yet the fossil evidence indicates that representatives of many of the major groups of mammals survived the end-Cretaceous extinction and began to diversify and flourish in the early Cenozoic era. In particular, the roots of the four major superorders of placental mammals (Figure 5.8) were probably present at the beginning of the Paleocene epoch.

One of these was the Euarchontoglires, which consisted of the ancestors of primates, flying lemurs, tree shrews, rodents, and rabbits. A critical feature that probably distinguished the early stem primates were their limbs and specifically their feet, whose digits lacked real claws. Instead, they had fingernails and a fifth and opposable digit that was ideal for grasping branches and climbing trees. In facial aspects, the first primates did not look especially different from the other early placentals, but they may have been particularly small. All that would prove most distinctive in primate faces along the LHLP path was yet to come.

From First Primates to the Anthropoid Primates

The specific changes that led eventually to the modern human face appear to have taken place largely in two stages separated by something on the order of 40 million to 50 million years. The first set of changes lay the foundations for the later ones and took place with the advent of the anthropoid primates, perhaps occurring only 5 million to 8 million years after the end of the Cretaceous. The shared facial characteristics of modern anthropoid primates included closely spaced eyes, naked (furless) faces, and reduced muzzles, which presumably date to the stem anthropoids. With those changes alone, a major set of steps toward the basic configuration of the human face had been taken. While the eye and muzzle characteristics were probably selected by a shift away from feeding on insects and other small animals and toward consuming fruit, the loss of fur on the face was probably a sexually selected trait in females initially, which later became expressed in males as well. A crucial point is that all three changes, whatever the initial reasons for their selection, would eventually favor one-to-one social interactions, with furless faces and reduced muzzles improving the visibility of facial expressivity and hence the "reading" of those expressions.

The first anthropoid primates were almost certainly small animals like the earliest primates. These early anthropoids were probably no larger than present-day mouse lemurs in size, weighing a few ounces.

Correspondingly, a major trend within the anthropoid primates involved an increase in size as they expanded and diversified. Those increases took place to various extents in the different lineages, and with some of those lineages there was undoubtedly a reversal of the trend of smaller animals. The general trend toward larger size, however, was particularly evident in the lineages that gave rise to the modern great apes, in the period lasting from 30 million to 20 million years ago during the Miocene epoch. In highly visual animals, an increase in head size would have favored an increase in eye size, which would have then permitted even better acute, close-up vision. That property would have been valuable for picking fruit or finding other food items but later would prove advantageous for inspecting the faces and expressions of conspecifics during social encounters.

The Rise of the Hominins and Eventually *Homo sapiens*

The second key phase in the evolution of the human face took place with the evolution of the hominins beginning about 7 million to 6 million years ago as the hominins branched off from the chimpanzee lineage. This, however, largely involved a continuation of the trends of the earlier changes. These changes included increasing loss of fur to yield the "naked" human body, increased size (from the australopithicines through the various *Homo* lineages toward modern humans), a further reduction in the muzzle, smaller canine teeth, and, in general, diminishing sexual dimorphism between males and females. The distinctive increase in brain size, which would eventually yield the human forehead and rounder head, began with the advent of the genus *Homo* and culminated in the Neanderthals and modern humans, 200,000 to 300,000 years ago. Behind the face, however, major changes in brain circuitry had evolved, allowing greater control of facial expression, presumably improved reading of subtler facial expressions, and, not least, the neural capacity for complex language.

Stepping Back to See the Big Picture

One can think of the whole half-billion-year sequence of events as a series of key innovations along the LHLP that ultimately contributed to making the human face what it is today. These are summarized in Figure 10.1. As I have stressed, this scheme does not imply inevitability in any sense; in-

facial / neural trait	taxonomic group	time/period of origin (probable)
basic facial structure	Craniates (stem)	510-500 mya Cambrian
jaws	Gnathostomes (stem)	420-410 mya Silurian/Devonian
simpler jaws specialized teeth two sets of teeth beginnings of hair and milk production first facial mimetic muscles (?)	Synapsida (stem) ↓ Cynodonta	320-220 mya mid-Carboniferous ↓ Triassic
fur-covered faces mammalian jaws (and ears)	Cynodonta ↓ Mammaliaformes	220-190 mya Triassic ↓ end-Jurassic
beginnings of primate vision (colour, visual acuity?)	Primates (stem)	82-80 mya Paleocene
reduction in muzzle loss of facial fur dentition changes for fruit eating closer set eyes for acute close-up vision full colour vision more mimetic muscles finer neural control of mimetic muscles (?) early cortical control of vocalizations new neural linkages connecting facial expressions and vocalizations	Anthropoid primates ↓ great apes	56-25 mya Eocene ↓ Miocene
loss of muzzle evolution of chin further modifications of teeth enlargement of cerebral cortex reduction of iris relative to sclera in visible part of the eye evolution of true language and requisite neural circuitry neural linkages coordinating facial expression and nascent language ability reduced sexual dimorphism in over-all body size and teeth loss of most body hair (late, post speciation) juvenilization of facial features	Hominins ↓ *Homo sapiens*	7-6 mya → 200 kya Pliocene ↓ Late Pleistocene
genetic differentiation of populations toward main racial groups with differences of facial features and skin pigmentation	post-speciation changes	

FIGURE 10.1 Capsule summary of the sequence of evolution of critical facial features along the LHLP in the eventual making of the human face as reviewed in this chapter.

numerable other primate, mammalian, and vertebrate lineages were unfolding with different changes quite happily and independently. It was simply one of many paths in vertebrate evolution that can be traced backward but can be viewed as a forward progression. If we could reconstruct the complete history in terms of the actual sequence of animals from earliest craniate to first humans, we would probably find that the sequence involved many hundreds of distinct species—taking the term *species* to denote an animal population whose members mate predominantly among themselves. Even without detailed knowledge of many of the events, one can be certain that there were countless factors that influenced the evolution of this lineage, and of the constituent faces of its component species, over the 500 million years or so in which it unfolded.

It is therefore reasonable to ask whether there were any salient unifying features or themes in this long and complex sequence. It could be regarded simply as a lengthy story of happenstance, a picaresque biological adventure story with one unrelated and unpredictable change following another, with neither theme nor coherence, which simply happened to result in human beings and the human face. That view, however, feels unsatisfactory and inadequate. Cannot something more general and substantial be extracted from it? I think the answer is "yes," although, if so, it divides the whole history into two major but unequal segments, one of more than 400 million years, the other of about 50 million years.

From Food to Social Interactions: The Story of a Major Shift in the Selective Factors That Shaped the Face

I propose that there was a major shift in the selective forces shaping the face that began with the anthropoid primates. For the greater part of vertebrate history, starting with the first jawed fishes and running into the beginnings of the primates—a span of about 350 million years—most of the changes in the face (in all land vertebrate lineages) were driven by matters of food, both in its pursuit and its initial processing in the mouth. The kinds of food that an animal needs ultimately determine the kind of jaws and teeth it has; in turn, jaw shape and tooth patterns set much of the architecture of the lower part of the face. For four-footed mammals (as well as most other vertebrates), the jaws serve to seize the food; for predators, it is particular important that the jaws project forward in order to do so. Of course, the sensory apparatus for detecting dietary items, in particular the organs for vision and smell, are also critical for sensing the location of food items; the architecture of these organs, especially their placement, shape, and size, help determine the visible architecture of the face. Nevertheless, it was jaw shape and architecture—adapted to specific types of food item—that were the preeminent determinants of overall face shape in the vertebrates from the earliest gnathostomes to the appearance of the primates.

With the advent and rise of the anthropoid primates, however, a new selective factor in shaping the face gradually appeared: social interactions. Although each facial change was probably first selected in connection with some property other than sociality, all of them would have facilitated social interactions. If they promoted social cohesion, hence survival of the group, which in turn would have promoted survival of members of the group, they would have been selected for that property, thus contributing to the shaping of the physical features of the face.

One crucial alteration was the reduction in the projection of the jaws, beginning with the stem anthropoids. This was made possible by an increasing use of the forepaws to handle food, a general primate feature that became even more pronounced in the anthropoid primates. With increasing reliance on the forepaws for obtaining food, there would have been less need to use the jaws as the primary instrument for seizing food items, in particular insects, which were probably the main dietary item in the earliest primates. There is only one group of anthropoid primates, the baboons, that have prominent muzzles, but the baboon muzzle was almost certainly a reacquired trait. Thus, anthropoid primates in general have reduced jaws and correspondingly flatter faces. It also seems probable that a diet consisting largely of fruit would be served better by such a mouth, with the highly mobile lips of the anthropoid primates useful in manipulating fruit items as they are ingested.

This shift toward faces with reduced muzzles, however, also would have had social consequences: with the muzzle no longer dominating the front view of faces, they would have had the potential to be more expressive. Mouths, one of the two primary loci of expressiveness, would have had more capacity to show expressions via modulation of lip movements. The two other anthropoid facial innovations would have also facilitated the reading of those expressions. Acute close-up stereoscopic vision of anthropoid primates would have been well placed to register those changes of expression around mouths which would have been easier to see than in mammals whose eyes lie behind protruding jaws. Furless faces would also have markedly helped: the naked skin of the anthropoid primate face was probably initially a sexually selected trait but would have permitted conspecifics at close range to more readily perceive expressions of others at close range.

Greater expressiveness, in turn, would have facilitated more one-to-one social interactions with fellow members of the troupe. To the extent that these social interactions fostered internal cohesion of the group, they would have been an adaptive response that tended to promote greater survival of members of the group. Furthermore, a tighter web of social relationships would have created further selection pressures for both its maintenance and growth. Sociality begets sociality. And this self-reinforcing process would probably have increased with the emergence of the cattharine primates as judged from the generally higher degree of sociality and expression- making in the catarrhines relative to both the prosimians and the platyrrhines.

Of course, there is no way to directly validate this scenario, but it seems probable from what we know of living primates. Furthermore, even if we provisionally accept it, there is no way to reconstruct its applicability specifically to the early and now extinct hominins, but it seems likely that similar selective pressures would also have applied to those early hominins. These social properties and dynamics, however, are greater in modern humans, particularly in conjunction with speech, and it is probable that the earlier trend to more complex social interactions within the anthropoid primates would have continued in the hominins. All primate sociality is built on individual-to-individual, one-to-one, face-to-face interactions and these would therefore have been an intrinsic and essential element in this evolution of sociality.

If so, increased sociality would have had major consequences for brain evolution. Improvements in the making of facial expressions and, even more, in their automatic reading, would have demanded more intricate neural apparatus. As expressions became connected to vocalizations, more neural circuitry would have been required. And as vocalizations became more voluntary and expressive, there would have been selective pressures to add new elements of cortical connectivity. As proto-language gave way to true language, the selective pressures for ever-more elaborate neural circuitry to permit that expansion of linguistic complexity would have increased.

Beyond expression making and expression reading, however, sociality in all its complexity is cognitively and behaviorally demanding: keeping track of who is who in the group, remembering what past interactions with those individuals had been like, knowing one's place in the social hierarchy, behaving appropriately in accordance with those relative rankings, and understanding how to get help from others in various situations, perhaps the most important of all. To the extent that friendships and working coalitions (such as for hunting) were important, they also would have required new elements of cognition. The greater the number and variety of the social interactions, the greater the number of individuals involved, the greater the neural capacity for it must have been, which would require increasing connectivity between different areas. These would have specifically included the different areas of the sensory cortex, the amygdala, the motor cortex, all the various centers involved in memory, specific face-recognition areas in the temporal and parietal cortex, and the various association areas in the prefrontal cortex. In effect, becoming more sociable requires far more neural capacity and connectivity than is needed for solitary existence or even the level of sociality characteristic of the herbivores who gather in big herds. Much of this new circuitry would have been in the cerebral cortex and in new connections to it from other regions in the brain for coordinating movements, in particular the cerebellum.

Such reflections lead to a fundamental question about human evolution: might increasing sociality and its multiple and complex demands have been the key factor that drove brain evolution and growth in human evolution? That is the basic contention of the social brain hypothesis, which we come to next.

The Social Brain Hypothesis and Its Addendum, the Social Face Hypothesis

The great neurobiological puzzle of human evolution—indeed, the central question in human evolution—concerns the increase in brain size that took place during hominin evolution. Even remembering that the hominins were a phylogenetic "bush," not a tree with a main trunk that led to humans at the apex (as in Ernst Haeckel's famous depiction of evolution), there was a definite trend of increasing brain size from the first australopithecines to the earliest species of the genus *Homo* and from them to later-emerging species of *Homo* (Figure 6.4). The increase during the past million years that led to *H. sapiens* seems particularly steep. This fact had been known for decades but, judging from the literature, it seems to have been initially little puzzled over. After all, since humans are a cognitively superior species, becoming ever more proficient at tool use, hunting techniques, and much else, is not this exactly what we would have predicted for our prior evolutionary history? If that assumption was the basis for the relative neglect of this issue until the late twentieth century, it reveals a reading back into the past requirements for capacities that were yet to evolve because the increase in brain growth preceded much of that later sophisticated behavior. Evolution does not operate with foresight, provisioning lineages with properties that will only be valuable later.

The evolution of larger brains, however, *is* a particular puzzle because brains are expensive in terms of energy, a fact that only came to be fully appreciated in the late twentieth century. The adult human brain weighs approximately 2 percent of total body mass yet consumes on average about 20 percent of total energy. That makes it the most energy-intensive organ and site in the body. Its energetic cost has to be met by increased calorie uptake, so the brain must "earn its keep" in terms of its energy demands. In other words, there has to be some direct and immediate compensating benefit. Some of the thinking about this matter has stressed the possibility of new stratagems to make possible the more efficient energetic and nutritional requirements—such as the invention of cooking or increased reliance on meat and fish to supply particular nutrients—demanded by a larger brain. Yet while an energy-expensive organ like the brain would require behavioral and nutritional responses that would support a larger brain, it is difficult to argue that increasing brain size was driven by both selection for more complex behaviors and by nutritional opportunities that made such increase possible.[3]

The assumption had long been that the chief selection pressures were in some sense ecological, deriving from the need to find more and better food and hence to hunt more efficiently. Some tasks would certainly have required increased cognitive power—for example, the increased cooperativity needed to hunt big game on the African savannahs and its correlated need for manufacturing weapons (spears) that would bring down the animals. This, however, was probably a relatively late development, well after the trend toward increasing brain size in the hominin lineage was underway. Furthermore, while involving new requirements, it is not at all clear that it would have required the extent of brain development that the overall trend in brain size exhibits. Some other form of selective pressure, it seems, was needed to explain it. Beginning in the 1980s, it began to be suggested that the growth in complexity of social interactions was the key factor. This idea has come to be known as the *social brain hypothesis*.

First discussed by the Columbia University anthropologist Ralph Holloway, though not named as such by him, one version of it had acquired traction by the late 1980s in the form of the *Machiavellian brain hypothesis*, which emphasized the complicatedness of judging other conspecifics' intentions (requiring more brain capacity to do so). Then later, it was termed the *social intellect hypothesis*, but it has now settled into its current name. Its function was to explain the relatively large brain sizes of catarrhine primates—in particular, the threefold greater size of the human brain (relative to body size) to that of the chimpanzee brain. The idea was relatively simple: these larger brains had been selected to deal with the multiple complex cognitive tasks demanded by increased numbers and kinds of social interactions. Although frequently contrasted to the earlier ecological hypotheses of selection for larger brains, with their explicit external environmental pressures, it posits a different kind of environmental pressure, namely, that stemming from the ecology of the social environment.[4]

To test the idea, two critical questions were posed from the start and had to be addressed; one was technical, the other substantive. The first, the technical question, concerned the best way to measure brain size in order to put the hypothesis on a quantitative footing. Initially, building on earlier analyses from the early 1970s that primates have brains that are larger relative to body size than expected (in comparison to other mammals), the first measure chosen was the *encephalization quotient* (EQ), the ratio of brain size to body size characteristic of a given species. In nonprimate mammals, there is a nearly linear relationship; anything in excess of that ratio for a particular species indicates new or more complex mental function in that species. Early work had suggested that primates had higher EQs than most other mammals and that humans had a particularly high EQ.[5]

Yet the important key brain-size variable in the primates was soon found to be not total brain size but that of the cerebral cortex, whose relative increase in primates is the main source of brain size increase. Hence, the question becomes: what should the size of the cerebral cortex be normalized to? Total body size was soon found unsatisfactory as a denominator

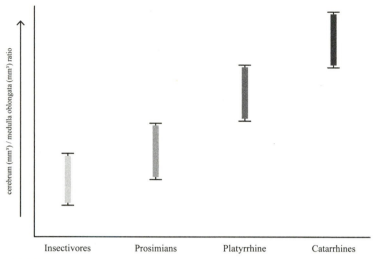

FIGURE 10.2 Cerebral cortex-to-medulla ratio in several groups of primates. The basic data are reported in Stephan et al. (1981) and the figure is modified from Dunbar (1997).

because it was too labile within certain closely related groups whose cerebral cortex dimensions were relatively stable in comparison. One possibility was another part of the brain, one whose size is far more fixed than the cerebral cortex—namely, the medulla, which is derived from the myencephalon of the hindbrain; its size is now the preferred measure. The results of one study of this ratio in several groups of primates are diagrammed in Figure 10.2, which shows that the apes have a larger cerebral cortex-to-medulla ratio than Old World monkeys, who, in turn, show larger ratios than the New World monkeys, who themselves exhibit greater values than the strepsirrhine primates. One must, of course, be cautious in identifying any evolutionary trend—which too often is in the eye of the beholder—and especially trends in a ratio (since such can be affected by independent changes in either the numerator or the denominator). Nevertheless, there is a strong suggestion of the existence of one trend here that is in line with measures of increasing social complexity in these primate groups from prosimians to advanced anthropoids. In particular, considering that humans have a cerebral cortex-to-medulla ratio twice as great as that of the largest value in the apes, it seems that there might well be a propensity toward such an increase, one that mirrors increasing degrees of sociality in these respective groups.

The second question was substantive, not technical, and lay at the heart of the matter: what are the critical *elements of sociality* that might actually have driven an increase in brain size? And how can these be measured comparatively? The two questions involve evolutionary inference at its most uncertain and demanding. One cannot, of course, directly interrogate the social history of the primates in the deep past; their fossils are silent on this matter, and they left no books or films to record their goings-on. Instead, one is forced to do correlations among traits within species living today. One strategy is to plot cerebral cortex size or the cerebral cortex-to-medulla ratio for different primate species against different variables—ecological and social—and determine which ones show some sort of statistically significant directional increase against cerebral cortex size. Any such variable that does shows such a correlation would be a candidate for being at least one of the selective factors that drove increased cerebral cortex (and brain) size. There is one oddity about this procedure, however: plotting any variable against cerebral cortex-to-medulla values and finding a linear relationship might make it look as if the brain property drove the behavioral one. In fact, it is the other way around: any such plot would suggest that it was the increasing values of the behavioral variable that provided the selection pressure for increased brain size—and complexity.

The earliest version of the social brain hypothesis postulated that it was the size of the social group in which the individual animal found itself. The central premise was that the greater the number of social partners, the more the brain would have to work to track individuals and specific interactions with them. In the first big study to investigate this, the cerebral cortex values for primate species from thirty-eight genera were tested against a range of variables, ecological and social. The results indicated that it was indeed this variable—social group number—that was the most reliable predictor of cerebral cortex increase. Thus, for the primates, this variable may have been the one driving increased brain size.[6]

Later tests extended to other kinds of social animals, especially ungulates and carnivores, however, did not support this simple relationship. The surprising critical variable that did seem to predict larger brain sizes for these groups was whether they formed lasting pair-bonds between males and females. Thus, species within these other mammalian groups that had permanent partners were found to have larger brains than related species in which mating partners were newly chosen each year. The implication is that such stable relationships are cognitively demanding in their own special ways. (Reflections on the requirements for successful human marriages can lead to the same conclusion.) Furthermore, among social mammals, those species in which coalitions are formed for hunting or protection also have larger cerebral cortices than related species that do not. These coalition-forming animals include elephants and some toothed whales whose brains substantially exceed human brains in absolute size. Indeed, for the successful workings of the social brain, it is absolute size that matters ultimately: the complex cognitive demands of intricate social arrangements require more absolute neural computing power, which can only be fulfilled (it seems) by increasing the amount of brain mass in the cerebral cortex.[7]

Now the most favored version of the social brain hypothesis is that the critical social variable affecting brain size is the *complexity* of social relationships. The greater that complexity, the greater would have been the selection pressure for larger brains. Yet an increase in that capacity in all probability would have increased the potential for more relationships as well. From this perspective, the increase in brain size in hominin evolution reflects the response to those selection pressures, as brains enlarged to deal with increasing social complexities. Correspondingly, an upward ratcheting of demands and new capabilities would have created brains with new potentialities, hence new selective opportunities for novel behaviors that, in turn, might have generated new selective pressures for the spread of those properties within the population. On the time scale of a normal human life, our natural temporal standard, all this would have proceeded at an imperceptible rate; on an evolutionary time scale, it could have been rapid.

As an explanation of human brain size, the social brain hypothesis has received much praise and much skepticism. It was developed, after all, to explain the outsized human brain as measured by EQ. While the kinds of social interaction variables in other animals correlated with increased brain size may help explain some of the increase in human brain size, are they sufficient to explain the extent of human brain increase relative to other primates? That increase might seem to require something unique, or at least highly distinctive, in humans, a thought that might render moot the kind of comparative studies with nonprimate mammalian species that have been done. As it happens, there is a natural candidate for such a unique human property: language, with its unprecedented demands on cognition.

Hence, there is a further refinement of the social brain hypothesis, though not usually labeled as such, which deserves consideration: that the increasing complexity of social relationships created a selection pressure for increasing use of language—specifically, complex language. In particular, the complexity of pair-bonding in long-term relationships, especially for raising human children, would demand communication skills of a very high order, the kind only supplied by language. This version of the social brain hypothesis can be diagrammed as:

complex social demands (long-term child rearing, in particular) → increasing
brain-neural complexity → increased language capacity → increased brain size
(where the arrows indicate selective pressures to generate the next step)

In this construction, it was the increased demands of complex language that drove the increase in brain size, but the former was itself driven by the complexity and uniqueness of human social arrangements, in particular the raising of children over time spans that are disproportionately long compared to other primates (and certainly most other mammals).[8]

In this discussion, it might seem that we have drifted away from our central topic of the face, but the digression is more apparent than real. In the line of thinking sketched in the preceding, the kind of highly mobile and expressive face that humans possess would have been one of the key instruments of the social brain—in effect, one of its essential enablers. Though the role of the face is not often discussed in the large literature on the social brain hypothesis, the social face would have been an indispensable adjunct to the evolution of the social brain in the anthropoid primates, becoming ever more versatile as its role expanded from the early anthropoid primates to the rise of the genus *Homo*.

To put these matters of brain and face into a larger evolutionary context, we can say that the anthropoid primates have a certain potential for creating circumstances that promote their survival. This general process in which the organism alters the environment to favor itself is termed **niche construction**, and it is an increasingly well-documented phenomenon, not only for animals but also among plants (where the latter largely create their niches through biochemistry rather than behavior). In the particular case of primates, especially the anthropoid primates, the niches constructed are social in nature. Part of the very nature of such social environments in animals with complex behaviors is that they possess the potential to enlarge and become more complex, though each such enlargement may initially generate new problems and potentially evolutionary pressures. We can visualize this process as a kind of evolutionary stumbling forward in a certain direction. It was, of course, not inevitable that our species would have arisen in this process—many primate lineages have formed and evolved without becoming human beings—but the *potential* for evolution toward ever-greater sociality must have been there to some degree from the beginning of "primate-dom" and then grown over evolutionary time in certain lineages, including ours. That expansion in social complexity, or at least diversity, is inherent in social niche construction in animals whose survival is highly dependent on group integrity.

With this line of thinking and returning to the role of the face, we can now confront a question that perhaps should have been addressed at the very start: does the vertebrate face truly have a major unitary function (such as the lungs or the heart) or is it just a spatially ordered collection of sensory elements in and around the mouth that operate largely independently but which evolution has happened to group in close proximity to one another? Another way to ask this would be: is the face more than the sum of its parts? For the great majority of animals whose faces function solely as the sensory headquarters of the individual animal, we could argue that it is not or slightly so at most because it presumably is somewhat advantageous when seeking food to have sites of three of the major senses grouped near the mouth and not far from the brain. Nevertheless, the face as such in those animals is still largely a collection of independent elements. For the primates, however, and especially the anthropoid primates, and for the hominins most of all, the social face clearly has a unitary and integrated function—namely, that for sending and receiving signals (often quite complex ones in humans) about social behavior between individuals. The whole face participates both in such signaling and as an identifier of the individual. Hence, in its communicative aspects, the face *is* definitely more than the sum of its parts. Indeed, one could argue that the face as an "organ of perfection" only truly came into its own as such with the rise of the anthropoid primates.

Evolution as a State of Continuing Dynamic Disequilibrium: The Interplay Between Selective Pressures and Selective Opportunities

The expression "stumbling forward" used in the preceding to describe the coevolution of the brain and hominin sociality, in which each change creates something new that then exerts a selective pressure for further change, may actually be quite apt. Earlier, I had suggested that this might be a general feature of the evolutionary process. Evolutionary biologists are highly familiar with one variant of such processes: where a change adds some selective advantage for the individuals showing it but is itself imperfect, thereby promoting selection for improvement or optimization of the new feature. Such optimization is a fundamental property of the evolutionary process. It is demanded by the theory and supported by much evidence.

What I suggest here, however, involves a somewhat different dynamic: that some of the changes triggered by an initial alteration may not always be simple optimizations but actually potentiate change in a new direction. In effect, the new change, while advantageous, creates some nonoptimality that opens up a new possibility for evolutionary progression down a new path. A classic example is the evolution of bipedal movement in the hominin lineage. Whatever its initial selective advantages were—a much-debated matter—its early stages would almost certainly have created selective pressures for further anatomical changes, especially in the pelvic girdle and the positioning of the head relative to the backbone. Those changes in themselves can be seen as both corrections and optimizations. At the same time, however, by liberating the forepaws from need for movement, bipedalism would have opened up new possibilities and hence new uses, especially for food gathering and handling. There would, in turn, have been selection for optimizing these new functions. Individuals with those new adaptations would have been favored to leave more offspring.

In effect, a sequence of evolutionary change can often be seen as a pattern of *dynamic disequilibrium* in which each change throws the organism's system off a little, and the attempt to equilibrate back to better function generates a need for or opens up a space for further changes, which may repeat the cycle. Whether there is selection for simple optimization or selection for something new (and its subsequent optimization) in each case will be determined by the properties of the organism, the precise nature of the perturbation from the first change, and the demands of the environmental system over time. Nevertheless, such dynamic disequilibrium as a propelling force in evolution may be far from rare.

Of course, there is nothing inevitable about evolutionary change. We know that evolution, at least morphological evolution, need not take place at all for long periods in long-surviving lineages, as witnessed by a so-called living fossil such as the horseshoe crab, whose morphology has stayed essentially the same for several hundred million years. Yet once a change occurs that alters something important in the form or function or physiology of an organism, the probable consequence is the need for further changes to maximize evolutionary fitness (the ability to leave larger numbers of offspring than those individuals who lack those alterations). The pattern of **punctuated equilibrium** is understandable in this light. First described by Niles Eldredge and Stephen Jay Gould in 1972, this phenomenon can be seen for many organisms in the geological record; a lineage exhibits stability of form (*stasis*) for many millions of years and then changes start occurring—often in a steady direction. Though previously observed by paleontologists but not named by them, this pattern was seen as simply an interesting phenomenon in the paleontological record rather than as a puzzle to be explained. Yet, as Gould and Eldredge stressed, it was not predicted by conventional evolutionary theory, and the pattern of change was not self-explanatory. They concentrated on what they saw as the anomaly of stasis, but the *punctuations* (the sudden changes followed by periods of continuing change, sometimes proceeding for millions of years) are equally mysterious since it is not immediately clear why selection pressures for change should be so continuous after long periods in which they had either been slight or had had no visible effect. If, however, an initial change can create some kind of suboptimal fit between the altered trait and other traits, creating selective pressure for further alterations that might then repeat the pattern, long periods of change following long periods of stasis may not be surprising. The evolutionary history of the human face in conjunction with the evolving human brain, especially during the last million years of hominin evolution, may well exemplify this dynamic "push–pull" character of the evolutionary process.[9]

Small versus Large Effects Revisited in the Light of Emergent Properties and the Matter of Costs in Evolutionary Change

There is no older debate in the history of evolutionary biology than the question of the magnitude of effects created by new genetic variations upon which natural selection acts to create changes in populations (see Box 6.2). Are the variations that actually contribute to evolutionary change of small phenotypic magnitude and cumulative in their effects or are there large, highly visible ones? Impressed by the ability of plant and animal breeders to detect minute variations and then amplify them over generations by always picking the offspring that showed the trait in slightly more magnified form and then breeding from them, Darwin argued that natural selection would be as efficacious as artificial selection and that small variations of this kind were the true source material for evolution. T. H. Huxley and others countered that such small variations in nature would not be comparably selected and that only variations of marked effect would work. "Small" and "large" are, of course, rather imprecise terms, but the lines of the debate were sufficiently clearly drawn to make sense to all participants.

The consensus in population genetics today, based on population genetic reasoning and calculations, is essentially in the middle: that the most probable genetic source material for selection leading to morphological change is of middling phenotypic effect—larger than Darwin imagined but smaller than Huxley's *saltations* (which would later be called *macromutations*). Nevertheless, it might be helpful to look at this matter not from the standpoint of theoretical genetics but using a test case, a lineage where many of the events are reasonably well known, and ask whether "small or medium" changes are sufficient to explain the main line of evolution or whether some "large" (salutatory) and inexplicable changes have to be invoked.[10]

Hominin evolution, in fact, provides such a good test case. After all, we now know a great deal of the probable sequence of events in the hominin lineage; when examined in detail, it is striking how few, if any, of the probable changes from the branch point with chimpanzees require sudden changes. Darwin's explanation of change by (relatively) small changes can explain most of the events. (In contrast, with much earlier events in the LHLP, such as the origins of the craniate head, the case is far from clear.)

Some of the most striking innovations that are central to human evolution—the advent of bipedality, the evolution of the human face (our subject), and the evolution of the human brain—can all be conceived of in terms of a series of changes of modest magnitude in processes that are fairly well understood. Even language, that most remarkable of human traits, whose origin involved both the face and the brain, can be put within a reasonable stepwise evolutionary framework without invoking saltations. The sequence could run from crude vocalizations to communication by lip smacking to lip smacking plus vocalizations to the development of proto-languages (essentially proto-sentences with just nouns and verbs) to true language.

Even if we accept that small- to medium-effect variations are the probable source material for much of evolution, however, the old question raised by Darwin's critics persists: can natural selection account for the origin of true novelties—examples such as bird flight, bat or cetacean sonar, spider web construction, and many more. For increasing numbers of such, the more we learn about the underlying genetics and developmental biology, the easier it is to fashion hypotheses that explain the basis of some of these dramatic new features. As discussed briefly in Chapter 6, at some point a critical threshold is passed or a tipping point is reached and a new quality or capacity appears. Such unexpected outcomes are said to be *emergent properties,* and many involve some of the most fundamental properties of the physical universe as well as special new properties of biological organisms.[11]

Nevertheless, there remains the matter of the human brain and its remarkable abilities. Those abilities were seen as perhaps the major challenge to classic Darwinism. As Alfred Russel Wallace pointed out, there was nothing conceivable in the Pleistocene environment that could have involved selection for the truly remarkable mental abilities displayed by (some) modern humans. What in that environment, it might be asked, could possibly have selected for the ability to play the violin, do calculus, design skyscrapers or rockets to the moon, or paint canvases like the old masters? For Wallace, the answer was obvious—"nothing could have"—and he therefore opted for the only alternative he could see: that the properties of the human brain had been created by some non-natural, divine source. Though his conclusion was a religious one, even if not derived from traditional religious belief, he had reached it on the basis of sound Darwinian principles.

Wallace's reasoning was based on a premise that is usually not stated but is nearly universally accepted: that natural selection only brings forth the *minimum* change needed. The idea that underpins this view is that anything more would involve unnecessary costs in terms of energy, materials, or perhaps time for some behaviors. And such unnecessary costs would reduce fitness and be selected against. Thus, by this reasoning, any trait that was more than what was strictly needed would be subjected to optimizing selection and reduced to its optimum value. Thus runs the reasoning, but is it true? Perhaps, for some traits, what seems to us to be excessive and hence costly is not sufficiently so to the organism itself that the continued maintenance of that trait would be precluded. An example is genome size in certain amphibians and flowers, where it is tens of times greater than that found in, for example, humans. It is hard, if not impossible, to explain such genome sizes in terms of adaptive advantage or as needed by the complexity of those organisms. Such DNA amounts reproduced in all body cells must entail some extra cost—relative to needs—in terms of biochemical source materials and energy, yet these organisms are living species that have evidently maintained themselves over exceptionally long time spans. Perhaps, some costs are more in the eye of the (human) beholder than experienced as such by the organisms themselves.[12]

Seen in this light, Wallace's argument might be answerable along these lines: that natural selection does not always deliver just what is needed but sometimes capacities that are more than what is needed. It cannot promote changes that are less than sufficient, but if ostensible costs are not always as deleterious as we might count them, it can in principle promote changes that are actually more than enough for the adaptive needs at hand. In the particular case of the brain, that could be the potential for abilities that were not demanded at the time those potentials were being selected. If the social brain hypothesis is correct, the demands of sociality—which we carry out with relatively little effort—demanded considerable brain power, enough to create a selective pressure for a much larger brain than our early hominin forebears required or possessed. In turn, those abilities that involve mental calculations (however unconscious), considerable memory capacity, capability of planning for the future, coordinating those plans with others, and possibly the beginnings of symbolic thought in connection with language could have been later adapted to such things as the invention of agriculture, developing new forms of communication, and creating all kinds of new technologies. Indeed, when we appreciate the probable relative facility with which new neural connectivities within the brain can form, linking regions in new ways or strengthening those connections, the brain looks like a device with a unique capability for expanding its activities. Wallace's argument was far from unreasonable, but he knew nothing of how brains construct themselves. Had more of this been known at the time, perhaps he would not have adopted a view that Darwin saw as threatening his whole conceptual edifice.

Behavior and Morphological Evolution: Justice for Lamarck?

As this story of the human face has progressed, the reader will have noticed that the focus has shifted from the morphology of the face per se to the ways that mental states may have indirectly but powerfully affected that morphology during evolution. Those mental properties or states have included perceptions of faces (both their identities and their expressions) and the emotional reactions evoked by those faces, the influence of aesthetic preferences for certain kinds of faces (in sexual selection), and the cognitive complexities entailed by living in social groups. Of course, mentalities by themselves cannot directly exert such influences. They can only do so indirectly through selection pressures that favor greater survival of individuals displaying certain behaviors triggered by the underlying mental states and the way those behaviors play out in the larger social groups to which the individuals belong. If those behaviors contribute to useful functioning of the group, then there would be pressures of social selection for their perpetuation and spread.

It is thus behavior that is or can be a factor in animal evolution and which, it has been argued here, was a major contributor to face evolution in the anthropoid primates and specifically in the hominins. A corollary of this proposition is that the animals with more complex behaviors reflecting their more complex mental operations (and more complex brains, which make such operations possible) are probably more affected by behavior in their morphological evolution than animals with simpler behaviors, mentalities, or brains. This thought gets relatively little consideration in evolutionary biology today, although as more attention is directed toward the phenomenon of niche construction by animals—which involves behaviors that feed back into evolutionary effects—it is receiving more. It is hardly, however, a new idea.

An early and important exposition of behavior and niche construction was given in *The Living Stream* by distinguished British zoologist Alister Hardy in 1965. There Hardy set forth the argument that individual animals will first explore a new niche with some new altered behavior. If it proves beneficial, not only will they continue to fill that niche but also in some cases their behavior will be copied by their conspecifics, something that has been seen with both birds and mammals, the phenomenon of *animal cultural evolution*. Thus, new beneficial behaviors are likely to persist and spread. Then if new mutations that alter physical structure in some way (via effects on their development, though Hardy did not stress this) to make the new behavior even more effective, those mutations will be selected. Such mutations, for example, might alter appendages used for obtaining food. Indeed, David Lack, the first twentieth-century zoologist to carefully study Darwin's finches, had suggested that the different kinds of beaks of the finches on the Galapagos had evolved in just this manner as different finch populations began to specialize in trying to obtain different kinds of foods on these islands that had emerged from the seas not long before. New mutations that facilitated access via particular beak forms to the new preferred food would be selected. He saw this process as the far more probable explanation than random mutations that first created new beak forms that then allowed the birds to go after new foods.[13]

Beaks, of course, are just one possible example. One can readily imagine extending this general mechanism to alterations in limbs or jaws or mouths or teeth. In the early 1980s, the pioneering molecular evolutionist Allan Wilson (1934–1991), who had done the first molecular dating of the chimpanzee and hominin lineages and who provided the first genetic proof of the out-of-Africa hypothesis, took up the idea. He and his colleagues analyzed the rates of morphological evolution in relationship to brain size and behavioral suites in birds and mammals. From their data, they argued that there were direct relationships between the rates of molecular and DNA sequence change and the complexity of behavior. Quite simply, their idea was that new behaviors would create selection pressures for genetic changes

that promoted morphological changes that, in turn, facilitated the new behaviors. In particular, and not least in this view, increasingly complex
behaviors would favor selection for increasingly complex and larger
brains.[14]

In essence, this idea of behavioral change paving the way for morphological change was the suggestion of J. Mark Baldwin (1861–1934), an
American psychologist, who proposed in 1896 the mechanism of evolutionary change later named after him (by G. G. Simpson) as the *Baldwin
effect* (introduced in Chapter 7). Baldwin had emphasized the role of new
behaviors in creating selection pressures for adaptations that altered animal morphology. That idea, in turn, would be taken up more than forty
years later by British developmental biologist C. H. Waddington (1905–
1975), who did not emphasize altered behavior but altered development.
(Waddington seems never to have cited Baldwin though he must have been
aware of him as an intellectual predecessor.) Waddington suggested that
if new, environmentally induced developmental alterations altered morphology in a way that met a new need and hence had adaptive value, there
would be selection pressures for mutations that made automatic those developmental alterations. He termed this process *genetic assimilation*.
Though their emphases were different, Baldwin and Waddington were
discussing the same general phenomenon: the appearance of a new induced (plastic) state with some advantage or advantages that would, if that
state persisted over generations, create selection pressures for mutations
that made it a fixed part of the organism's biology.[15]

Yet preceding all of these individuals in this line of thought was someone
who had been a highly influential biologist in the early nineteenth century.
Indeed, he is one of three individuals who independently coined the word
biology to denote the study of living things as a distinct branch of the
natural sciences. This was Jean-Baptiste Lamarck (1744–1829). Today Lamarck is famous, or rather infamous, for propagating the idea that physical traits acquired during an animal's lifetime could be inherited by the
progeny of that animal, an idea now denoted as Lamarckian and whose
defeat in the late nineteenth and early twentieth centuries is often credited
with helping facilitate the rise of modern genetics. As recent scholarship
shows, however, this characterization of Lamarck's thinking constitutes a
serious injustice. Lamarck did believe in the inheritance of acquired traits,
but he did not originate the idea, associate himself with it in particular, or

spend much time on it. He regarded it as obvious, and in any event he seems to have lacked any strong interest in modes of heredity. Furthermore, many scientists of his time and much later also believed in the inheritance of acquired characteristics. They included Charles Darwin, who likewise accepted it as a fact in his treatise on heredity. If to believe in the inheritance of acquired traits makes one a Lamarckian, then Darwin himself was a Lamarckian. And like Waddington with respect to Baldwin, he did not initially acknowledge Lamarck as any sort of scientific precursor, though he was fully aware of Lamarck's role as a pioneering advocate of the reality of evolution, a fact that he publicly recognized later.[16]

Lamarck's central interest, however, was the way that new behaviors could pave the way for the evolution of new structures, new physical features, in animals. His proposed mechanism for how this might work looks simply bizarre to today—making no connection with our ideas—and is not worth describing. Nor did he have the idea of natural selection or the concept of selection pressures for new adaptive features. (Thus, while Darwinian was Lamarckian in the sense we use the word today, the reverse is not true.) In his basic idea, however, which was elaborated in his most important books—that new behaviors could be an evolutionary force contributing to changes in animal form—Lamarck was a precursor of Baldwin, Lack, Hardy, Wilson, and others who much later adumbrated this idea in modern ideas and terminology. To acknowledge as much is simply to do justice to Lamarck.[17]

One cannot leave the subject of Jean-Baptiste Lamarck and his ideas, however, without mention of a topic that is now closely associated with his name: *epigenetic inheritance* across generations. Given Lamarck's relative lack of interest in matters of heredity, it is debatable whether it is appropriate to use the word *Lamarckian* for this phenomenon, but the association is often made in the literature. Regardless of the appropriateness or otherwise of labeling it Lamarckian, *epigenetic inheritance* refers to the (possible) transgenerational inheritance of chromosomal regulatory states—in other words, modifications of gene activity involving proteins bound to specific genes that can be passed to progeny through the reproductive cells. Abundant evidence for the transmission of such states through mitotic divisions in the body now exists. The question is whether such states commonly are transmitted to progeny and, if so, to successive generations derived from those progeny. If they can, and there is certainly some evidence that this can occur in plants, with weaker (but growing) evidence for its occurrence in animals, then in principle if the changes are truly stable enough, these epimutations should be subject to natural selection, just as true mutations (DNA sequence–based changes) are. The importance of this is twofold. First, it really would constitute a departure from traditional twentieth-century thinking about evolutionary mechanisms—which are solidly anchored in the idea of DNA sequence change as the source of hereditary variations; support for this idea would open up large new areas for investigation. Second, epimutations in body tissues can be induced much more readily than specific mutations that affect those traits. If epimutations are found to play a significant part in evolution—an idea still highly speculative and controversial—it might help account for the fact that some evolutionary changes seem to be far more rapid than one would expect if all hereditary variation waited on the occurrence of true mutations—namely, DNA sequence alteration, which occur with low frequencies.[18]

The potential relevance of these considerations specifically to the evolution of the face is its possible connection to the phenomenon of domestication, which is discussed in Chapter 8. The major experimental work on domestication of animals was initiated by a Russian scientist, Dmitry Belyaev (1917–1984), a pioneer of Russian genetics who began his work during the Stalinist era, when to study the mechanisms of inheritance was considered a potentially subversive act in the Soviet Union, one subject to imprisonment or exile and possibly death. Belyaev was fascinated by the phenomenon of the domestication syndrome (though he did not call it that) and began his experimental studies in 1960, the point when it had at last become possible to do real genetics in the USSR without substantial personal risk. Even in its early stages, the work of Belyaev and his coworkers showed how rapid domestication could be achieved via selective breeding of effectively wild animals by selecting for the tamer ones at each generation. His group initially carried out this out with mink, rats, otters, and silver foxes but ended up concentrating on the foxes. (Compare **Plates 16** and **17.**) To him, the process seemed too rapid to be accountable solely in terms of rare mutations in the farmed fox populations he was breeding from. His thinking about the mechanism therefore began to focus on epigenetic changes as initiating events because of their ease of induction.

To regard the phenomenon in this light does not exclude a major contribution from true genetic (DNA-based) variation existing in the wild (fox farm) progenitors, which then contributed, during the selection for tameness, to the domesticated phenotype. Belyaev, however, postulated that such classic genetic variations were not the whole story. Indeed, one of the genetic changes that did appear early, which caused a white patch on the domesticated foxes' heads that he named "Star," has more the character of an epimutation than a mutation in its frequent rate of occurrence and its rate of *reversion*—that is, its loss in stocks showing it. "Star" was one of the first traits to appear in the breeding for tamer, domesticated foxes, but it was soon followed by changes in the face—in particular, the reduction of the jaws and the acquisition of a more juvenile face in the adult domesticated animals.[19]

The relationship between epigenetic changes and domestication remains a highly speculative subject, yet Belyaev's thinking about these matters is not unreasonable and deserves consideration. If forthcoming evidence supports it, then this connection might have some relevance to the evolution of the human face, particularly if humans are, as suggested already, a *self-domesticated* species. Furthermore, if epimutations can be stabilized via selection by true mutations, analogously to the Baldwin effect, it may turn out that the evolution of the human juvenilized ape face owes something to epimutations. This possibility is hypothetical, of course, and the prior question that needs answering is whether humans are truly self-domesticated. Nevertheless, it would be an interesting quirk of scientific history if pioneering twentieth-century genetic work carried out in Russia on the domestication of animals, after being initially suppressed in the name of an ideology hostile to modern genetics, eventually shed light on something fundamental in human genetics, namely the evolutionary history of the human face and, more generally, the properties that have made our species so distinctive.

CODA: THREE JOURNEYS

In this book, I have given an account of the evolutionary events that led to the formation of the human face. It is, of course, just one possible version of the story. Undoubtedly, other authors would have told the tale somewhat differently. Furthermore, it is inevitably incomplete. There remain major gaps in our knowledge and some of the ideas ventured here will surely need rethinking. Those flaws, however, are inherent in the nature of science and its communication: scientific understanding is always provisional, imperfect, and incomplete.

Whatever deficiencies there might be in this specific treatment, a basic fact remains: the evolutionary genesis of the human face is an intriguing and important story, one highly relevant to our character as a species—indeed, to the nature of "human nature." One view of that evolutionary history is to see it as a rather remarkable journey through time and space and many incarnations, stretching over half a billion years and countless generations of animal life and many, many species. That journey was not predestined—at least as far as science can tell—but it resulted in something quite extraordinary. Its end point, the human face, is both an efficient collector of information about the world for the individual (as all animal faces are) and a remarkably good transmitter of information about our individual selves, feelings, and intentions to our fellows.

The description of that evolutionary journey in this book, however, has provided the basis for a second journey, that of the reader. The reading of any book indeed involves an excursion, a mental one, at the end of which, if the book has had any impact on the reader, she or he is not quite the same person. Something in her or his knowledge and views will have changed. That journey need not and should not end with the reader nodding in perfect agreement with the author's conclusions. It is far better if the reader is provoked either to disagree at some points or moved to examine some of the ideas further. A truly useful book, after all, does more than provide information; it prompts reactions and further thinking.

This book has also involved a third journey—that of the author. No matter what intentions and prior knowledge an author has at the beginning of a book project, by the end that book will inevitably be somewhat different than the one envisaged at the beginning. Correspondingly, the author is somewhat altered, too. Scouring sources for this book and continually mulling over the material I was finding, I became far more impressed than I had ever been before with what extraordinarily cooperative, social creatures we human beings are. Our whole existence is intimately and nearly continuously tied up with our sociality, not simply in our overt activities, but in our imaginations, goals, dreams, hopes. Of course, there are highly solitary humans who are worlds unto themselves— at the extreme would be psychopaths and sociopaths—but such individuals are the outliers.

The idea that humans are exceptionally sociable creatures is hardly new. There is, however, a difference between thinking something, particularly a conventional thought like that, and, actually *feeling* it with all the emotional resonances that feeling evokes. Furthermore, we can embrace this particular idea without denying that human beings often do cruel, even horrifying, things to one another. The propensity for such is every bit as much a part of human nature as the capacities for cooperation, friendship, and love. Nevertheless, our nature as intensely social beings is intrinsic even to much of the evil we do. The worst human actions are carried out in groups relying on group identity and drawing on our fundamental impulses to cooperate and conform, perhaps the essential elements of human social nature. For example, the most horrific activity that humans engage in—genocide—is for the perpetrators an intensely social activity, as shocking as this may sound. It is socially organized and depends utterly on identification with one's own social group, in effect one's tribal identity. In drawing on those social bonds, genocide reflects the darkest side of human sociality, not its absence.

Nevertheless, despite the occasional intrusion of such gloomy thoughts, the experience of writing this book has made me more optimistic about our species' future. The key to our flourishing, perhaps to our survival, will lie in harnessing our strong social natures to positive ends while preventing them from being captured by the destructive obverse side of our social nature: tribal bonding. That, in turn, will mean that many more people will have to start seeing the whole "human race"—a delightful old-fashioned phrase that deserves revival—and not their immediate group, as their real "tribe." There is no social engineering scheme that can achieve this but, to some extent, through globalized shared media experiences and increasing intermarriage between members of different ethnic groups, the forging of this much larger sense of human solidarity is already underway and actually has been proceeding for some time.

And what will be the role of our face in the future of our species? The human face, I maintain, has not been rendered obsolete as a communications medium either by print or all the various forms of electronic media. Face-to-face conversation will remain an essential part of human existence, and the face will continue to play a vital and stabilizing role in how we interact with one another. It evolved, after all, as a facilitator of social interactions. As such, I predict it will remain the handmaiden of the "better angels of our nature" for as long as human beings exist.[1]

NOTES

ACRONYMS

GLOSSARY

BIBLIOGRAPHY

ACKNOWLEDGMENTS

INDEX

NOTES

1. THINKING ABOUT THE HUMAN
FACE AS A PRODUCT OF EVOLUTION

1 Immediately behind the face lie the aural openings, the paired organs for the fourth primary sense—hearing. Though the external ears in humans and many mammals visually frame the face, most vertebrates lack external ears, and this feature is therefore not a universal part of the vertebrate face. Nevertheless, four of the five main senses are tightly clustered in and around the face. The fifth, sensitivity to touch, is dispersed around the body within the skin. For discussion of the idea that the original adaptive function of the face was to allow animals to orient toward food, and specifically predators toward prey, see Gans and Northcutt (1983).

2 Enlow (1968), p. 168.

3 The corollary of the fact that the facial expressions often serve to reinforce or amplify the meanings of the words in a conversation is that if people deliberately want to mislead others in an exchange, then they must consciously control their facial expressions. Acting ability requires just this, matching one's expressions to the scripted words. "Method acting" goes further, requiring the actor to "feel" those feelings deeply, not just simulate them, to ensure that the spoken words are in the correct emotional register. This visual layer of information is missing, of course, in e-mails or letters, a fact that explains why it is easier to succeed in lying in writing than in spoken conversation. It also helps explain why written communications so often lead to misunderstandings; they are missing the facially expressive content. Adding a smile "emoticon" in an e-mail is an attempt to express a positive feeling that one suspects is not conveyed by the words alone.

4 Despite the diversity of schools of physiognomic thought, there have been two main divisions: one stresses the physical features of the face as clues to personality, the second emphasizes expressions. The first germinated in the late eighteenth century with a Swiss pastor, Johan Kastor Lavater (1741–1801), and culminated in the twentieth-century race theories of the Third Reich in Germany; this branch of physiognomy essentially self-destructed along with Hitler's regime (Gray 2004). Nevertheless, a version minus the racism may be undergoing a small revival (Chapter 9). The other kind of physiognomy, which emphasizes expressions as a key to personality, lives on as a general intuitive belief. Abraham Lincoln once said that, by age forty, every man has the face he deserves. (The same statement is also attributed to George Orwell, though he gave the figure as fifty, an age he did not reach.) There is much truth in this statement: habitual expressions, whether of happiness, sadness, or anger, reveal dominant strains in an individual's personality, and eventually become imprinted in the lines of the face, a consequence of the aging of the skin.

5 The principal interest in facial expressions in the first half of the twentieth century was in the question of whether human expressions were universally shared, as Darwin believed, or a product of culture. The history is reviewed in Russell (1994). The extent of influence of culture in modulating possibly universal human expressions remains an area of controversy. With respect to the evolutionary roots of human facial expressions, De Waal (2003) reviews the history of investigation of primate expressions, initiated by Darwin, as it was revived in the twentieth century.

6 Gregory's book needs to be understood in the context of its time in American culture. It was written in the aftermath of the famous "Scopes monkey trial," which took

place in 1926 in Tennessee. That trial was intended by the state to enforce a law pro-
hibiting the teaching of evolution in Tennessee schools. In the first pages of the
book, Gregory emphasizes that the structure of the human face can only be understood from an evolutionary perspective—that, for example, the basic structural similarity of shark and human faces is not accidental—and he makes clear, in sly,
humorous language that his larger aim is to combat antievolutionary ideas.

7 The earlier term denoting the human-specific lineage within the taxonomic order of
the primates was *hominid*. This term reflected the belief that humans constituted a
distinct family within the Linnaean taxonomic system (see Chapter 5), one that excluded the great ape**s**. As a result of growing understanding of the genetic and hence
evolutionary relationships of humans to chimpanzees and bonobos ("pygmy chimpanzees"), however, humans are no longer placed in their own distinct Linnaean
family but are now included in that which includes the great apes, the **Hominidae**—
orangutans, gorillas, chimpanzees, bonobos, *and* humans. The present term *hominin* designates all those species that arose within the branch of the Hominidae that
gave rise to humans eventually as they split off from the chimpanzee-bonobo lineage.

8 An example of how starting with a small quantitative question can make it possible
to tackle a larger qualitative one concerns the immune system. The intriguing big
question in the late 1940s, "How does the mammalian immune system produce such
a vast repertoire of different antibodies?" provided no entry point to its solution. Only
when a smaller question was asked, "Is an individual antibody-producing cell capable
of producing all antibodies or only a few or even only one type?" could the larger
question be investigated. (The answer was that each antibody cell produces only one
type of antibody, the whole repertoire of different antibodies in an individual being
due to millions of different cells, each with its own specific capability.)

9 As will be discussed in Chapter 6, there are two main divisions of the primates: the
anthropoid primates and the more primitive (or "less diverged," that is, from ancestral primate stocks) prosimians.

10 Recognizing that there was a succession of different kinds of fossils in a sequence of
geological strata did not necessarily require the observer to believe this reflected the
operation of evolution. The great nineteenth-century paleontologist Baron Georges
Cuvier rejected so-called species transmutation (evolution) and postulated that different fossil forms in successive strata reflected local extinctions of existing forms at
different times, followed by movements into new territories by animals that had previously lived elsewhere.

11 For a history of the often fraught relationship between developmental biology and
evolutionary biology, see Wilkins 2002, chapter 4. Today, a branch of evolutionary
biology is devoted to their intersection—*evolutionary developmental biology* or "evo-
devo"—which emerged in the late 1980s.

12 Comparative biology as a method for reconstructing the evolutionary past from living
related species is described in depth by Harvey and Pagel (1991).

13 The conservation of genetic machinery for development actually extends far more
widely than the vertebrates, existing throughout the various bilaterally symmetric
animals, the so-called Bilateria. See Carroll et al. (2001), chapter 2, and Wilkins
(2002), chapter 5.

14 For the details of this work, see Abzhanov et al. (2006).

15 The similarities of six basic expressions in humans and chimpanzees are described
in Parr and Waller (2006). The difficulties of interpreting such comparisons, in the

light of the physical differences between human and chimpanzee faces, are discussed in Waller et al. (2007).

2. HOW THE FACE DEVELOPS

1 This picture also embodied the view of evolution as an inherently progressive process—in this case, one that proceeded inexorably from ape via ever more advanced stages to human. It is not surprising that this aspect came in for spoofing in the late twentieth century, for example, in versions that showed the end product as an obese couch potato watching television. There was, however, something else wrong with the traditional picture: it conveys the mistaken idea that the evolutionary process in this case was unidirectional, inevitably heading toward modern humans. In reality, hominin evolution from apelike precursors involved many branchings and different species, of which only one survives today—our own (as will be discussed in Chapter 6).

2 The high degree of structural complexity that characterizes the head and the face are described in the first two chapters of Lieberman (2011).

3 Intriguingly, the mitochondria are descendants of bacteria-like cells that colonized the early ancestors of the kind of nucleated cells that make up our bodies. There is a good argument that the extra energy supplies that these symbionts provided was crucial for permitting the size of **eukaryotic cells** to increase over their microbial cell ancestors, ultimately making possible the evolution of complex multicellular organisms, plants, and animals (Lane and Martin, 2010).

4 For an accounting and description of the different types of human cells, and the estimate of >400 cell types, see Vickaryous and Hall (2006).

5 The neural crest cells were first identified in 1868 by a major nineteenth-century embryologist, William His, Jr., who intuited but could not prove their importance in vertebrate embryo development. Their role in giving rise to many different specific cell types was only firmly established in the 1940s and 1950s. For the idea of neural crest cells as the "fourth germ layer," see Hall (2000). Some objections to this classification, however, and an argument that the term *germ layer* is outmoded in terms of molecular criteria, can be found in Sarnat and Flores-Sarnat (2005). A good general reference to neural crest cells is that by Le Douarin and Kalcheim (1999).

6 For a full account of human embryonic development, an excellent source is the classic textbook of Larsen (2009).

7 The word *ventricle* denotes an enclosed space within a tissue or organ. The ventricles of the heart are the kind of ventricle familiar to most people from high school biology. Within the developing brain, there are many ventricles. The cells that line these cavities are often termed the *ventricle layer*, and the cells just above them are denoted (perhaps somewhat confusingly) the *subventricular layer*.

8 The evidence that cranial neural crest cells are essential for forebrain development is provided in several important papers, but see especially Etchevers et al. (1999), Creuzet, Martinez, and Douarin (2006), and Creuzet (2009).

9 The dorsal side of the telencephalon will eventually become the *cerebral cortex* (or *cerebrum*), which comes to enclose the brain, while the ventral side will later develop into essential brain structures termed the *basal ganglia*; both will be discussed later.

10 A good description of the FEZ and its role in patterning the upper face is that of Hu and Marcucio (2009), while the essential role of SHH in supporting the development of the lower jaw from the mandibulary arches is discussed in Le Douarin, Brito, and Creuzet (2007).

11 The pharyngeal arches are evolutionary derivatives of the anteriormost gill arches of ancestral fishes. In mammals, there are five pairs of these structures, the other four situated behind the first pair, and these give rise to the pharynx, several neck cartilages, the anterior section (*foregut*) of the digestive tube, and several muscles. The first branchial arches will not only form the mandibulary facial prominences but also contribute to both the hinge region that connects the upper and lower jaws and the lower (ventral) part of the upper jaws. The role of SHH, secreted from the anterior endoderm, in permitting development of the mandibulary arches is given in Le Douarin, Brito, and Creuzet (2007).

12 These findings on embryonic face shape and their different trajectories of change are described in Young et al. (2014). As noted in the text, the bird beak develops from the premaxillary bone; it is anterior-posterior growth in beak extension that requires calmodulin , as discussed in Chapter 1.

13 See Brugmann, Kim, and Helms (2006).

14 For details, see Boughner et al. (2006) and Parsons et al. (2011).

15 It was the father of taxonomy, Linnaeus, who first emphasized the importance of the head in the identity of vertebrates and referred to them as *craniates*, a usage that has been revived by Janvier (1996) and others.

16 Until the 1980s, it was generally believed that the complement of neurons in the brain is fixed from early childhood, when net brain growth ceases, and that the only change in their numbers is decline during aging. This rather depressing perspective, however, turns out to be false: there exist many neural brain stem cells that can generate new neurons. Some new neurons can be generated on demand as it were, with appropriate learning regimes (e.g., violin playing or learning immense amounts of local geography as London taxi cab drivers must).

17 See Lieberman, McBratney, and Krovitz (2002) and Lieberman (2009). One selective factor that may have contributed to the special roundedness of the *Homo sapiens* head, unknown in other hominins, is that this shape contributes ideally to balance during running. The idea that long-distance, endurance running involved in chasing wounded game to ground, contributed to shaping the head, along with other features of the skeleton, is discussed in Bramble and Lieberman (2004).

18 For a discussion of the role of physical factors in particular tensile stresses ("mechanical load"), see Lieberman (2011), pp. 44–51, while the role of different nutritional systems in affecting the size and shape of the mandible is described in Cramon-Taubadel (2011).

19 On the elements that go into recognition of gender differences in faces, see Burton, Bruce, and Dench (1993) and Brown and Perrett (1993).

20 Evidence that testosterone levels, including those experienced in the womb, can exert a masculinizing effect on facial features (effects that are perceived as such by women) and other secondary sexual characteristics is reported in Fink et al. (2005) and Mareckova et al. (2011).

21 For the role of sexual selection in shaping the characters of men's and women's faces, see Weston, Friday, and Lio (2007) and Windhajer, Schaefer, and Fink (2011).

3. THE GENETIC FOUNDATIONS OF THE FACE

1 The height of genetic deterministic thinking was reached during the lobbying for financing of the Human Genome Project in the late 1980s. During this period, several eminent scientists claimed that knowing the full human genome sequence, the

"Book of Life," would uncover human biology to its depths, revealing what it means to be human, and so forth. Such rhetoric became rarer, however, as the project neared completion in the early 2000s and it became apparent that the complete human genome sequence could not deliver such understanding. This conclusion, in fact, had been reached much earlier and from basic considerations; see, for example, Wilkins (1986), pp. 484–486. For a full dissection of the fallacy of genetic determinism, however, see Dennis Noble's *The Music of Life* (2006).

2   A fine exposition of the role of environmental factors in altering pathways of development and thus the properties of organisms is that of Lewontin (2000).

3   A classic explanation of the story of how genes give rise to proteins is provided in James Watson's *The Molecular Biology of the Gene* (1970). For a recent short treatment, see chapters 2 and 3 of Fields and Johnston (2010). The landmark history of the discoveries that created the molecular biology revolution of the 1950s and 1960s is that of Judson (1979). Preceding that revolution was the classic era of genetics; an excellent account of its history comes from a great twentieth-century geneticist, Alfred Sturtevant (1965).

4   A recent and penetrating review of noncoding RNAs and their multiple functions is that of Cech and Steitz (2014).

5   The work was published in 1961 and is known as the *lac operon model*. A full description is in Jacob and Monod (1961). For histories of this work, see Judson (1979), chapter 7, and Carroll (2014), chapter 28. For a review of the work that established the reality and importance of "positive control" in gene regulation, see Hahn (2014).

6   See Hofmeyr (2007) on the cell as a "self-fabricating" entity.

7   To distinguish gene names from their corresponding protein products, genes are written in italics and proteins in roman type. To further distinguish human genes and their products from those of their counterparts in other vertebrates, the former are written in capitals and the latter in lowercase letters. Hence, FGF8 designates that specific human protein, *FGF8* the corresponding human gene, and *fgf8* the mouse (or other vertebrate) gene. Since most of the molecular biology was initially worked out in experimental organisms, most of the gene names given here will be lowercase. To readily distinguish protein products from their genes, all protein names will be given in capital, nonitalic letters.

8   The name sonic hedgehog should be explained. The story starts with the original *hedgehog* gene in the fruit fly *Drosophila*. The gene was named for the appearance of the larvae in the original mutant. Insect larvae consist of an array of body segments, with the ventral side of the abdomen showing thin bands of bristles, one per segment. In the mutant, these bristles cover the whole ventral surface, giving a hedgehog-like appearance. When the fruit fly gene was used to purify genes with a similar DNA sequence from vertebrate genomes, three genes were found. The first two were given the names of actual hedgehog species, *Indian hedgehog* and *desert hedgehog*. The third, however, was discovered by a graduate student with a sense of humor, who named his gene after a video game character, Sonic Hedgehog (Riddle et al. 1993). The name has stuck. There—is, however, a nonhumorous side effect of the name; this involves clinical situations where doctors have to explain to parents the nature of the defect in their children when this gene is mutant.

9   See Wolpert (1969).

10 For a review of limb bud development, see Towers and Tickle (2009). A good treatment of the FEZ, including a comparison of differences between the FEZ in chick

and mouse embryos, is that of Hu and Marcucio (2009). The laboratories that have played a major role in elucidating facial development are those of Ralph Marcucio, Jill Helms, and Benedict Hallgrimmson.

11 For Fgf8 in limb development, see Crossley et al. (1996); for its roles in face development, see Brugmann et al. (2006).

12 For the roles of *Shh* in bat wing growth, see Hockman et al. (2008). For loss of *Shh* expression in hind-limb buds in dolphins, see Thewissen et al. (2006). The role of SHH in regulating the size of the midface as a whole is documented in Young et al. (2011) and in that of individual facial primordia in Young et al. (2014). The relationships between SHH pathway deficiencies and holoprosencephaly are described in Cordero, Tapadia, and Helms (2005).

13 For a critical evaluation of the significance of retinoic acid, see Lewandoski and Machem (2009).

14 For the role of the Wnt pathway in some of the details of facial patterning, see Brugmann et al. (2007) and for the BMPs and the FGF8 pathway, Brugmann et al. (2006). BMP-SHH interactions in limb development are described in Bastida et al. (2009), and synergistic reactions between SHH and FGF8 pathways in face development are described in Abzhanov and Tabin (2004). These papers help explain the complicated relationships between the different pathways involved in limb and face development.

15 For *Dlx* genes in limb and face development, see Kraus and Lufkin (2006). For *HAND2* roles in development, see Liu et al. (2009).

16 The name of the *Hox* genes should be explained. They were first discovered in fruit flies and initially defined by their striking mutant phenotypes, the replacement of one segment type by seemingly another, a phenomenon termed *homeosis.* An example of such homeotic mutants is *bithorax,* in which the small metathoric segment of the fly is replaced by a large mesothoracic-like segment, leading to a four-winged fly (instead of the normal two wings.) When the genes were cloned in the 1980s, many were found to share a region encoding a sixty amino acid "motif" that allows binding of the entire protein to a region of DNA. These genes encode transcription factors, and the region of the gene encoding the binding region of the protein was termed the *homeobox,* with *Hox* becoming the acronym of choice for the genes themselves.

17 See Wilkins (2014).

18 The evidence that different modules can be recruited for different developmental processes is given with examples in Davidson (2006); a genetic-evolutionary framework for this phenomenon is provided in Wilkins (2007a).

19 There is a large literature on the resemblances between fin and limb development. Most of the same molecular players involved in the early steps—*Hox* genes for the initial specification of the positions of the limb buds, so-called T-box transcription factors for differentiating forelimbs and hind limbs, and the various signal transduction systems (SHH, FGFs, WNTs, BMPs)—are used in largely similar fashion. It is the late steps involving the development of the autopods in limbs, where the major genetic-molecular differences become apparent. For another example of a *shared core module,* this one involving eyes and muscles, see Heanue et al. (1998).

20 See Tucker et al. (1998) for the evidence of shared genetic machinery between the mandibular primordia and the limb buds.

21 Darwin recognized but did not emphasize that evolution works with the materials to hand in the origination of new structures, not from scratch as an engineer might. The modern simile that compares evolution to tinkering comes from Francois Jacob,

whose role in elucidating the basic facts of gene regulation was discussed earlier. Jacob's argument that evolution works as a tinkerer rather than as an engineer was intended to rebut the basic tenet of evolutionary biologists that evolution is an ever-optimizing process. While natural selection acts to improve existing traits with time, the initial steps by which new features arise *cannot* be optimizations. Rather, they have the character of someone reaching for something ready to hand that will do the job—the response of a tinkerer, not an engineer. See Jacob (1977), Jacob (1982), and Duboule and Wilkins (1998).

22 See Rakic (1995).

23 For a clear, if detailed, explanation of how this might take place, see Fish et al (2008). For the genes identified in humans that create this condition, see Montgomery et al. (2010); for a description of the effects of these mutations on cell divisions, see Pulvers et al. (2010).

24 See Bastida et al. (2009).

25 The idea of canalization is described well in Waddington (1957). For one aspect of its basis, see Wilkins (1997).

26 A good discussion of the general point about developmental flexibility and evolvability can be found in Lieberman (2011), pp. 10–12. This issue has also been discussed by Deacon (2000) while "developmental accommodation" is described in West-Eberhard (2003).

27 See Coleman (1964), chapter 7.

28 Charles Darwin's classic treatise *Variations in Animals and Plants Under Domestication* (1868) remains a fine survey of the hereditary variations found in domesticated animals and plants and the results of their breeding. For a recent look at the extents to which selective breeding for domesticates can unexpectedly alter the genetic material and physical traits of breeds without altering their species' identities, see Glazko et al. (2014).

4. THE GENETIC BASIS OF FACIAL DIVERSITY

1 The surprising extent of human facial diversity has long been appreciated, as seen in this comment by Sir Thomas Browne (1605–1682), a seventeenth-century polymath— doctor, writer, philosopher—and physiognomist: "It is the common wonder of all men, how among so many millions of faces, there should be none alike" (Browne, 1645). His comment can only be faulted for understatement: today, we would speak of billions of different faces.

2 In considering the influence of relatedness on the degree of facial resemblance between individuals, it is worth noting that the word *unrelated* is ambiguous. The members of one species are all related to each another, only to different degrees; in effect, every species consists primarily of a vast set of cousins (of different "removes"). Furthermore, our species is a relatively new one, probably no more than 200,000 years old (Chapter 6) or approximately 10,000 generations, a relatively small number for the age of a species. Hence, humans are far more closely related than members of most other animal species.

3 The degrees of facial difference within the species of the four kinds of great apes (chimpanzees, bonobos, gorillas, orangutans) can be seen in two fine compilations of photographic portraits of apes: *James and Other Apes* (Mollison 2005) and *Menschenaffen Wie Wir (Apes Like Us)* (Hof and Sommer, 2010). Other primate species also have detectable degrees of facial diversity, as reported in particular

by field zoologists, but to human eyes facial variability in the great apes is much more striking.

4   The term *wild-type allele* predates the molecular era, reflecting a time in which the properties of a gene could only be deduced from the appearance of a trait that the gene affected, its phenotype. It was assumed then that if a trait looked identical between individuals, then the genes underlying that trait must be, too. Today, given the analytical power of gene sequencing, it is apparent that a lot of hidden DNA sequence variability does not affect phenotypes, hence the idea of the wild-type allele is a fiction, embracing all those alleles that essentially give a fully functional gene activity. As shorthand for all those normal versions of a gene, however, it is still useful.

5   Today, it is known that the *R* gene (designated by its wild-type allele) specifies an enzyme of starch metabolism. The *r* form (allele) , however, has a bit of exogenous DNA sequence inserted into it, which disrupts the protein chain, making it incapable of carrying out its normal function. The *rr* plants thus lack the functional enzyme and the subsequent interference with starch metabolism, triggers other metabolic effects, ultimately changing the water content of the pea and leading to the wrinkled pea form. For a description of modern knowledge of the seven garden pea genes studied by Mendel, see Reid and Ross (2011).

6   In recent years, a phenomenon has been discovered that requires a slight modification of the idea that all individuals have the same gene set (apart from the X–Y chromosome difference.) It concerns something that occurs during the first cell division of meiosis (meiosis I) in reproductive cells during the production of the gametes—namely, sperm in males and eggs in females. This process is responsible for creating some differences in total gene content between individuals. In meiosis I, the two copies of each chromosome, the two *homologues,* line up and undergo switching of segments through the process of genetic crossing over (see note 4 in Chapter 3.) Sometimes this process is not precise and leads to the production of *copy number variants* (CNVs) in which a small number of genes (per reproductive cell) undergo an increase in number through the error of *unequal crossing over*, and leading to duplicated genes in tandem. It has become increasingly apparent that many CNVs have effects, including some major clinical effects, in the individuals that carry them in their genomes. These meiotic errors, however, are comparatively rare and the gametes can effectively be regarded as having the same gene content consisting of one *haploid* set of genes.

7   We can demonstrate these combinations graphically in what is known as a Punnett square, named after the early twentieth-century geneticist R. C. Punnett, who invented this method. One first writes the gamete (haploid) genotypes among the eggs in a horizontal straight line and then the same genotypes from the sperm running down a line perpendicularly. Each pairwise combination then generates a box, with each square occupied by the resulting diploid genotype. This allows a ready graphical representation of the set of relationships between gamete and fertilized egg (diploid) genotypes (Edwards 2012). A simple Punnett square is shown in Figure 4.2.

8   There are, in fact, two distinct forms of gene *shuffling*—that is, the mixing up of the original contributions of the maternal and paternal gene sets: that of random segregation in the first division of meiosis of the members of each pair of chromosomes, and the phenomenon of **genetic recombination**. The latter involves the exchange of segments of chromosomes in a process of breakage and the highly accurate rejoining of chromosomes during meiosis.

9   See Brugmann and colleagues' discussion in the article titled "Looking Different: Understanding Diversity in Facial Form" (Brugmann, Kim, and Helms 2006).

10 The experiments and observations are described in Roberts et al. (2011).

11 A general treatment of enhancers as modular sites for changing patterns of a gene's transcription is that of Davidson (2006), while a recent review is that of Schyueva, Stampfel, and Stark (2014). Not all *cis*-regulatory sites are enhancers, however; other kinds exist that act via different mechanisms and are probably significant in evolution (Alonso and Wilkins, 2005). A good review of the evolutionary significance of *cis*-regulatory sites in evolution , permitting localized effects of pleiotropically employed genes, is that of Stern and Orgozogo (2008).

12 *Cis*-regulatory sites are, of course, not the only sources of mutations significant in developmental evolution. Another route involves the structural organization of proteins, specifically the fact that many are organized into separately folding domains, each with a discrete biochemical activity. Because of the frequently complex employment of different domains within a protein in different developmental roles, mutations in one domain can often affect one function with slight or nil effects on others (Alonso and Wilkins, 2005; Wagner and Lynch, 2008) Lynch and Wagner 2008; Wilkins, 2007a).

13 This phenomenon is discussed in Wilkins (1997).

14 For a discussion of gene expression without function, see Yanai et al. (2006).

15 The two studies are those of Feng et al. (2009) and Attanasio et al. (2013).

16 See Hill (2012).

5. HISTORY OF THE FACE I

1   That two eyes and a mouth instantly suggest a face is readily illustrated by the smiley face symbol. More intriguingly, the front views of automobiles, with their paired headlights, are often perceived as faces, with different car fronts conveying the sense of different personalities (Windhajer et al. 2011).

2   The Cambrian period was named by nineteenth-century biologist Adam Sedgwick, a senior colleague of Darwin's who found the first fossils of this period in Wales, a region whose earlier name was *Cambria*.

3   Stephen Jay Gould's book on the life of the Cambrian era, *Wonderful Life* (1991), in particular the large, strange, soft-bodied animals of the Lagerstaette of the Burgess Shale in British Columbia, remains a classic description of those forms. It is slightly marred, however, by a thesis that is probably wrong: that the Cambrian saw an explosion of new animal phyla, most of which went extinct. Today, most of the very strange fossils of the Burgess Shale and comparable sites in China are believed to be the remains of early, *stem* representatives of modern phyla (Budd, 1998). For a scholarly and well-illustrated account of the animals of the Burgess Shale, see Briggs et al. (1994).

4   For a review of pre-Darwinian attempts to show relationships of both living things and languages using treelike imagery, see Sutrop (2012).

5   All three bilaterian supraphyla were named in reference to particular developmental characteristics shared by the phyla within each subdivision. Thus, the Ecdysozoa were named for their shared trait of molting their outer cuticles under the influence of hormones of the ecdysone family. The Lophotrochoa were designated on the basis of certain shared traits in their larval forms, while the Deuterostomia share a pattern of embryonic development in which two openings, "mouths," characterize the embryos. See Adoutte et al. (2000).

6   The two papers that originally set forth the "new head" hypothesis and the data and reasoning a behind it are Northcutt and Gans (1983) and Gans and Northcutt (1983). For a revised and more up-to-date discussion, see Northcutt (2005).

7   The new phylogenetic results are in Philippe, Lartillot, and Brinkmann (2005) and Delsuc et al. (2006).

8   For details, see Yu (2010) and Simoes-Costa and Bronner (2013).

9   The findings are described in Jeffrey, Strickler, and Yamamoto (2004) and some current caveats given in Simoes-Costa and Bronner (2013). Another insightful, though earlier, review is that of Donoghue, Graham and Kelsh (2008).

10 For this argument about the origins of neural crest cell migratory ability, see Kerosuo and Bronner-Fraser (2012). A good review of the phenomenon of gene recruitment, also termed *gene co-option*, is that of True and Carroll (2002); the idea was anticipated, though not named, in the celebrated paper on evolutionary "tinkering" of Jacob (1977).

11 On the evolution of the neural placodes, in relationship to that of the neural crest cells, see Schlosser (2005, 2008).

12 For a discussion of these ideas, see Retaux et al. (2014).

13 The first investigation was that of Meulmans and Bronner-Fraser (2007), the second that of Jandzik et al. (2015).

14 For the two putative Cambrian craniates, see Shu et al. (1999). A thorough and still valuable, if slightly dated, general treatment of early vertebrate evolution is that of Janvier (1996).

15 For a comparison of the two ideas, see Mallatt (2008).

16 See Cerny et al. (2010).

17 For example: "The basic craniate pattern, which evolved in the Ordovician or even in the Cambrian, has remained unchanged for some 450 to 500 million years. The developmental and anatomical complexity attained by gnathostomes has been relatively stable at least since the Silurian." (Maisey 1996, p. 68).

18 For a detailed account of the emergence of the tetrapods, the first land-living vertebrates, see Clack (2012).

19 The exception to this general rule is the group of frogs known as *direct developers*. In these species, the young hatch directly into tiny froglets within the body of the mother and are released through her cloaca, already breathing air. These frogs, however, still require nearby aqueous environments. Direct development in amphibians, however, is a relatively late evolutionary development.

20 The term *mammal-like reptile* has been given up by evolutionary biologists. Indeed, while *reptiles* survives as a taxonomic category, it is no longer considered an evolutionary one. (The problems of combining taxonomic categories with evolutionary ones will be discussed in Chapter 6.) With respect to the original defining skull features of diapsids and synapsids, those also have proved malleable in evolution, with, for example, snakes and turtles having respectively no and one temporal fenestra and, even more strikingly, some mammals having none. It is the *ancestral* skull traits of these groups of animals that define them, respectively, as diapsids and synapsids.

21 The causes and consequences of the great end-Permian mass extinction are detailed in Stow (2012). A full description of the large changes in terrestrial ecosystems during the Triassic is given in Sues and Fraser (2010).

22 See Rubidge and Sidor (2001) for descriptions of the early forms and stages in synapsid evolution. A full history and description of the synapsids up to and including the first mammals is given in Kemp (2005).

23 The newer findings concerning mammalian diversification in the Age of Dinosaurs are discussed in Luo (2007).

24 The lineage that would eventually give rise to the modern egg-laying mammals, the monotremes, split off from what would give rise to the therians in the early Jurassic (circa 190 mya), as judged primarily from molecular clock data. The roots of the marsupial and placental lineages probably extend back to the mid-Jurassic (circa 160 mya). The question that vexes students of mammalian evolution is how ancient the evolutionary roots of the modern *crown group* mammals are, as will be discussed briefly later in this chapter. (The terms *crown group* and *stem group* will be defined more explicitly in the next chapter.)

25 For recent estimates and discussions of evolutionary history of the Mammalia, and in particular the placental mammals, see Wibble et al. (2007), Luo et al. (2011), Meredith et al. (2011) and dos Reis et al. (2012). The latter two articles present the base case for roots of the modern mammalian orders stemming mostly from the late Cretaceous.

26 See O'Leary et al. (2013) for the full analysis leading to these conclusions—and for a rebuttal, dos Reis et al. (2014).

27 The issue need not forever be unresolved, however. When the great twentieth-century evolutionary biologist J. B. S. Haldane was asked if there were anything that could convince him that the theory of evolution was not true, he is said to have replied (without a pause), "Finding a rabbit [fossil] in the Ordovician." Similarly, the finding of one true mammalian crown group fossil dated to the late Cretaceous would disprove the central conclusion of O'Leary et al.

28 Not surprisingly, the idea of tooth specification by a *Dlx* code based on nested gene expression is too simple to explain the entire complexity of tooth diversity, though it is probably part of the explanation. A detailed evaluation is in Depew et al. (2005).

29 For a detailed discussion of the evidence relevant to hair evolution and the two hypotheses, see Alibardi (2012).

30 For the differences in genetic regulation between the trunk and craniofacial muscles, see Sambasivan, Kuritani, and Tajbakhsh (2011). Discussions of how genetic networks in development can evolve are in Wilkins (2007a, 2007b, 2014).

31 See Oftedal (2012).

32 I am grateful to Olaf Oftedal for the suggestion that the reduced muzzles of mammalian young is probably related to their need to nurse and to Michael Depew for reminding me that the young of mammalian species that make a rapid transition to adult diets, such as herbivores, are born with larger muzzles than those that nurse exclusively for their nutrition for longer periods.

6. HISTORY OF THE FACE II

1 Orangutans, Asia's only great apes, are usually regarded as the exception to primate sociality, being highly solitary creatures throughout much of their range. Yet one group, those found in the alluvial forests of western Sumatra, live in groups and are highly sociable (van Schaik, 2004). With respect to the human propensity to interact socially with all our conspecifics, this trait is probably shared with only one other animal species, namely, our "best friend," the domestic dog.

2 For a general description of the different kinds of primates, and their evolutionary history, see the text by Fleagle (1999). Though dated in some details and lacking

molecular information about relationships, it remains a good general source on the anatomy, ecological relationships, adaptive features, fossil history and evolution of the primates. For a short and up-to-date account of primate evolution, see Martin (2010).

3   For *Teihardinia,* see Beard (2008); the probable relationships of the plesiadapforms to the crown group primates are discussed in Bloch et al. (2007).

4   The first analysis of this kind was carried out by Tavare et al. (2002). A later and more detailed analysis that produced estimates of the ranges of dates for the origins of both primates as a whole and subgroups is that of Wilkinson et al. (2011). Martin (2006) provides a good, clear, and nontechnical discussion of these issues.

5   The analyses and the reasoning are given in Steiper and Seiffert (2012), but also see Bromham (2011) for a general discussion of the inverse relationships between life-history traits and the running of the molecular clock. The estimates of primate origins in Steiper and Seiffert fall closer to the end-Cretaceous boundary than the analyses of Tavare et al. (2002) and Wilkinson et al. (2011) but still within the Cretaceous.

6   See Steiper and Seiffert (2012) for this discussion of the ancestral primate but also Gebo (2004), who argues for the likelihood of an even smaller ancestral primate.

7   See Martin (1990) and Soligo and Martin (2006).

8   As a point of historical interest, it should be noted that the first published phylogenetic tree of a group of living animals was of the primates and appeared in 1865. It was constructed by St. George Mivart (1827–1900), Darwin's most important contemporary critic, a Catholic who evidently struggled to reconcile his religious faith with his conviction of the reality of evolution. (His basic argument with Darwin was over the issue of whether slow increments of change could create truly novel features when significant function would not be expected to accrue incrementally but would only appear when the change is complete.) While Darwin formulated the idea of the phylogenetic tree as a picture of evolutionary change over time and drew the first one—the frontispiece and sole figure in *The Origin of Species*—it was a wholly abstract conception, neither based on nor referring to any actual organisms. Mivart took Darwin's concept one step further by constructing a phylogenetic tree specifically for the primates. It was based solely on comparative anatomy and morphology—there was effectively no fossil evidence at the time and trees based on molecular sequence data were a century in the future—yet it foreshadows the modern versions in showing the division into *prosimians* and *simians* (see text), a division anticipated by Linnaeus. See Bigoni and Barsanti (2011).

9   This older idea, which goes back to Aristotle, denoted the "Great Chain of Being" and is a linear, hierarchical ordering of major forms of life from plants at the bottom to humans (or, in some later versions, angels) at the top. Its influence was still felt in the nineteenth century and early twentieth in the belief shared by many scientists that humans were the predestined end product of evolution; for this history, see chapter 9 in Bowler (1986). The history of this idea is described in a classic of biological history by Lovejoy (1936).

10 The difference almost certainly reflects a difference in embryonic development in the extent to which the two mandibular facial primordial fuse. One may think of the strepsirrhine split-lip condition as a form of natural, developmentally programmed "cleft lip." In these animals, the two halves of the upper limit are anchored just beneath the nose in an internal strip of soft tissue termed the *philtrum* (a term that is also used to denote the visible outer groove in the upper human lip, a uniquely human feature.)

11 The three traits named, however, are not the ones that anatomists and taxonomists focus on. For them, the definitive haplorhine trait is *postorbital closure*, completion of the bony ring from around the rim of the eye to surround the whole eye. A further ten relatively subtle cranial and dental traits collectively distinguish haplorhines from the strepsirrhines as listed in Table 1.1 of Miller, Gunnell, and Martin (2005).

12 For a thorough discussion of anthropoid origins, especially its biogeography, see Miller, Gunnell, and Martin (2005). For a more current if less detailed treatment, see Williams, Kay, and Kirk (2010).

13 For these datings, see Steiper and Young (2006) and Wilkinson et al. (2011).

14 The name *Proconsul* derives from that of a famous show-business chimpanzee, Consul, who performed at the Folies Bergère in Paris in the early 1900s. The name *Proconsul* was coined when its first fossils were found in the early 1930s, indicating that this was a species that existed before chimpanzees. Its genus is the best characterized of the Miocene apes.

15 For a discussion of early ape evolution and diversification and spread out of Africa into Asia, see Andrews and Kelley (2007).

16 This issue became the focus of a legendary debate held in the lecture hall in what is now the Pitt-Rivers Museum of the University of Oxford in 1860 between an opponent of Darwinian evolution, Bishop Samuel Wilberforce (dubbed "Soapy" Wilberforce by his detractors for his oratorical style), and Darwin's most vocal champion, Thomas Henry Huxley. The debate was intended to be a broad-ranging discussion of the merits (or lack thereof) of Darwin's theory. From eyewitness accounts, it seems that the key elements of Darwinian evolution were largely neglected in the debate, which is remembered today chiefly for Huxley's brief but witty putdown of Wilberforce's pomposity and ignorance.

17 The term *the missing link* seems to have been coined only in the 1890s, but the objection to Darwinism that it embodied—the absence of evidence for transitional forms between apelike creatures and humans—was invoked in many of the early criticisms of Darwinism.

18 Dubois's story is recounted in Shipman (2001). Histories of the revival of Darwinian theory during the 1920s and 1930s, initiated primarily by new genetic findings and analysis, are in Provine (1971) and Wilkins (2008).

19 Apart from the lack of good fossil evidence, there is the vexing problem that defining extinct species, *paleospecies*, is inherently difficult. Modern animal species are defined in terms of their breeding with one another exclusively (though there are species whose members clearly sometimes stray to have relations beyond their species' normal boundaries), but that operational definition is useless for extinct animals. Hence, judgments of species membership from fossils are based on morphological resemblance and rough temporal concordance. Morphology, however, is an uncertain guide, with many mammalian species showing some variability in their traits. The specific difficulties and ambiguities in deducing the early hominin phylogenetic tree are discussed in Wood and Harrison (2011).

20 A highly readable account of much of the history of discovery of fossil australopithecines and early *Homo* species in Africa, though centered on one particularly important set of discoverers, Louis Leakey and his family, is that of Morell (1995).

21 A recent review of the early history of the genus *Homo* is that of Anton, Potts, and Aiella (2014). These authors regard *H. habilis* and *H. rudolfensis* not as discrete species but as species groups, as they do *H. erectus* itself. Perhaps even more important,

they relate the variability in body size and other physical features within these populations to findings indicating highly variable climates and landscapes during the past 2 million years in Africa, essentially the period of the Pleistocene, which was characterized by multiple ice ages in the Northern Hemisphere and worldwide climatic effects.

22 For a thorough discussion of the data, issues, and ambiguities of later hominin evolution, see Lieberman (2011), chapter 13.

23 A full set of these reconstructions of hominins as they might have looked in flesh and blood are depicted in the book of Sawyer and Deak (2007). The methods of forensic reconstruction of extinct hominins are described in Sawyer and Deak's appendix, and the history of how early humans have been visualized is given in a second appendix by Richard Milner. The reconstructions reprinted in this book (Plates 9–13) were made by John Gurche.

24 The terms *muzzle* and *snout* are often used interchangeably, but there is a difference: *snout* refers to the part of the muzzle that bears the nostrils—namely, the upper jaw— while *muzzle* refers to the ensemble of the two jaws.

25 The roles of certain BMPs in muzzle development are explored in Brugmann, Kim, and Helms (2006). A specific involvement of BMP3 in head and jaw shape has been found in dogs (reviewed in Schoenebeck and Ostrander 2013). For the synergistic effects of *Fgf8* and *Shh* in facial bone development, hence muzzle development, see Abzhanov and Tabin (2004). The role of Wnts in facial patterning are described in Brugmann et al. (2007).

26 In 2004, an intriguing report indicated that the extent of muzzle outgrowth in dogs was a function of certain genetic changes in the gene *Runx2*, a gene for a transcription factor expressed in and required for muzzle development (Fondon and Garner 2004). These had to do with the numbers of a small repeated DNA sequence in the coding region of this gene; the greater the number of those repeats, the larger the jaws. Subsequently, however, this correlation was found to hold solely for members of the order Carnivora and not for other mammals (Pointer et al., 2012). Evidently, there need not be only one mechanism for setting muzzle size and length in mammals.

27 For the evidence on the accelerated evolution of the *ASPM* gene in the hominoid lineage, see Kouprina et al. (2004). These authors discuss ways in which this gene product might have evolved to extend the period of neural cell progenitor divisions. Those mechanisms, however, are less direct than extended expression of the microcephalic genes during cortical development.

28 For good reviews, see O'Leary, Chou, and Sahara (2007) and Pierani and Wassef (2009).

29 For further information on *FoxG1*, see Hanishima et al. (2002); for DUF1220 domains, see Dumas and Sikela (2009) and O'Bleness et al. (2012).

30 Incidentally, Darwin made his case for evolution by increments in human evolution in *The Descent of Man*—not from fossil evidence, which did not exist then, but primarily on the evolutionary continuities between certain mental states and behaviors in animals, particularly apes and humans. His argument was that with the roots of those qualities so obvious in our closest animal kin, it took no stretch of the imagination to believe that quantitative enhancement of those qualities could produce their human equivalent. (His fundamental belief in the general importance of variants of small effect as the source of evolutionary change, however, was grounded in his observations of how animal and plant breeders worked.) As will be discussed, such extrapolation is probably not a sufficient argument.

31 The reversal toward a smaller brain occurred within the islands of Indonesia, resulting in the relative dwarf species *Homo florensis,* probably a descendant of *H. erectus,* found on the island of Flores. Animal evolution on islands frequently shows size reduction, and *H. florensis* may be an example.

32 A particularly good look at the phenomenon of emergence has been provided by Harold Morowitz in *The Emergence of Everything,* in which he follows a sequence of twenty-eight emergent events that can be traced from the Big Bang to the arrival of the human mind (Morowitz 2002). This approach, too, is a *hominocentric* one because the sequence leads to and ends with our species.

33 Darwin's reasoning is given in *The Descent of Man* and elaborated in Richard Leakey's *The Origin of Humankind* (1994). In addition to the fact that his hypothesis is not supported by the evidence, it ignored the fact that apes use their hands extensively, which presumably was true of early hominins as well.

34 An early but informative review on the use of gestures, both facial and hand, by apes is that of De Waal (2003). The recent work on wild chimpanzees is described in Hobaiter and Byrne (2011), who have described sixty-six gestures, each apparently associated with a specific communicative intent.

35 See Senghas, Kita, and Ozyuerek (2004).

36 A full account of the hand-to-mouth theory is that of Corballis (2002).

37 Gentilucci and Corballis (2006) describe one scenario for the transition from a primarily gesture-based communication system to a vocalized one. Another, and more detailed, "co-evolution" scenario, emphasizing the underlying neural evolution, is that of Garcia et al. (2014).

38 A full analysis of the requirements for human speech and how its mechanics might have evolved is found in Fitch (2010), chapters 8–10.

39 A detailed, informative, and provocative book-length account of the probable evolutionary origins of human symbolic thought capacity is given in Deacon (1997).

40 See Ghazanfar and Takahashi (2014) for a review of this work and its evolutionary implications. The first suggestion that rhythmic facial movements in catarrhine primates were the precursors to speech is that of MacNeilage (1998).

41 For *Teilhardina's* impressive migrations, see Smith, Rose, and Gingerich (2006). The fact that Earth has experienced overall climate change of +10 °C without wiping out all complex animal life is good news for those who fear it could. Nevertheless, the rise in ocean levels and other changes, should this occur again, would play havoc with human life in many ways.

42 For discussion of climatic variability during the Pleistocene and some of its consequences for human evolution, see Anton, Potts, and Aiella (2014). For the roles of meat eating, and even more important, the adoption and use of fire in cooking, which may have greatly increased nutritional and caloric value of foods and profoundly influenced human evolution, see Wrangham (2009).

43 An early, formal argument for social selection and its possible ground state of docility is that of Simon (1992). A good book-length treatment of the origins of language within the context of social cooperativity is that of Tomasello (2008). A more general argument case for social selection in shaping cultures and its influence on human evolution is that of Richerson and Boyd (2005).

44 See Tomasello et al. (2007) on ascertaining gaze direction from the eyes and its social functions in humans. Wolves use light irises, framing dark pupils, against a darker background of fur surrounding the eyes to detect the gaze of conspecifics; wild,

nonsocial canids of other types do not have these visual clues and pay less attention to the gaze of their fellows (Ueda et al., 2014. Domestic dogs, descended from wolves and famously sociable, clearly inspect their owners' gaze for clues as to what is going on.

7. BRAIN AND FACE COEVOLUTION

1 Traditionally, the emotions are seen as clouding rational judgment. In fact, they are often an essential guide in sensibly evaluating situations. This appreciation, however, is relatively new historically. For still useful if somewhat older general accounts of how the emotions are involved in rational decision making, see Damasio (1994) and Panksepp (1998).

2 For example, see Krubitzer (2009), Krubitzer and Stolzenberg (2014), Aboitz (2011), and Finlay and Uchiyama (2015).

3 See Krubitzer (2009).

4 For a detailed but readable account of human vision, see Kandel (2012), pp. 225–303.

5 For stereopsis and its roles in primate brain evolution, see Barton (2004).

6 One careful evaluation of this idea is in Regan et al. (2001).

7 The snake-detection hypothesis is that of Isbell (2006, 2009). Her account of the neurobiology that might be involved in rapid snake detection is intriguing. Nevertheless, the idea remains controversial. For example, see Wheeler, Bradley, and Kamiliar (2011), although their critique focuses on evidence of contemporary anthropoid primates and snakes, not the initial circumstances when anthropoid primates first arose.

8 There is substantial literature on the FFA and facial recognition. A good review is that of Kanwisher and Yovel (2006), though it neglects the roles of the STS and the OFA. For the importance of number of neurons devoted to particular faces in the FFA, see Kandel (2012), p. 296, and Viskontas et al. (2009).

9 A good and relatively recent review of what is known about the entire nexus of neurobiological dynamics of face processing, including but not limited to face-identity recognition, is that of Atkinson and Adolphs (2011).

10 Though a high level of individual facial recognition is a general human trait, some people are better at it than others; at the extreme are super-recognizers who are extraordinarily good at it (Bennhold 2015). Furthermore, identical twins are more similar in their facial-recognition capacities than same-sex fraternal twins, indicating a strong genetic component to this ability (Wilmer et al. 2010).

11 A thorough review of the capacities of different animals to recognize faces as a distinct category of object and to identify individual faces is that by Leopold and Rhodes (2010). The references for many of the statements made in this section can be found there.

12 See Tibbetts (2002).

13 For the comparative aspects of *Polistes* face-recognition abilities and the potential of species that do not normally show individual face recognition to be trained to learn some degree of face discrimination, see Sheehan and Tibbetts (2011).

14 See Reiss (2001) for the bottlenose dolphin work; for elephants, see Plotnik, de Waal, and Reiss (2006).

15 It would be of interest to know whether the monotremes and the marsupials also possess face-recognition capacities, but they also have not yet been tested. Should they be found to have conspecific face-recognition capacity, especially if centered in the ventral temporal lobe, its evolutionary beginnings of this ability would be pushed back to the origins of the mammals in the Jurassic, or even earlier in the history of the synapsids.

16 The work on sheep is that of Keith Kendrick and his colleagues. An early but still useful description of the general properties of the sheep face-recognition system can be found in Kendrick (1991). For a relatively recent review, see Tate et al. (2006).

17 See Freiwald, Tsao, and Livingstone (2009).

18 For the learning element in the "other race" effect, see Anzures et al. (2013).

19 For the measurement and comparison of chimpanzee and human expressions, see Vick et al. (2007) and Parr et al. (2007).

20 The comparisons of mimetic muscles for different kinds of mammals are in Diogo et al. (2009); prosimian mimetic muscles are examined in Burrows and Smith (2003); chimpanzee facial muscles are described in Burrows et al. (2006); those of gibbons in Burrows et al. (2011).

21 Dobson and Sherwood (2011) review the earlier literature on the presumed lower expressivity of platyrrhine primates relative to catarrhines and produce data relevant to a possible neural difference (see the next section). Barton (2006) suggests that platyrrhines are less reliant on vision than olfaction, which could be related to a lower degree of facial expressivity. Marmoset facial expressions within the context of marmoset sociality are described in Kemp and Kaplan (2013).

22 The lesser robustness of the facial muscles in gibbons is described in Burrows et al. (2011) and the comparisons between different catarrhines by Dobson and Sherwood (2011).

23 For a discussion of the relationship between body size and facial expressivity, see Dobson and Sherwood (2011).

24 For the data on cranial nerve VII and a discussion of their significance, see Sherwood (2005). For the case of birds, see Striedter (1994).

25 For discussions on cultural differences in facial expressions, see Matsumoto (1992) and Elfenbein et al. (2007).

26 The brain regions involved in face reading are discussed in Kawasaki et al. (2012).

27 For the integration of faces and vocalization in the VLPFC, see Romanski (2012). For voice-and-face association studies, see Sliwa et al. (2011).

28 For discussions of the earlier dates of possible proto-languages, see Deacon (1997) and Corballis (2011). The late estimate for at least "true" language is given by Klein and Edgar (2002).

29 The two learned societies to ban the subject were the Linguistic Society of Paris and the Philological Society of London (Corballis, 2011, p. 19).

30 For contemporary views of the true complexity of the neural architecture underlying language, see Aboitiz (2012); Garcia, Zamorano, and Aboitiz (2014); and Hagoort (2013, 2014).

31 For the concept of the *phonological loop,* see Aboitiz, Aboitz, and Garcia (2010) and Aboitiz (2012). For the argument that recursion is *the* defining property of true language, see Hauser, Chomsky, and Fitch (2002) and Corballis (2011). For an argument against this claim, however, see Bickerton (2009), chapter 9.

32 A good review on axon guidance molecules is that of Kolodkin and Tessier-Lavigne (2011). A comparable review of how the main connections in the brain are established and the role of *pioneer axons* is by Chedotal and Richards (2010).

33 The factors and processes involved in axonal pruning and neuronal apoptosis are described in Vanderhaeghen and Cheng (2010).

34 For an early Darwinian perspective on the shaping of neural connections see Edelman (1987). Later, Terence Deacon placed these aspects of the development of neural

connections within a Darwinian framework of variation in growth patterns followed by selection for those that generate a new, adaptive property (Deacon 2000). He emphasizes the role of slight changes in *bias* in outgrowth or pruning of connections (or cells) (Deacon 1997, chapter 7).

35 Selemon, 2013.

36 See Jablonka and Raz (2009) and Heard and Martienssons (2014) for discussions of epigenetic inheritance and epimutations; we return to this subject at the end of chapter 10.

8. "POSTSPECIATION"

1 As remarked previously, one other mammalian species exhibits extensive physical diversity, indeed far more than in humans and involving many body characteristics, including those in its faces—namely, the domestic dog. Dogs, however, are creatures of our making, generated through selective breeding for particular traits over centuries, although the great majority of breeds originated only within the past two centuries. The visible differences in breeds of other species of domesticated animals have the same provenance: selective breeding by humans of different varieties.

2 His comment on the multiplicity of ideas concerning the number of "races" was this: "Man has been studied more carefully than any other organic being, and yet there is the greatest possible diversity amongst capable judges whether he should be classified as a single species or race, or as two (Virey), as three (Jacquinot), as four (Kant), five (Blumenbach), six (Buffon), seven (Hunter), eight (Agassiz), eleven (Pickering), fifteen (Bory St. Vincent), sixteen (Desmoulins), twenty-two (Morton), sixty (Crawford), or as sixty-three, according to Burke." From Darwin (1871), vol. 1, p. 226.

3 The earliest study to show there is far more genetic diversity within individual "races" than between them is from Lewontin (1972), who used fifteen readily assayable proteins, each from a separate human gene. Later studies employed the DNA sequence markers termed *microsatellites,* which cover the entire genome at reasonable resolution; to a high degree, these studies confirmed Lewontin's estimates, which were based on a much smaller data set (Barbujani et al. 1997). The logical demonstration that groups with somewhat separate evolutionary histories will develop characteristic constellations of allele frequencies for different genes is given by Edwards (2003). Risch et al. (2002) present evidence that "racial" groups differ in just this way and discuss the implications for diagnostics of genetic predispositions toward various clinical conditions that can differ between the different groups.

4 The simplest demonstration that behaviors and abilities are largely determined by cultural environments concerns the rapidity and completeness of acculturation of young humans to new societies. One of the first descriptions of this process was that of Charles Darwin. He did so in his first book, *Journal of Researches* (published originally in 1838 and later titled *The Voyage of the Beagle*.) In it, he discusses the three young Fuegians taken from Tierra del Fuego three years previously to live in England and who were then returned to their homeland via *The Beagle* in 1833. Darwin was struck by how English they had become in their speech, dress, and manners, a set of changes that had taken place during their short stay in England. Such adaptability showed there could be no deep biological differences between "savages" and civilized Englishmen and revealed the importance of culture and upbringing on behavior. For modern commentaries on rapid acculturation between different racial groups, see Diamond (1997) and Richerson and Boyd (2005), chapter 3.

5 A history of these early debates about the evolution significance of different human "racial" groups is given in Bowler (1986), though he takes the history only to 1944.

6 The original modern statement of the MRE hypothesis is that of Wolpoof et al. (1984). The earliest version, which is not frequently cited today, was Coon's *The Origin of Races* (Coon 1963).

7 See Howells (1976) for the first version of the out-of-Africa hypothesis. The most influential early statement, however, is that of Stringer and Andrews (1988).

8 The work came out of the laboratory of Allan C. Wilson, who had pioneered the molecular dating of the divergence of humans and chimpanzees at 7 mya to 6 mya. For the "mitochondrial Eve" work, see Cann, Stoneking, and Wilson (1987).

9 For a recent review of the DNA evidence confirming an African origin of *Homo sapiens,* though prior to the discovery of archaic sequences in the genome, see Relethford (2008). The description of the nearly 200,000-year-old modern *H. sapiens* cranium is in McDougall, Brown, and Fleagle (2005).

10 The picture of human migratory paths following the exit from Africa presented in the following pages is largely based on two accounts by Stephen Oppenheimer (2003, 2012). I have borrowed his apt phrase, "the peopling of the world," for the phenomenon. Also, see Mellars et al. (2013).

11 See Kitter, Kayser, and Stoneking (2003).

12 For discussions on how to estimate ancestral population size from the genetic diversity of contemporary populations, see Takahata, Satta, and Klein (1995) and McEvoy et al. (2011).

13 This increased dependence on meat might have depended on two innovations, one evolutionary, the other cultural. The evolutionary trait was the development of the capacity for endurance running, which would have permitted the hunting down of large game (Bramble and Lieberman 2004). The cultural innovation was the invention of cooking, which greatly increases the energy values obtained from food, both meat and plants; see Carmody and Wrangham (2009). Wrangham (2009) speculates that cooking might have started with *H. erectus,* 1.8 mya, based on several indirect inferences.

14 For one early discussion on the probable effects of increasing human populations in Africa in the late Pleistocene, see McBrearty and Brooks (2000); for the possible influence of this factor on both the out-of-Africa exodus and further human movements into Asia, see Mellars (2006) and Henn, Cavalli Sforza, and Feldman (2012).

15 See Oppenheimer (2003), pp. 124–125.

16 See Richards et al. (2006).

17 Although much of Beringia is now submerged, hence much of the archaeological evidence of human habitation and transit lost forever, there are grounds for thinking that it was largely a grasse steppe and, correspondingly, a relatively hospitable environment during much of the Last Glacial Maximum (28,000 to 17,000 years ago). There was a "pause" somewhere within this interval in the eastward migration of the peoples who would later give rise to Native Americans, of perhaps as much as 10,000 years (Pringle, 2014). There was probably a physical barrier to further movement, but its nature, whether glaciers or something else, is unknown.

18 For a review of the evidence and the debates on what caused the late Pleistocene mass extinctions of large mammals, see Stuart (1991).

19 A popular account of the wanderlust hypothesis, underlain by alleles favoring risk taking, is given in Dobbs (2013).

20 The general scenario presented here is consistent with the genetic data of present-day populations, especially that derived from the analysis of mitochondrial DNA sequences. It is not universally agreed, however, and some have argued for both multiple migrations of H. sapiens out of Africa including an initial, much earlier one on the basis of other genetic data and cranial measurements (Reyes-Centeno et al. 2014).

21 The word *mutations* is usually taken to mean base-pair substitutions in DNA—for instance, the replacement of an AT base pair by a GC. Small deletions or insertions of one or more base pairs can also occur, these being called *indels*. In coding regions, the latter tend to have severe effects—they throw off the translational reading frame downstream of the mutation to give greatly altered and often shortened polypeptide chains. Hence, in general most mutations in coding regions tend to be the less deleterious base-pair substitutions. In regulatory regions, indels are probably more common.

22 The key papers are Green, Krause, and Briggs (2010) and Reich et al. (2010), which deal, respectively, with Neanderthal and Denisovian genome reconstructions.

23 See Bowler (1986), pp. 75–111, on the historical controversies surrounding the possible evolutionary relationships of Neanderthals and *Homo sapiens*. For the current assessment of Neanderthal history from fossil findings, see Hublin (2009).

24 It is ironic and rather satisfying given the history of racism that people of European descent have some ancestry from the long-derided and caricatured Neanderthals—in contrast to modern Africans, who do not.

25 The paper that describes these last findings is that of Vernot and Akey (2014).

26 The two analyses are presented respectively in Yang et al. (2012) and Sankararaman et al. (2012).

27 These findings are reported in Vernot and Akey (2014) and Sankararaman et al. (2014).

28 Charles Darwin (1871), vol. II, pp. 370–371.

29 A special twist on this idea of the liking for exaggerated traits has come out of neurobiology and the discovery that mammals when taught to prefer one shape to another tend to like disproportionately exaggerated versions of those shapes—caricatures, if you will. See Kandel (2012), pp. 295–300. A population genetic analysis that supports the idea that cultural preferences, including those involving sexual esthetics as proposed by Darwin, can drive rapid physical evolution is that of Laland (1994). A relatively recent defense of Darwin's idea for the molding power of sexual selection with respect to skin color is that of Aoki (2002).

30 See Fisher (1915) and Anderson (1994).

31 In reality, the word *naked* in connection with human furlessness is misleading. Humans have nearly the same density of hair follicles on their skin as other primates; in the naked body areas, the hairs are much thinner and grow more slowly than hairs on the head, the armpits, and the pubic areas.

32 Perhaps the most extreme application of the idea of sexual selection influencing a human trait pertains to one of the most important and distinctive of such traits: our large brain size. It was argued by Miller (2000) that the outsized human brain is the ultimate sexually selected trait, whose complexity was required by the demands put on males to successfully court females. For any man who has had to work hard to win his female mate, this notion may resonate, but today it is a minority view. If true, however, this trait too would have to have become gender equalized over time because males and females have similar-sized brains when corrected for average body size differences. Other explanations for the size of the human brain will be discussed in Chapter 10.

33 The relationships of skin pigmentation to various aspects of physiology are described in Jablonski (2012), pp. 24–46, with the folic acid hypothesis described on pp. 30–34. If valid, then human skin must be a major body repository for folate, but that has not yet been shown.

34 A still worthwhile if dated history of these ideas is in Gould (1977), chapter 10. For a more detailed and recent treatment, see Deacon (1997), chapter 5, where the consequences of prolonged brain growth on human brain size are evaluated in detail.

35 For a detailed accounting of the brain differences between domesticated and wild mammals, see Kruska (2005).

36 This experimental work will be described in more detail in chapter 10. For a good review, see Trut, Oksina, and Kharmalova (2009).

37 A good discussion of the domesticated nature of humans is that of Leach (2003).

38 Brüne (2007).

39 The full explanations are given in Bagehot (1872) and Boas (1938).

40 The comments are scattered throughout the final two chapters of volume II of *The Descent of Man* in dealing with sexual selection in humans. For a good précis of his conflicts on this issue, see Brüne (2007).

41 See Hare, Wobber, and Wrangham (2012) for the full description and exposition of bonobos as self-domesticated apes. A recent discussion of the idea of humans as self-domesticated is in Francis (2015), chapters 13 and 14.

42 See Etchevers et al. (1999), Creuzet (2009), and Creuzet, Martinez, and Le Douarin (2006).

43 The hypothesis and the evidence are described in detail in Wilkins, Wrangham, and Fitch (2014). For recent confirmation that domestication is associated with mutations in neural crest cell genes, see Carneiro et al. (2014) and Montague et al. (2014). These authors also find many other genetic changes in domesticated breeds relative to their wild progenitors, as would be expected given the extensive selection most domestic animal breeds have experienced for various traits.

44 These behavioral aspects of the domestication syndrome are discussed in Wilkins, Wrangham and Fitch (2014), pp. 802–803.

45 The data and interpretations of changed skull sizes are given in Cieri et al. (2014). A much earlier discussion of the possible role of hormonal changes in promoting the kinds of cooperative behavior that are part of domestication is that of Holloway (1981). Before Cieri et al., Mellars (2006) suggested that growing population sizes were a key element in the early peopling of Africa by *Homo sapiens* and that the resulting population pressures may have contributed to the out-of-Africa emigration.

9. ON FACE CONSCIOUSNESS AND THE FUTURE OF THE FACE

1 The case that human artistic activity, however difficult to define precisely, has a long history, from the earliest days of *Homo sapiens* and perhaps even prior to that, is made by McBrearty and Brooks (2000), Oppenheimer (2003), pp. 89–123, and Morriss-Kay (2010). This view contradicts the position that there was a sudden human revolution in cognition as signified by the Paleolithic cave paintings (discussed in this chapter) and as proposed by Klein and Edgar (2002). Rather, as the authors noted above argue, the uniqueness of the Upper Paleolithic cave art almost certainly reflects a conjunction of special circumstances: culturally unique facets of those populations and geological opportunity (the caves themselves and their walls as surfaces for image making) rather than a unique artistic and cognitive ability.

2   For an excellent accounting and analysis of early art during the Paleolithic throughout the world, see White (2002).

3   White (2002) is particularly strong in dismissing the art-for-art's sake motive (see pp. 40–54). For the probable importance of religious observance of some kind in connection with the Paleolithic cave art, see Lewis-Williams's *The Mind in the Cave* (2002).

4   See Garlake (1995).

5   One site, however, has etchings full of human figures and human faces: the Paleolithic cave of La Marche in southern France. It dates to 15,000 to 14,000 years ago and is thus more recent than the major Paleolithic cave painting sites. The nature, genuineness, and significance of this art are controversial, but if validated by subsequent work it will provide an exception to the thesis given here that human faces were of little general interest to late Paleolithic humans. It raises the significant question of major cultural differences between different art-creating Paleolithic cultures. Face painting or tattooing as seen in some non-Western tribal peoples is, of course, another form of face consciousness but not one directed at bringing out individual differences of look or expression.

6   The oldest Venus figure is that of the Hohenfels venus, which has been dated to 40,000 years ago (Morriss-Kay 2014).

7   Bagehot (1872) discusses the need for strong rule to initially establish civilizations, but then the civilizations so established would have inevitably fostered the emergence of individual differences of opinion. This would have entailed an ability to perceive individuals as such. These ideas seem to have received little attention in recent times but are perhaps worth a new look.

8   For a recent picture essay showing and discussing this phenomenon along with the growing acceptance of being "mixed race" in the United States, see Fundberg (2013). A short report on the same phenomenon in the United Kingdom appeared in *The Economist* on February 8, 2014, pp. 55–56.

9   Ironically, in view of the fact that geometric morphometrics can now be put to use to study genetic questions, as discussed here, Thompson himself did not have much use for either genetics or Darwinian evolution when he first published *On Growth and Form* in 1917. This was, however, a period when genetics was just developing and still embroiled in a major debate about whether Mendelian genetics underlay heredity generally (Provine 1971). Also at this time, Darwin's theory of evolution was still generally held in low repute. More surprising, however, is that Thompson similarly neglected genetics and evolution twenty-five years later in the second edition, even though both fields had made major advances in that interval. Those lacunae, however, do not detract from *On Growth and Form*'s standing as a scientific prose masterpiece.

10 For a thorough description of geometric morphometrics, see Zelditch et al. (2004).

11 For an older but still useful review of the history of measuring craniofacial differences and their genetic foundations, see Kohn (1991).

12 The great majority of these SNPs are in noncoding regions; even in gene regulatory regions, many seem not to have large deleterious effects and are thus selectively neutral or nearly so. On the other hand, a significant fraction of those SNPs associated with a genetic difference causing a difference in facial phenotype are in such gene regulatory regions.

13 The first study discussed is that of Paternoster et al. (2012), the second is Boehringer et al. (2012), and the third, Liu et al. (2012). The earlier study that showed an important role for *Pax3* in setting normal mouse facial morphology is that of Asher et al. (1996).

14 See Adhikari et al. (2016).

15 The work is that of William DeMyer and his colleagues (1964).

16 Physiognomic ideas came close to derailing one of the great scientific careers before it had even begun. Captain Fitzroy of *The Beagle* nearly turned down Charles Darwin as his naturalist companion because he felt that Darwin's small nose indicated lack of character. Fortunately, his more considered and rational judgment—that Darwin in fact had the right personal qualities for the trip—overrode his physiognomic prejudices.

17 "Ancient and Modern Physiognomy," pp. 220–240 in Gregory (1929).

18 A particularly interesting gene that is missing in the Williams syndrome deletion is a transcription factor that is active in neural crest cells (as well as other cell and tissue types). Tests of complete deficiency of this gene in the experimental test animal, the frog *Xenopus*, indicate various developmental deficiencies, several of which can be related to Williams syndrome in humans (Barnett et al. 2012).

19 The work is described and discussed in Sheehan and Nachmann (2014).

20 Two recent news story headlines give a hint of what is to come: "An Eye That Picks You Out of Any Crowd" and "Banking on Facial Recognition Software, NSA Stores Millions of Images"; see Singer (2014) and Risen and Poitras (2014).

10. SOCIAL SELECTION IN THE SHAPING OF THE HUMAN FACE

1 Darwin begins this passage with the following oft-quoted words, which appear to throw down the gauntlet to his own theory: "To suppose that the eye, with all its inimitable contrivances for adjusting the focus to different distances, for admitting different amounts of light, and for the correction of spherical and chromatic aberrations, could have been formed by natural selection, seems, I freely confess, absurd in the highest degree" Darwin (1859, p. 152). He then goes on to explain why it is really not absurd at all. A more convincing analysis, however, complete with numbers that supported Darwin's view, was given much later by Nilsson and Pelger (1994). That visual organs— "eyes" in the broad sense, though most are far simpler than those of vertebrates—have evolved independently in many animal lineages shows that the initial steps in eye formation are not in the least improbable (see Salvini-Plawen and Mayr 1977).

2 A good popular book on the emergence of tetrapods is Shubin (2012). A more scholarly and detailed treatment is that by Clack (2012).

3 For two different nutritional arguments, see Wrangham (2009) and Cunnane and Crawford (2014).

4 Probably the first important statement of the hypothesis is that of Holloway (1981). A later multiple-author edited volume by Byrne and Whiten (1988) provided the first extended treatment of the idea that social environments require high cognitive abilities in the service of the "Machiavellian brain". Much of the further development of the social brain hypothesis has been carried out by R. I. M. Dunbar and his colleagues; see Dunbar (1992, 1998) for two early important discussions.

5 See Jerison (1973) for the first results pointing out the increased size of primate brains as assessed by the EQ. In recent years, EQ in the popular psychological literature has come to stand for "emotional quotient," the emotional equivalent of IQ, but the abbreviation used here is only in Jerison's sense.

6   See Dunbar (1992) for a clear description of how the variables were sorted through.

7   Discussions of the kinds of social complexity that might have been crucial in driving selection for larger cerebral cortices are Dunbar and Schutz (2007) and Dunbar (2009).

8   See chapter 12 ("Symbolic Origins") in Deacon (1997) and chapter 6 ("The LCA: Our Last Common Ancestor with Chimpanzees") in Fitch (2010) on the special characteristics of human social structures, especially those related to the demands of child-rearing, and the resulting selection pressures for the evolution of complex language.

9   For the classic, and clearest, statements on punctuated equilibrium, see Eldredge and Gould (1972) and Gould and Eldredge (1977).

10 The present-day population genetic view can be found in Orr (1998). In the past, the discussion has all too frequently been bedeviled by unintentionally conflating the size of the effect with the physical extent of the mutation in terms of the amount of DNA sequence affected. There is no such correlation. Many mutations changing a single base pair can be lethal, while some large deletions of certain chromosomal regions are without visible or only minor phenotypic effect in many organisms. The small versus large debate is only about the relative magnitude of the observed phenotypic effects.

11 Morowitz (2002).

12 I am indebted to Daniel Milo of L'École des Hautes Études (Paris) for many fascinating discussions of just this point, including how tight are the costs for evolutionary innovations and whether the actual results of selection can yield results far in excess of what seemed to be demanded.

13 See Hardy (1965) and Lack (1947) for the general point and specific example, respectively.

14 See Wiles, Kunkel, and Wilson (1983). A full account of animal cultures and their evolutionary significance is in Avital and Jablonka (2000).

15 To compare their ideas and their contrasting emphases, see Baldwin (1896) and Waddington (1961).

16 Darwin attempted to do full justice to his predecessors with an accessory sketch appended to the third edition of *The Origin of Species*. He expanded it later to include thirty-four individuals in the fourth edition published in 1866. The story of some of the more important figures in that history, along with Darwin's sketch, is in Stott (2012).

17 For this history, see Burkhardt (2013).

18 Transgenerational epigenetic inheritance is reviewed in Jablonka and Raz (2009) and Heard and Martienssen (2014).

19 See Belyaev (1979) for an early statement and Trut, Oksina, and Kharmalova (2009) for a recent review of the domesticated fox work and the epigenetic hypothesis of domestication.

CODA: THREE JOURNEYS

1   The quoted phrase is from Abraham Lincoln's second inaugural address, but it is also the title of a recent book by Stephen Pinker, a book that makes the counterintuitive case, with much statistical documentation and cogent argument, that given the right circumstances humans are a far more sociable and cooperative species than commonly believed and have been becoming more so in recent history, whatever the horrors and disasters of the twentieth and early twenty-first centuries might seem to say to the contrary (Pinker 2011).

ACRONYMS

AER—apical ectodermal ridge
ATP—adenosine triphosphate
AU—action unit
BMP—bone morphogenetic protein
BRM—basal metabolism
CNV—copy number variant
CP—cortical plate
DNA—deoxyribonucleic acid
EQ—encephalization quotient
FACS—facial action coding system
FEZ—frontonasal ectodermal zone
FFA—fusiform facial area
FGF—fibroblast growth factor
fMRI—functional magnetic resonance imaging
fWHR—facial width-to-height ratio
GRN—genetic regulatory network
GWAS—genome-wide association study
Hox—homeobox
IFL—inner fiber layer
ISVZ—inner subventricular zone
kbp—thousands of base pairs
kya—thousands of years ago
LGN—lateral geniculate nucleus
LHLP—long hominocentric lineage pathway
lncRNA—long noncoding RNA
M—maternal
mbp—million base pairs
MRE—multiregional hypothesis
mya—millions of years ago
MZ—marginal zone
NCC—neural crest cells
NFDS—negative frequency-dependent selection
OFA—occipital facial area
OFL—outer fiber layer
OMIM—Online Mendelian Inheritance in Man

OSVZ—outer ventricular zone
P—paternal
PETM—Paleocene-Eocene Thermal Maximum
RAO—recent African origin
RNA—ribonucleic acid
SC—superior colliculus
SHH—sonic hedgehog
SNP—single nucleotide polymorphism
SP—subplate
STS—superior temporal sulcus
SVZ—subventricular zone
TSS—transcriptional start site
UVRB—ultraviolet radiation B
VLPFC—ventrolateral prefrontal cortex
VZ—ventricular zone
ZPA—zone of polarizing activity

GLOSSARY

adaptive radiation The rapid diversification of a group of organisms that can occur when exposed to new ecological conditions; frequently accompanied by a geographical spreading of the new species

Afrotheria The superorder of placental mammals that includes such different-looking mammals as elephants, hyraxes, tenrecs, and aardvarks that are shown to be related by their DNA sequences; all have African origins

agnathans All the jawless fishes, both extinct and those still living, the latter consisting today solely of the lampreys and hagfishes

allele A variant form of a particular gene

amino acids The small organic molecules containing carbon, nitrogen, and hydrogen atoms that are the basic structural unit of proteins; the universally shared set of these consists of twenty members

amnion *See* amniote.

amniote Any vertebrate species whose embryo develops within an amniotic sac

amniotic sac The fluid-containing sac that contains the developing embryos of the fully terrestrial four-legged vertebrate classes (reptiles, birds, and mammals)

anthropoid primates The larger of the two suborders of Primates, consisting of the more "advanced" groups, namely, the monkeys, apes, and hominins

apical ectodermal ridge (AER) The cap of cells under an epidermal sheath at the tip of a developing limb bud, which is the main center of cell division and growth of the growing limb

apoptosis Developmentally "programmed" cell death that serves to eliminate surplus cells, thereby contributing to the functioning and often shaping of tissues and organs

Archaic DNA In principle, any sequence of DNA from earlier life forms found in a contemporary animal species; in practice, most commonly denotes DNA sequences from earlier hominins found in the genomes of living members of *Homo sapiens*

Archaeon A prokaryotic cell belonging to any of the species of the kingdom of prokaryotes known as the Archaea; the eukaryotic kingdom may derive from an archael cell that swallowed a eubacterial endosymbiont, which eventually evolved into the mitochondrion

arealization The provisional demarcation of the developing cerebral cortex in the developing vertebrate embryo, by key gene activities, into distinct areas that will acquire distinct functions

Asberger's syndrome One condition of autism, whose bearers are often distinguished by a combination of high analytical intelligence and poor social skills and interactions

asymmetric cell division A cell division yielding two daughter cells of different fate and often of slightly different size

autopod The terminal part of a limb bud that gives rise to the terminal part of the limb such as the hand or foot in primates

autapomorphy A trait unique to a branch on a cladogram, hence one that arose early in the history of that group

axon The long process emanating from a nerve cell that makes contact with the shorter cell processes termed *dendrites* of other neurons

basal forebrain nuclei Earlier called the *basal ganglia*, the nodulelike structures beneath the cerebrum that are derived from the telencephalon

basicranium The base or foundation of the skull, to which the axial skeleton attaches and which lies just above the face

Bilateria The superphylum of the Animal Kingdom consisting of all those phyla that exhibit bilaterally symmetrical forms at one or all stages of the life cycle

bone morphogenetic proteins A group of diffusible proteins related by sequence that play numerous roles in growth and development

branchial arches Developing from the cartilaginous pharyngeal arches in the embryo, these are the bony loops that support the gills in fishes; in mammals, the pharyngeal arches contribute to the jaws, the larynx, thyroid gland, and small bones of the middle ear; also called *pharyngeal arches*

Broca's area A specific region of the inferior frontal lobes, localized to Brodmann areas 44 and 45, required for speech production but also involved with language ability in several other ways as well

Brodmann's areas A system of dividing the cerebral cortex into numbered smaller areas, each defined by its location and cytoarchitecture; devised by anatomist Korbinian Brodmann

calmodulin A small protein (148 amino acids long) that binds calcium and is involved as a chemical messenger in many processes that are modulated by calcium

Cambrian period The geological time period lasting from 542 to 485 million years ago, which saw the initial flourishing and diversification of animals; it marks the start of the Phanerozoic era

cartilage The flexible form of connective tissue consisting of several different types found in many parts of the body such as bone joints, the ears, and the intervertebral discs; it is also the precursor for many of the bones of the skeleton and the base of the skull

catarrhine primates The group of the anthropoid or haplorhine primates consisting of the Old World monkeys and the hominoids (the apes and the hominins)

cell The tiny basic unit of biological organization of all living things above the level of viruses, consisting of a nucleus and a complex enveloping cytoplasm, all contained within a cell membrane. There are two kinds of basic cell organization—*prokaryotic* and *eukaryotic*—the former restricted to bacteria-like cells and lacking a true (membrane-bounded) nucleus.

Cenozoic The last major biological era of the Phanerozoic eon, extending from approximately 65 million years ago to 11,000 years ago, the start of the Recent era, and featuring the dominance of mammals in the food chain

central nervous system (CNS) The major part of the nervous system, consisting of the brain and spinal cord, that receives and processes all the information received by the body

cerebellum The region of the brain derived from the hindbrain of the early embryo; it is situated under the cerebrum, which plays a crucial role in motor control but also participates in some cognitive functions

cerebral cortex The region of the mammalian brain, derived from the forebrain of the early embryo, that carries out most of the sensory interpretation and cognitive functions of the brain; also known as the *neocortex* and the *isocortex*

cingulate cortex A central and medial part of the cerebral cortex, lying above the corpos callosum, which connects the two hemispheres. It is involved in the processing of emotions, learning, and memory, and it has been implicated in motivation and responses to emotionally charged situations, and, from its anterior portion, control of expressive vocalizations.

***cis*-regulation** Regulatory control over one or more genes exerted by a protein (or in some cases an RNA molecule) adjacent to the binding site of that molecule

Chordata One of the thirty or so animal phyla distinguished by the possession of a notochord and a postanal tail; it includes the subphylum of the Vertebrata

chromatids The two genetically identical subunits of a replicated chromosome as produced by the replication of that chromosome

chromosomes The long fibers that contain DNA, the hereditary material, which is often coiled to various extents within the nucleus

cladistics The current name of the system originally called *phylogenetic systematics*. It is devoted to the reconstruction of evolutionary events, which can be diagrammed as sequences of bifurcating branches.

Cnidaria One of the nonbilaterian phyla of the Animal Kingdom, hence one of the first to arise in evolution; characterized by radial symmetry, hence a member of the superphylum Radiata

coevolution The phenomenon in which two (or sometimes more) structures or properties influence each other's evolution

combinatorics All the permutations of the combinations of elements in a set, e.g. the set of 64 possible 3-base combinations for the genetic code of protein-coding sequences

comparative biology The set of methods involved in reconstructing likely aspects of evolutionary history through the comparison of related living organisms

conservation (evolutionary) The perpetuation through evolution of long-lasting features, such as structural properties or molecules, relative to others

conspecifics Fellow individual members of the same species

cranial sutures The long, thin gaps between neighboring bony plates in the skull. In early development, they are open, but with time and aging the plates tend to come together and fuse, erasing the sutures.

craniates The subdivision of the chordates containing true heads with complex brains; since all members of this group today also feature backbones consisting of vertebrae, a synonym for the group is the *Vertebrata*.

craniosyntoses Clinical conditions in which skull plates prematurely fuse across a cranial suture, usually resulting in some channeling of brain growth, with subsequent distortion of skull and brain shape

crown group The entire set of a taxonomic group of organisms consisting of those living today and all the ancestors that specifically gave rise to members of that group

cynodonts A group of synapsids originating during the Triassic, one of whose branches led to the true mammals

cystoskeleton The proteinaceous inner fretwork of a cell that gives shape and strength. The key components of the cytoskeleton are microtubules, actin filaments, and intermediate filaments.

cytoplasm The colloidal gel surrounding the nucleus, bounded by the cell membrane, and which usually occupies the greater part of a cell's volume

dendrites The short extensions from neurons that receive informational input from the axons of other neurons

deoxyribonucleic acid (DNA) The long polymeric molecule, usually in the form of a double helix, that carries the genetic information (via the sequence of "base pairs" that project into the center of the molecule from the two helical strands)

dermal bones *See* **membranous bones**.

Deuterostomia One of the three superphyla that make up the Bilateria, the largest division of the Animal Kingdom, whose signature trait is the presence of two openings—one oral, one anal—in their embryos; consists of the Chordata and the Echinodermata

development All the processes that contribute to change in form, shape, and size during the lifetime of an organism, most particularly plants and animals from their inception as fertilized eggs

developmental plasticity The ability of organisms to develop into somewhat altered or variant forms in response to different environmental conditions

diapsids The amniotic vertebrates distinguished by two openings (*fenestrae*) in their temporal bones, most famously the dinosaurs and their living descendants, the birds

diencephalon The posterior division of the forebrain or *prosencephalon* of the early vertebrate embryo that will give rise to several structures, including the thalamus and hypothalamus

diploid *See* diploidy.

diploidy The genetic property of a cell designating the possession of two sets of chromosomes, as is true of most somatic (nongerm line) cells

DNA *See* deoxyribonucleic acid.

dominant The genetic condition in which when there are two different alleles of a gene; one is expressed and its effects visible in the appearance (phenotype) of the cell or organism

domestication syndrome The suite of traits commonly seen in domesticated but not wild strains of animals. In mammals, it often includes such traits as shorter muzzles, smaller forebrains, a prolongation of juvenile forms and behaviors, floppy ears, and development of lighter-colored patches within the fur coat.

Ecdysozoa One of the three superphyla of the Bilateria of the Animal Kingdom, in which larval forms shed their outer coats in response to ecdysoid hormones. It includes the Arthropododa and the Nematoda, the two most species-rich animal phyla.

ectoderm One of the three classic *germ layers* of animal embryos; gives rise to the outer skin and the nervous system

ectodermal placodes *See* **neural ectodermal placodes**.

Ediacaran fauna The earliest undisputed complex forms of animal life that arose during the final period of the Proterozoic eon, the Vendian, and which are known solely from their fossil remains

effective population size (Ne) The estimated size of a population that gave rise to descendants, a number that is always less, sometimes substantially so, from the estimated maximum size of that population in its prime

endochrondral bones Bones that are formed in and around a prior cartilage *model*, hence taking the shape of that precursor structure

endoderm One of the three classic *germ layers* of the embryo; gives rise to the gastrointestinal tract, lung tissues, and various endocrine glands

enhancer A relatively short sequence, often 200 to 300 base pairs long, that serves as the binding site for transcription factors that govern the transcription of a gene; enhancers may be contiguous to or quite distant from the genes so regulated.

epimutations Heritable states of a gene that can be passed through a mitotic division to daughter cells or, in some cases, through a meiotic division to a gamete (a sperm or egg) and then to the next generation

epistasis The blocking or hiding of expression of a mutation in one gene by a second mutation in a different gene

epithelial-mesenchymal transition The process whereby an epithelial tissue with a fixed location turns into motile mesenchymal cells—for example, the neural crest cells.

epithelial tissue A single layer of covering tissue in which the cells are tied together by special "tight junctions" that serves as a barrier to the external environment such as skin or gastrointestinal surfaces.

Euarchonta The superorder of mammals consisting of the Dermoptera (flying lemurs), Scandentia (tree shrews), and Primates

Euarchontoglira The placental mammalian superorder that includes the Euarchonta, Rodentia (rodents), and Lagomorpha (rabbits)

eukaryotic cells Those kinds of cells that possess a true (membrane-bounded) nucleus that contains genetic material

evolutionary lineage The sequence of organismal types that led to a particular kind of organism, for example, the hominin lineage, which began with australopithecines and led to the genus *Homo*

evolutionary synthesis *See* neo-Darwinism.

evolutionary novelty A biological structure with one or more novel properties that did not exist in ancestral forms of the organism in which it is detected—for example, a bat's wing

facial primordia / facial prominences The tissue nodules that are the precursor structures for different parts of the developing face in the embryo

faunal association The fact that in fossil deposits, certain animal forms occur together, indicating they lived at the same time

fibroblast A precursor cell for connective tissues

fibroblast growth factor 8 (FGF8) A particular member of the fibroblast growth factor family of messenger proteins that is essential for both brain and limb development in mammals

fitness The measure of relative reproductive success for a particular genotype versus other genotypes in the population

foramen magnum The major opening in the occipital bone of a vertebrate skull through which the spinal cord passes along with the major vertebrate arteries

frontonasal ectodermal zone (FEZ) The region of the embryo from which all the upper facial primordia develop

frontonasal process (FNP) The structure produced by fusion of the medial and nasal prominences and from which the central face will develop

frugivores Animals that rely predominantly or exclusively on fruit for their nutrition

functional magnetic resonance imaging (fMRI) A method for detecting regions of the brain experiencing increased blood flow, a sign of heightened neural actvity

fusiform facial area (FFA) A region of the temporal lobes important in the recognition of faces, as distinct from other entities

gastrulation The infolding of a central region of the early embryo that leads to the development of the mesoderm between the endoderm and the ectoderm

gene A stretch of DNA that directly specifies one or more related RNA or protein products

gene activity The outcome of expression of a gene—for example, for a traditional protein-encoding gene, the production of the active form of that protein

gene family A group of sequence-related genes, derived ultimately from rounds of gene duplication in ancestral organisms

gene pool The potential total set of all copies of a gene within the gametes of a population of organisms, including all allelic variants

gene recruitment The employment of a preexisting gene for a new function during evolution

gene regulation The modulation of expression of a particular gene to affect the amount of its active product

genetic background The set of alleles in different genes in a genome that can alter the expression of a particular allele of a given gene

genetic drift The changes in allelic composition of a gene pool produced by chance in a small subset of a population that had been separated from the main population

genetic recombination The process or outcome of the swapping of identical segments of homologous chromosomes, usually during the process of meiosis but also sometimes occurring in mitosis between *non-sister chromatids*

genetic pathway The sequence of gene activities, one regulating the next, that results in turning on or off a gene activity that has a distinctive phenotypic effect

genetic regulatory network (GRN) An ensemble of gene activities, often decomposable into linked genetic pathways, that collectively has a major phenotypic outcome—for example, the production of segments in an insect embryo or of a limb in a vertebrate embryo

genetic variance The measure of the extent of genetic difference (variation) within a population for a particular gene that contributes to the total phenotypic variation within that population for a trait affected by that gene

genome The sum total of the physical genetic material that defines the genotype of an organism, namely all the DNA of a DNA-based organism as found in a cell of that organism or, for an RNA virus, the set of RNA instructions

genotype The genetic composition of an individual with respect to the existing alleles for one or more genes; for diploids, this will include both alleles for each gene in the set of interest.

geological time scale The system for dividing and apportioning temporal segments of Earth's history, involving (in order of decreasing magnitude) eons, eras, geological periods, and epochs

geometric morphometrics A set of methods for quantitating shape differences in organisms or parts of organisms—for example, faces

germ layers (or primary germ layers) The three distinctive regions of an early animal embryo that are the proximate source of all major tissue types in that individual as it develops; the three classic germ layers are termed the *ectoderm*, *endoderm*, and *mesoderm*.

glial cells The non-neural cells that exist in tight association and provide support, protection, and nutrition for neurons; also known as *neuroglia* or simply *the glia*

gnathostomes All vertebrate animals that possess jaws

Gondwanaland The southern supercontinent that existed from about 300 to 180 million years ago and contained the present-day continents of Africa,

South America, Australia, and Antarctica. During the Jurassic period (about 180 million years ago), it began to merge with the northern supercontinent, Laurasia, to form one major land mass, Pangaea.

Hadean The first eon within the geological time scale, when newly formed Earth lacked any life forms; from about 4.6 to 4 billion years ago

haploid *See* haploidy.

haploidy The existence of only one full set of chromosomes in a cell, which is characteristic of animal gametes (sperm and egg cells)

haplo-insufficiency The production of a phenotypic effect when there is only one active copy of a wild-type gene in a diploid organism, signifying the need for two full copies to get proper development of the trait in question

haplorhines The clade of primates consisting of the tarsiers and all the simian primates (the monkeys and apes); coextensive with the group of primates known as the *anthropoid primates*, except for the tarsiers

haplotype Either the complete genotype of a haploid set of chromosomes, inherited from one parent, or the genotype of a specific set of tightly linked genes

herbivores Animals that eat only plant material

heterozygous The diploid condition for a gene in which the two alleles of a gene differ; for example, the *Rr* condition for round-seed peas

holoprosencephaly A clinical condition in which development of two separate brain hemispheres fails to occur, leading to fusion of two or more regions of the brain derived from the forebrain, with either lethal effects or, in cases of live births, severe mental deficiencies

homeothermy The maintenance of a steady body temperature independently of external temperatures; the property of warm-bloodedness

hominin lineage The lineage that split off from that of the chimpanzees to form the hominins, the branch that gave rise to modern humans

Hominidae The primate family consisting of the four genera that comprise the great apes: *Pongo* (orangutans), *Gorilla*, *Pan* (chimpanzees), and *Homo* (hominins, including modern humans).

homologous *See* homology.

homology The possession of similar characteristics due to shared ancestry; homologous features can range from the molecular to those of major parts of the body—for example, forelimbs

homoplasy The possession of similar characteristics by two distantly related organisms as an accidental consequence of independent evolutionary acquisition

homozygous The genetic condition of a diploid genotype having two copies of the same allele for a gene; the cell or organism is said to be *homozygous* for that gene; for example, the *RR* genotype in peas for the dominant round-seed allele

Hox **genes** The set of genes that helps set axial patterning along the antero-posterior axis in a developing bilaterian animal embryo

housekeeping genes The genes that specify the essential enzymes for general cellular metabolism or other cellular functions shared among all cells of an organism

hypertelorism The condition of wider heads, signaled by more widely spaced eyes than usual, the result of greater than average sonic hedgehog signaling during early head development in the embryo

hypotelorism The reverse of hypertelorism: a shortening of head width due to below-average Sonic hedge-hog signaling; at its extreme, there is a loss of head bilateral symmetry and the production of embryos with a single, central eye

intermediate filament proteins A group of proteins that are key constituents of the cytoskeleton, defined as a group by being in between the filament widths of actin microfilaments and microtubules; some are also vital constituents of other structures, such as the keratins in hairs.

introgression The passage of the genes of one species into the genotype of another created by (usually) rare acts of hybridization

keratins One essential member of the group of *intermediate filament proteins,* used both for the cytoskeleton of eukaryotic cells and the production of hair, fingernails, and hooves

key innovation An evolutionary novelty that permitted an adaptive radiation of the organisms that carried it

lateral geniculate nucleus (LGN) The small nodular brain structures derived from the diencephalon that provide an essential way station in neural transmission of visual information from the retina to the brain centers, in particular the V1 region; required for processing such information

Laurasia The northern supercontinent that existed roughly from 300 million years ago to 180 million years ago before eventually merging with Gondwanaland, the southern supercontinent to create Pangea, a single supercontinental landmass

Laurasiatheria One of the four major placental mammalian superorders, including the Carnivora (dogs, cats), Cetartiodactyla (cows, whales), Chiroptera (bats).

ligand A small molecule that binds to a specific receptor in a cell triggering a distinct biochemical interaction followed by some change in cellular properties

limb bud The initial bud of ectodermal and mesenchymal tissue that will develop into a full limb during embryonic development

lineage See **evolutionary lineage**.

long hominocentric lineage path (LHLP) The sequence of lineages, beginning with the first vertebrates, that can be retrospectively traced as leading eventually to the modern human species, *Homo sapiens*

Lophotrochozoa One of the three superphyla that together make up the Bilateria in the Animal Kingdom; defined by their shared and unique possession of a certain kind of larval form, the *trochophore larva*, distinguished by two midbody bands of cilia

loss-of-function mutations Mutations in a gene that are characterized by reductions in the amount of active gene product(s) specified by that gene

mammaliaform synapsids Groups or species of synapsids that showed many but not all mammalian characteristics, hence precursors to or related to precursors of the true mammals

mammals The vertebrate class characterized by the production of milk for feeding the young, the presence of hair, and three kinds of middle-ear bones not found in other vertebrates

marsupials Mammals whose young are born at what amounts to an early fetal stage and who obtain their mother's milk in special pouches that carry them

matrilineal inheritance The situation in which a particular cellular element is inherited only through the maternal germline and hence not carried in male gametes; the archetypal example is the mitochondrion.

meiosis The special form of cell division, involving two stages, in the generation of haploid gametes from diploid precursor cells; it thus involves the reduction of two chromosome sets to a single one per cell via the separation of homologous chromosomes from each other in the first meiotic division.

membranous (or dermal) bones The kind of bones produced by mesenchymal precursor cells that transform into bone-secreting cells, usually in long, flattened regions, and without a prior cartilage model; the bones of the cranial vault are classic membranous bones.

mesencephalon The part of the early embryonic brain situated between the forebrain and the hindbrain that will give rise to midbrain structures; its derivatives are located centrally in the mammalian brain in mammals under the cerebral cortex and above the structures of the hindbrain

mesenchymal cells Motile cells that often differentiate into particular, but collectively diverse, cell types after reaching different destinations; they are derived from multipotent cells designated as MSCs, initially a designation for *mesenchymal stem cells* but today for *mesenchymal stromal cells.*

mesoderm The third classic *germ layer* of the embryo, along with the ectoderm and endoderm; typically situated physically between those two other germ layers, where it gives rise to blood and muscle and various kinds of connective tissues (from tendons to bones).

Mesozoic The second era of the Phanerozoic eon, following the Paleozoic era and extending from 252 mya to 66 mya; classically seen as the "Age of Reptiles," it was the period when nonavian dinosaurs dominated the terrestrial food chain.

messenger RNA (mRNA) The kind of RNA molecule whose base sequence directly specifies a polypeptide chain; the mRNA is thus the intermediary molecule between the gene and the proteinaceous chain specified by that gene.

microtubules The relatively thick filaments that make up a major element of the cytoskeleton and, more particularly, the mitotic and meiotic apparatus for separating chromatids and chromosomes, respectively

mimetic (facial) muscles The set of muscles, unique to mammals, that attach directly to the skin (of the face) and are responsible for facial expressions

mitochondria The cellular organelles, derived from a bacterial cell at the evolutionary origins of eukaryotic cells, responsible for the production of adenosine triphosphate (ATP), the energy currency of the cells

mitosis The process of cell division in eukaryotic cells by which a cell gives rise to two "daughter" cells, each of which contains the complete set of chromosomes possessed by the initial ("mother") cell

model systems Organisms used in experimental research that are particularly useful for investigating one or more basic genetic, cellular, neural or developmental properties

modern synthesis *See* neo-Darwinism.

module Any component structural unit of a larger entity; in this book, the main reference is to discernible pieces of genetic regulatory networks that carry out distinct actions within those networks

molecular clock (1) The process of roughly steady accumulation of mutations in evolutionarily "neutral" sites; (2) the set of analytical approaches used to estimate time points at which lineages diverged, as based on the presumptive steady accumulation of base-pair changes with time

molecular phylogenetics The set of techniques used to reconstruct evolutionary history from detailed comparisons of DNA sequences of living organisms

monomorphic The situation in which a population of organisms effectively has just one allele for a particular gene such that (nearly) all individuals are homozygous for that allele

monophyletic The situation in which a taxon of organisms is deemed to have originated from one ancestral type

monotremes The egg-bearing mammals, today consisting just of the platypus and several species of echidna

morphology The detailed shape and form of organisms, particularly complex, multicellular organisms such as animals and plants

mosaic evolution The process denoting the piecemeal acquisition during evolution of traits that characterize a taxon today

motoneurons Neurons that contact muscles and activate contractions of those muscles

motor responses Movements initiated by motoneurons

mRNA *See* messenger RNA.

multiregional hypothesis (MRE) The idea that modern *Homo sapiens* originated independently from *Homo erectus* in several places in Africa and Eurasia

mutation Any change in the genetic material, although the term generally refers to changes of relatively small physical extent within the DNA, from single base pairs up to several contiguous ones, the latter referred to as *indels*

natural selection The process by which particular genetic changes are either favored to spread within a population because they promote some favorable or adaptive change or are eliminated from that population because they have deleterious consequences

neocortex The part of the mammalian brain derived from the forebrain of the embryo that covers the rest and which is responsible for so much of the processing of sensory information and of the distinctive behaviors of the animal; also known as the *isocortex* or cerebral cortex

neo-Darwinism The body of modern thought that evolved in the first half of the twentieth century and constitutes the core of the main contemporary theory of biological evolution

neoteny The evolved state in which juvenile body features (morphological or behavioral or both) are retained in sexually mature animals

neural crest cells The multipotent cells derived from the dorsal part of the neural tube in the developing vertebrate embryo that migrate to various parts of that embryo and, in those new locations, give rise to many different kinds of cell types—for example, neural sensory cells, pigment cells, connective tissue, and endocrine cells

neural ectodermal placodes Thickened regions of the outer ectoderm in early vertebrate embryos, derived from the anterior part of the neural tube, that give rise to many cranial sensory structures during embryonic development

neural tube The tubular structure that develops in the dorsal region of the early vertebrate and that will give rise to the central nervous system (brain and spinal cord) and the neural crest cells

neurocranium That part of the skull (the cranium) that covers and houses the brain

neurotrophic substances Particular molecules that promote survival and/or growth of neural tissue

niche construction The ability of organisms to modify some aspect of the environment in ways that promote their survival

nucleotides The basic structural unit of all nucleic acids, both RNA and DNA, each consisting of a phosphate molecule linked to a sugar molecule that is linked itself to a small nitrogen-containing molecule (a *nitrogenous base*)

nucleus The small, essentially spherical, membrane-bounded organelle within cells that contains the chromosomes, hence the genetic material

Online Mendelian Inheritance in Man (OMIM) database Electronic database that compiles all the information about hereditary conditions in humans, both the phenotypes and the genes known to be involved in such conditions

orders The taxonomic category in the Linnaean taxonomic system that lies between classes and families—for example, the order Rodentia within the class Mammalia

paleoanthropology The science of deducing the properties of earlier hominins from their fossils, the surrounding matrix of those fossils, and (when present) the artifacts left behind by those hominins

Paleozoic era The earliest and longest geological era of the Phanerozoic eon, stretching from 543 mya to 252 mya

parallel evolution The independent acquisition of particular properties during evolution of distinct lineages within a larger taxon; "convergent evolution"

patrilineal inheritance Something that is inherited solely from the male parental line; for example, the Y chromosome

pattern formation The phenomenon in which stereotypic spatial organizations of elements come into being during development in the same way within organisms

phalanges The separate bony units that constitute the digits as seen, for example, in the fingers of mammalian forepaws or hands

Phanerozoic Within the geological time scale, the eon of "recent life," stretching from 542 million years ago to the present

pharyngeal arches The paired mesodermal pouches that arise in the anterior part of the vertebrate embryo that will give rise to various structures; the nature of the structures depends on the kind of vertebrate; thus, the gill arches in fishes but structures such as the lower jaw and the pharynx in mammals; see also *branchial arches*

phenotype The visible appearance of a genetically determined or influenced trait, in contradistinction to the "genotype," the underlying genetic information involved in development of that trait

philtrum The shallow groove on the face that runs between the bottom of the nose and the top of the upper lip; among living animals, found only in humans

phonemes The basic units of sound that distinguish one word from another

phonological loop The circuitry underlying memory of the sequence of sounds permitting spoken formulation of sentences

phyla The largest taxonomic unit within the Animal and Plant Kingdoms; for example, the phylum Chordata within the Animal Kingdom (though, in recent times, using DNA sequence evidence, phyla are now often grouped in *superphyla*)

phylogeography The study of the patterns of geographical distribution of organisms with respect to their genotypes and of the historical processes that generate these patterns.

phylogeny The reconstructed evolutionary history of the relationships and origins of a group of organisms

phylogenetic tree The graphical form of representation of a phylogeny, often given today in the form of a *cladogram* that emphasizes the points of bifurcation (splittings) within that history

placentals All mammalian species whose developing fetuses receive their nourishment within the mother's womb from a placenta

Placozoa The phylum including the simplest and most primitive (*basal*) forms of animal life, consisting of tiny round flattened circles, with two tissue surfaces, the lower one serving to take in particles of food

platyrrhine primates A major branch of the Haplorrhini (the anthropoid primates), consisting of the so-called New World monkeys

pleiotropy Multiple kinds of usage (two or more) of a particular gene product, usually deduced from the occurrence of two or more independent phenotypic effects of a mutation

plesiadiforms A group of primate-like animals deduced solely from fossil evidence of the Paleoeocene, once thought to be stem primates, now believed to be a sister group of the primates

plesiomorphy A trait shared by all branches on a cladogram, hence a primitive trait for that phylogenetic group

polymorphism The existence of two or more common allelic forms of a gene within a population; opposite of monomorphism, where there is one overwhelmingly predominant allelic form

polypeptide chain A string of linked amino acids encoded by a DNA sequence; the simplest and basic form of protein organization

pons The anterior part of the hindbrain, which acts as a relay station from the forebrain for signals going to the cerebellum

population bottleneck A great reduction in size of a population such that all descendants derive from a relatively small number of the original members of that population

Porifera The phylum of the sponges, among the most primitive forms of animal life, lacking body symmetry and organs

prefrontal cortex That part of the cerebrum consisting of the anterior part of the frontal lobes, concerned with "executive functions", inhibiting impulsive behaviors, and several important cognitive functions

primary germ layers *See* germ layers.

primary neural progenitor cells The stem cells in the developing brain of the early mammalian embryo that are the initial precursor cells for the neocortex

Primates The order of placental mammals now grouped within the superorder Euarchontaglira, which includes lemurs, monkeys, apes, and humans

prosencephalon The embryonic forebrain in its initial stage.

prosimians The kinds of primates (lemurs, lorises, and tarsiers) considered the most primitive

prosopagnosia The inability to recognize faces

protein domain A distinct, often globular region of a protein that carries out a particular function

proteins Complex polymeric molecules consisting of polypeptide chains, which provide most of the structure and carry out most of the activities of cells

Proterozoic The eon in the geological time scale in which simple life forms, predominantly unicellular forms, became common; it extended from 2.5 billion years ago to 542 million years ago.

proto-languages Hypothesized initial language forms in early human history, probably predating *Homo sapiens,* which lacked complex syntax and consisted mostly of nouns and verbs

punctuated equilibrium The evolutionary phenomenon in which long periods of nonchange (*stasis*) in an evolutionary lineage are interrupted by periods of (often) steady change, those being the so-called punctuations

Quaternary Earlier denoted as an era following the Cenozoic, now primarily used to denote a geological period encompassing the Pleistocene (2.6 mya to 11 kya) and the Holocene (11 kya to the present)

radial unit The basic structural 6-layered unit of the neocortex

radiometric dating Methods for assigning absolute temporal dates to events in the past based on the rates of decay of specific radioisotopes

recent African origin (RAO) hypothesis The idea that modern *Homo sapiens* arose in Africa, about 200 kya, and subsequently (from no earlier than 72 kya) spread out to other continents

receptor A larger molecule that binds a specific ligand or family of ligands, triggering a biochemical and then a cellular response

recessive The condition in which a particular allele does not show a phenotypic effect if paired with a dominant allele; most recessives are loss-of-function alleles

recursion In language, the process of referring back to something within a sentence

regulatory gene A catchall term, in decreasing use, for a gene that controls the expression of one or more other genes

rhombencephalon The early hindbrain of a vertebrate embryo

ribonucleic acid (RNA) The major class of nucleic acids in which the sugar moiety of the nucleotide units is ribose; in contrast to deoxyribonucleic acid (DNA), in which the sugar moiety lacks a particular oxygen atom, hence deoxyribose

ribosomes The protein and RNA containing particles in the cell on which protein chains are synthesized

RNA *See* ribonucleic acid.

sauropsids The predominant group of the diapsids from which the dinosaurs arose

secondary neural progenitor cells Progeny cells of the primary neural progenitor cells; they are the immediate precursors of the neurons that come to populate the layers of the neocortex.

secondary sexual characteristics All those characteristics, apart from the actual reproductive organs, that characterize and visibly distinguish the two sexes of a species—for example, the antlers of male deer

sex chromosomes The chromosome pair that distinguishes the cells of males and females in many animals—for example, the X and Y chromosomes of humans

sexual dimorphism The visibly different appearances of the two sexes in many animal species due to their secondary sexual characteristics

sexual selection The evolution of differences that contribute to sexual dimorphism, based on appeal of those features in one sex to members of the other

signal transduction cascade A sequence of molecular and biochemical events that transmit and *transduce* some initial external signal into some molecular or biochemical result at the end of the sequence that has a particular functional consequence

single nucleotide polymorphism An altered base pair in a DNA sequence that distinguishes some members of a population from the majority of the members of that population

sister group A lineage that acquires a separate identity from its closest related lineage at an evolutionary bifurcation

social selection The selection of characteristics that favor social cohesion and hence survival of the individuals with those traits as they interact within their social groups

sonic hedgehog A protein that is the initial signal in the signal transduction cascade that bears its name; that cascade is crucial for many developmental processes, including the development of the face

stem groups The earliest arising members of a taxon; they lack one or more of the defining characteristics of the modern (*crown group*) taxon

stereopsis The perception of depth and three dimensional structure, which is particularly facilitated by binocular vision where both eyes can take in the same image, albeit with slightly different angles

strepsirrhines The "wet-nosed," more basal groups of primates—namely, the lemurs and the lorises

structural gene An older term denoting genes that encoded proteins that directly carried out general cellular functions such as those of metabolism

superior colliculus A part of the midbrain that is particularly important in directing eye movement; also called the *tectum*

superior temporal gyrus The uppermost fold (gyrus) in the temporal lobe, involved in the processing of auditory information and in particular language comprehension but also in the visual processing of facial expressions

superordinal (or supraordinal) group A set of related taxonomic orders within a class, e.g. the Afrotheria of the placental mammals

synapomorphies A trait shared by a sister group on a cladogram, hence one that arose in their common ancestor

synapsids The group of amniotes that arose early in the history of those vertebrates; characterized by a single opening (*fenestra*) in the temporal bone

taxons Units of taxonomic classification; for example, order, family, genus

telencephalon The anterior part of the embryonic forebrain; in mammals, the precursor of the cerebral cortex and the basal forebrain nuclei

tetrapods All vertebrates having four limbs; hence, the amphibia, reptiles, birds, and reptiles

thalamus A medially located structure of the brain beneath the cerebral cortex and above the midbrain; it is the main derivative of the embryonic diencephalon and is primarily concerned with relaying sensory and motor information to the cerebral cortex but also performs necessary functions in consciousness.

therapsids An early major group of the synapsids, from which the cynodonts and later the true mammals emerged

therioanthropes An artistic image of a creature with a human body and an animal head and face

tissues An ensemble of cells of one kind that serves a particular function; an intermediate level of biological organization between that of cells and organs

transcription The copying of one strand of a DNA molecule into a complementary RNA molecule

transcription factor A protein that carries out transcription of one or more specific genes

translation The act of synthesizing a polypeptide chain from a messenger RNA

transposable elements Discrete DNA sequences that have the capacity to move from one location in the genome to another; when they do so, they often alter the transcriptional activity of nearby genes in their new location.

unconscious selection The selection of new traits as an accidental accompaniment of selection by a breeder for certain other desired traits

Upper Paleolithic The final part of the Paleolithic, or "Stone Age", roughly from 40,000 to 10,000 years ago.

vertebrates All animals possessing backbones

viscerocranium The collective set of facial bones of the skull

wild-type The presumptive most common allele of a gene within a population, hence the *normal* allele of that gene

Xenarthra One of the four placental mammalian superorders, consisting of the sloths, anteaters, and armadillos

BIBLIOGRAPHY

Aboitiz, F. 2011. "Genetic and Developmental Homology in Amniote Brains: Toward Conciliating Radical Views of Brain Evolution." *Brain Research Bulletin* 84: 125–136.

———. 2012. "Gestures, Vocalizations, and Memory in Language Origins." *Frontiers in Evolutionary Neuroscience* 4; doi: 10.3389/fneuro 2012.00002.

Aboitiz, F., S. Aboitiz, and R. R. Garcia. 2010. "The Phonological Loop: A Key Innovation in Human Evolution." *Current Anthropology* 51: S55–S56.

Abzhanov, A., W. P. Kuo, C. Hartmann, B. M. Grant, and P. R. Grant. 2006. "The Calmodulin Pathway and Evolution of Elongated Beak Morphology in Darwin's Finches." *Nature* 442(3): 563–567.

Abzhanov, A., and C. J. Tabin. 2004. "Shh and Fgf8 Act Synergistically to Drive Cartilage Outgrowth During Cranial Development." *Developmental Biology* 273: 134–148.

Adhikari, K., M. Fuentes-Guajardo, M. Quinto-Sanchez, J. Mendoza-Revilla, J. C. Chacon-Duque, et al. 2016. "A Genome-wide Association Scan Implicates DCHS2, RUNX2, GLI3, PAX1 and EDAR in Human Facial Variation" *Nature Communications.* doi: 10.1038/ncomms 11616.

Adoutte, A., G. Balavoine, N. Lartillot, O. Lespinet, B. Prud'homme, and R. de Rosa. 2000. "The New Animal Phylogeny: Reliability and Implications." *Proceedings of the National Academy of Sciences USA* 97: 4453–4456.

Ahlgren, R. C. 2002. "Sonic Hedgehog Rescues Cranial Neural Crest from Cell Death Induced by Ethanol." *Proceedings of the National Academy of Science USA* 99: 10476–10481.

Alibardi, L. 2012. "Perspectives on Hair Evolution Based on Some Comparative Studies on Vertebrate Cornification." *Journal of Experimental Zoology Part B: Molecular and Developmental Evolution* 318B: 325–343.

Alonso, C. R., and A. S. Wilkins. 2005. "The Molecular Elements That Underlie Developmental Evolution." *Nature Reviews Genetics* 6: 709–715.

Anderson, M. 1994. *Sexual Selection.* Princeton, NJ: Princeton University Press.

Andrews, P., and J. Kelley. 2007. "Middle Miocene Dispersals of Apes." *Folia Primatologica* 78: 328–343.

Anton, S. C., R. Potts, and L. C. Aiella. 2014. "Evolution of Early *Homo:* An Integrated Biological Perspective." *Science* 345; doi: 10.1126 Science.1236828.

Anzures, G., P. C. Quinn, O. Pascalis, A. M. Slater, J. W. Tanaka, and K. Lee. 2013. "Developmental Origins of the Other-Race Effect." *Current Directions in Psychological Science* 22: 173–178.

Aoki, K. 2002. "Sexual Selection as a Cause of Human Skin Colour Variation: Darwin's Hypothesis Revisited." *Annals of Human Biology* 29: 589–608.

Asher, J. H., R. W. Harrison, R. Morell, M. L. Carey, and T. B. Friedman 1996. "Effects of Pax3 Modifier Genes on Craniofacial Morphology, Pigmentation and Viability: A Murine Model of Waardenburg Syndrome Variation." *Genomics* 34: 285–298.

Atkinson, A. P., and R. Adolphs. 2011. "The Neuropsychology of Face Perception: Beyond Simple Dissociations and Functional Selectivity." *Philosophical Transactions of the Royal Society B* 366: 1726–1738.

Attanasio, C., A. S. Nord, Y. Zhu, M. J. Blow, Z. Li, et al. 2013. "Fine-Tuning of Craniofacial Morphology by Distant-Acting Enhancers." *Science* 342: 1241006. doi: 10.1126.

Avise, J. C. 2000. *Phylogeography.* Cambridge, MA: Harvard University Press.

Avital, E., and E. Jablonka. 2000. *Animal Traditions: Behavioral Inheritance in Evolution.* Cambridge, UK: Cambridge University Press.

Bagehot, W. 1872. *Physics and Politics, or Thoughts on the Application of the Principles of Natural Selection and "Inheritance" to Political Society.* London: Henry S. King.

Baldwin, J. M. 1896. "A New Factor in Evolution." *The American Naturalist* 30: 441–451.

Bajpal, S., R. F. Kay, B. A. Williams, D. P. Das, V. V. Kapur, and B. N. Tiwari. 2008. "The Oldest Asian Record of Anthropoidea." *Proceedings of the National Academy of Sciences USA* 105: 11093–11098.

Barbujani, G., A. Magagni, E. Minch, and L. Luca Cavalli-Sforza. 1997. "An Apportionment of Human DNA Diversity." *Proceedings of the National Academy of Sciences USA* 94: 4516–4519.

Barnett, C., O. Yazgan, H. C. Kuo, S. Malabar, T. T. Fitzgerald, et al. 2012. "Williams Syndrome Transcription Factor Is Critical for Neural Crest Cell Function in *Xenopus laevis*." *Mechanisms of Development* 129: 324–338.

Barton, R. A. 2004. "Binocularity and Brain Evolution in Primates." *Proceedings of the National Academy of Sciences USA* 101: 10113–10115.

———. 2006. "Olfactory Evolution and Behavioral Ecology in Primates." *American Journal of Primatology* 68: 545–558.

Bastida, M. F., R. Sheth, and M. A. Ros. 2009. "A BMP-Shh-Negative Feedback Loop Restricts Shh Expression During Limb Development." *Development* 136: 3779–3789.

Beard, K. C. 2008. "The Oldest North American Primate and Mammalian Biogeography During the Paleocene-Eocene Thermal Maximum." *Proceedings of the National Academy of Sciences USA* 105: 3815–3819.

Bearn, A. G. 1993. *Archibald Garrod and the Individuality of Man.* Oxford, UK: Clarendon Press.

Belloni, E., M. Muenke, E. Roessler, G. Traverso, J. Siegel-Burtelt, et al. 1996. "Identification of Sonic Hedgehog as a Candidate Gene Responsible for Holoprosencephaly." *Nature Genetics* 14: 353–356.

Belyaev, D. 1979. "Destabilizing Selection as a Factor in Domestication." *Journal of Heredity* 70: 301–308.

Bennhold, K. 2015. "Policing with a Facebook of the Mind." *International New York Times,* October 10–11, p. 1.

Bickerton, D. 2009. *Adam's Tongue: How Humans Made Language, How Language Made Humans.* New York: Hill & Wang.

Bigoni, F., and G. Barsanti. 2011. "Evolutionary Trees and the Rise of Modern Primatology: The Forgotten Contribution of St. George Mivart." *Journal of Anthropogical Sciences* 89: 1–15.

Bloch, J. I. 2007. "New Paleoeocene Skeletons and the Relationship of Plesiadapiforms to Crown-Clade Primates." *Proceedings of the National Academy of Sciences USA* 104: 1159–1164.

Boas, F. 1938. *The Mind of Primitive Man.* New York: Macmillan.

Boehringer, S., F. van der Lijn, F. Liu, M. Guenther, and S. Sinigerova. 2011. "Genetic Determination of Human Facial Morphology: Links Between Cleft Lips and Normal Variation." *European Journal of Human Genetics* 19: 1192–1197.

Boughner, J. C., S. Wat, V. M. Diewert, N. M. Young, L. W. Browder, and B. Hall-grimsson. 2008. "Short-Faced Mice and Developmental Interactions Between the Brain and the Face." *Journal of Anatomy* 213: 646–662.

Boveri, T. 1902. "Ueber mehrpolige Mitosen als Mittel zur Analyse des Zellkerns." *Verhandlungen Physikalisch-Medizinische Gesellschaft Würzburg* 35: 67–90.

Bowler, P. J. 1983. *The Eclipse of Darwinism.* Baltimore: Johns Hopkins University Press.

———. 1986. *Theories of Human Evolution: A Century of Debate 1844–1944.* Oxford, UK: Basil Blackwell.

Bramble, D. M., and D. Lieberman. 2004. "Endurance Running and the Evolution of *Homo.*" *Nature* 432: 345–352.

Briggs, D. E. G., D. H. Irwin, and F. J. Collier. 1994. *The Fossils of the Burgess Shale.* Washington, DC: Smithsonian Institution Press.

Britten, R. J., and D. Kohne. 1968. "Repeated Sequences in DNA." *Science* 161: 529–540.

Bromham, L. 2011. "The Genome as a Life-History Character: Why Rate of Molecular Variation Varies between Mammal Species." *Philosophical Transactions of the Royal Society B* 366: 2503–2513.

Brown, E., and D. I. Perrett. 1993. "What Gives a Face Its Gender?" *Perception* 22: 829–840.

Browne, T. 1645. *Religio Medici,* Part 2, Section 2, paragraph 2. London: Andrew Crooke.

Brüne, M. 2007. "On Human Self-Domestication, Psychiatry, and Eugenics." *Philosophy, Ethics, and Humanities in Medicine* 2: 21–29.

Brugmann, S. A., L. H. Goodnough, A. Gregorieff, P. Leucht, D. ten Berge, et al. 2007. "Wnt Signaling Mediates Regional Signaling in the Vertebrate Face." *Development* 134: 3283–3295.

Brugmann, S. A., J. Kim, and J. A. Helms. 2006. "Looking Different: Understanding Diversity in Facial Form." *American Journal of Medical Genetics Part A* 140A: 2521–2529.

Budd, G. E. 1998. "Arthropod Body-Plan Evolution in the Cambrian With an Example from Anomolocaridid Muscle." *Lethaia* 3: 197–210.

Burkhardt, R. W., Jr. 2013. "Lamarck, Evolution, and the Inheritance of Acquired Characters." *Genetics* 194: 793–805.

Burrows, A. M., R. Diogo, B. M. Waller, C. J. Bonar, and K. Liebal. 2011. "Evolution of the Muscles of Facial Expression in a Monogamous Ape: Evaluating the Relative Influences of Ecological and Phylogenetic Factors in Hylobatids." *The Anatomical Record* 294: 645–663.

Burrows, A. M., and T. D. Smith. 2003. "Muscles of Facial Expression in *Otolemur* with a Comparison to Lemuroidea." *The Anatomical Record* 274A: 827–836.

Burrows, A. M., B. M. Waller, L. A. Parr, and C. J. Bonar. 2006. "Muscles of Facial Expression in the Chimpanzee (*Pan troglodytes*): Descriptive, Comparative, and Phylogenetic Contexts." *Journal of Anatomy* 208: 153–167.

Burton, A. M., V. Bruce, and N. Dench. 1993. "What's the Difference between Men and Women? Evidence from Facial measurement." *Perception* 22: 153–176.

Byrne, R. W., and A. Whiten (Eds.). 1988. *Machiavellian Intelligence: Social Expertise and the Evolution of Intelligence in Monkeys, Apes, and Humans.* Oxford, UK: Oxford University Press.

Cann, R. L., M. Stoneking, and A. C. Wilson. 1987. "Mitochondrial DNA and Human Evolution." *Nature* 325: 31–36.

Carmody, R. N., and R. W. Wrangham. 2009. "The Energetic Significance of Cooking." *Journal of Human Evolution* 57: 379–391.

Carneiro, M., C-J. Rubin, F. Di Palma, F. W. Albert, and J. Alfoeldi. 2014. "Rabbit Genome Analysis Reveals a Polygenic Basis for Phenotypic Change During Domestication." *Science* 345: 1074–1079.

Carroll, S. B. 2013. *Brave Genius: A Scientist, A Philosopher, and Their Daring Adventures from the French Resistance to the Nobel Prize.* New York: Broadway Books.

Carroll, S. B., J. K. Grenier, and S. D. Weatherbee. 2001. *From DNA to Diversity: Molecular Genetics and the Evolution of Animal Design.* Malden, MA: Blackwell Publishing.

Cech, T., and J. A. Steitz. 2014. "The Non-Coding RNA Revolution: Trashing Old Rules to Forge New Ones." *Cell* 157: 77–94.

Cerny, R., M. Cattell, T. S. Spengler, M. Bronner-Fraser, F. Yu, and D. M. Medeiws. 2010. "Evidence for the Prepattern / Cooption Model of Vertebrate Jaw Evolution." *Proceedings of the National Academy of Sciences USA* 107: 17262–17267.

Chan, J. A., S. Balasubramanian, R. M. Witt, K. J. Nazeria, Y. Choi, et al. 2009. "Proteoglycan Interactions with Sonic Hedgehog Specify Mitogenic Responses." *Nature Neuroscience* 12: 409–417.

Chedotal, A., and L. J. Richards. 2010. "Wiring the Brain: The Biology of Neural Guidance." *Cold Spring Harbor Perspectives in Biology* 2: a001917.

Cieri, R. L., S. E. Churchill, R. G. Franciscus, J. Tan, and B. Hare. 2014. "Craniofacial Feminization, Social Tolerance, and the Origins of Behavioral Modernity." *Current Anthropology* 55: 419–443.

Clack, J. A. 2012. *Gaining Ground: The Origin and Evolution of Tetrapods.* Bloomington: Indiana University Press.

Cloutman, L. L. 2012. "Interaction between Dorsal and Ventral Processing Streams: Where, When, and How?" *Brain and Language* 127: 251–263.

Coleman, W. 1964. *Georges Cuvier, Zoologist: A Study in the History of Evolution Today.* Cambridge, MA: Harvard University Press.

Coon, C. S. 1963. *The Origin of Races.* London: Jonathan Cape.

Corballis, M. 2002. *From Hand to Mouth: The Origins of Language.* Princeton, NJ: Princeton University Press.

———. 2011. *The Recursive Mind: The Origins of Human Language, Thought, and Civilization.* Princeton, NJ: Princeton University Press.

Cordero, D. R., S. Brugmann, Y. Chu, R. Bajpai, M. Jame, and J. A. Helms. 2011. "Cranial Neural Crest Cells on the Move: Their Roles in Craniofacial Development." *American Journal of Medical Genetics Part A* 155(2): 270–279.

Cordero, D. R., M. Tapadia, and J. A. Helms. 2005. "Sonic Hedgehog Signaling in Craniofacial Development." Pp. 166–189 in *Shh and Gli Signalling and Development,* edited by S. Howie and C. Fisher. Georgetown, TX: Eurekah Publishing.

Cramon-Taubadel, N. 2011. "Global Human Mandibular Variation Reflects Differences in Agricultural and Hunter-Gatherer Subsistence Strategies." *Proceedings of the National Academy of Sciences USA* 108: 19546–19551.

Creuzet, S. 2009. "Regulation of Pre-Otic Brain Development by the Cephalic Neural Crest." *Proceedings of the National Academy of Sciences USA* 106: 15774–15779.

Creuzet, S., S. Martinez, and N. M. Le Douarin. 2006. "The Cephalic Neural Crest Exerts a Cortical Effect on Forebrain and Mid-Brain Development." *Proceedings of the National Academy of Sciences USA* 103: 14033–14038.

Crossley, P. H., G. Minowada, C. A. MacArthur, and G. R. Martin. 1996. "Roles for Fgf8 in the Induction, Initiation, and Maintenance of Chick Limb Development." *Cell* 84: 127–136.

Cunnane, S. C., and M. A. Crawford. 2014. "Energetic and Nutritional Constraints on Infant Brain Development: Implications for Brain Expansion During Human Evolution." *Journal of Human Evolution* 77: 88–98.

Damasio, A. R. 1994. *Descartes' Error: Emotion, Reason, and the Human Brain.* New York: G. P. Putnam's Sons.

Darwin, C. 1859. *On the Origin of Species by Means of Natural Selection; or the Preservation of Favoured Races in the Struggle for Life.* London: John Murray and Sons. Reprinted edition 1996. Oxford, UK: Oxford University Press.

———. 1868. *The Variations of Animals and Plants Under Domestication.* London: John Murray and Sons.

———. 1871. *The Descent of Man, and Selection in Relation to Sex.* London: John Murray and Sons.

———. 1872. *The Expression of the Emotions in Man and the Animals.* London: John Murray and Sons.

———. 1890. *Journal of Researches,* 2nd ed. Edinburgh: T. Nelson and Sons.

Davidson, E. H. 2006. *The Regulatory Genome.* San Diego: Academic Press.

Deacon, T. W. 1997. *The Symbolic Species: The Co-Evolution of Language and the Brain.* New York: W. W. Norton.

———. 2000. "Evolutionary Perspectives on Language and Brain Plasticity." *Journal of Communication Disorders* 33: 271–290.

Dehay, C., and H. Kennedy. 2007. "Cell-Cycle Control and Cortical Development." *Nature Reviews Neuroscience* 8: 438–450.

Delsuc, F., H. Brinkmann, D. Chourrot, and H. Phillippe. 2006. "Tunicates and Not Cephalochordates Are the Closest Living Relatives of Vertebrates." *Nature* 439: 965–968.

DeMyer, W., W. Zeman, and C. G. Palmer. 1964. "The Face Predicts the Brain: Diagnostic Significance of Medial Facial Anomalies for Holoprosencephaly (Arhinencephaly)." *Pediatrics* 34(2): 256–263.

Depew, M. J., C. A. Simpson, M. Morasso, and J. L. Rubenstein. 2005. "Reassessing the Dlx Code: The Genetic Regulation of Branchial Arch Skeletal Pattern and Development." *Journal of Anatomy* 207: 501–561.

De Waal, F. B. 2003. "Darwin's Legacy and the Study of Primate Visual Communication." *Annals of the New York Academy of Sciences* 1000: 7–31.

Diamond, J. 1997. *Guns, Germs, and Steel: A Short History of Everybody for the Last 13,000 Years.* New York: Random House.

Diogo, R., B. A. Wood, M. A. Aziz, and A. Burrows. 2009. "On the Origins, Homologies, and Evolution of Primate Facial Muscles, with a Particular Focus on Hominoids and a Suggested Unifying Nomenclature for the Facial Muscles of the Mammalia." *Journal of Anatomy* 215: 300–319.

Dobbs, D. 2013. "Restless Genes." *National Geographic* 223: 44–57.

Dobson, S. D. 2009. "Allometry of Facial Mobility in Anthropoid Primates: Implication for the Evolution of Facial Expression." *American Journal of Physical Anthropology* 138: 70–81.

Dobson, S. D., and C. C. Sherwood. 2011. "Correlated Evolution of Brain Regions Involved in Producing and Processing Facial Expressions in Anthropoid Primates." *Biology Letters* 7: 86–88.

Donoghue, P. C., A. Graham, and R. N. Kelsh. 2008. "The Origin and Evolution of the Neural Cord." *BioEssays* 30: 530–541.

dos Reis, M., J. Inoue, M. Hasegawa, R. J. Asher, P. C. Donoghue, and Z. Yang. 2012. "Phylogenomic Data Sets Provide Both Precision and Accuracy in Estimating the Time Scales of Placental Mammalian Phylogeny." *Proceedings of the Royal Society B* 279: 3491–3500.

dos Reis, M., P. C. Donoghue, and Z. Yang. 2014. "Neither Phylogenomic Nor Paleontological Data Support a Paleogene Origin of Placental Mammals." *Biological Letters* 10(1): 20131003.

Dover, G. 2000. *Dear Mr Darwin: Letters on the Evolution of Life and Human Nature.* Berkeley: University of California Press.

Drummond, A. J., S. Y. W. Ho, M. J. Phillips, and A. Rambaut. 2006. "Relaxed Phylogenetics and Dating with Confidence." *PLOS Biology* 4(5): e88. doi: 10.1371.

Duboule, D., and A. S. Wilkins. 1998. "The Evolution of Bricolage." *Trends in Genetics* 14: 54–59.

Dumas, L., and J. M. Sikela. 2009. "DUF1220 Domains, Cognitive Disease, and Human Brain Evolution." *Cold Spring Harbor Symposia on Quantitative Biology* 74: 1–8.

Dunbar, R. I. M. 1992. "Neocortex Size as a Constraint on Group Size in Primates." *Journal of Human Evolution* 20: 469–493.

———. 1998. "The Social Brain Hypothesis." *Evolutionary Anthropolology* 6: 178–190.

———. 2009. "The Social Brain Hypothesis and Its Implications for Social Evolution." *Annals of Human Biology* 36: 562–572.

Dunbar, R. I. M., and S. Shultz. "Evolution in Social Brain," *Science* 317 (2007): 1344–1347.

Economist, The. 2014. "Into the Melting Pot," February 8, pp. 55–56.

Edelman, G. 1987. *Neural Darwinism: The Theory of Neuronal Group Selection.* New York: Basic Books.

Edwards, A. W. F. 2003. "Lewontin's Fallacy." *BioEssays* 25: 798–801.

———. 2012. "Reginald Crundall Punnett: First Arthur Balfour Professor of Genetics, Cambridge, 1912." *Genetics* 192: 3–13.

Ekman, P., W. V. Friesen, and J. C. Hager. 2002. *The Facial Action Coding System CD-ROM.* Salt Lake City: Research Nexus.

Eldredge, N., and S. J. Gould. 1972. "Punctuated Equilibria: An Alternative to Phyletic Gradualism." Pp. 82–115 in *Models in Paleontology,* edited by T. J. M. Schopf. San Francisco: Freeman, Cooper.

Elfenbein, H. A., M. Beaupre, M. Levesque, and U. Hess. 2007. "Toward a Dialect Theory: Cultural Differences in the Expression and Recognition of Posed Facial Expressions." *Emotion* 7: 131–146.

Enard, W. 2011. "FoxP2 and the Role of Cortico-Basal Ganglia Circuits in Speech and Language Evolution." *Current Opinion in Neurobiology* 21: 415–424.

Enlow, D. H. 1968. *The Human Face: An Account of the Postnatal Growth and Development of the Craniofacial Skeleton.* New York: Harper & Row.

Etchevers, H. C., G. Couly, C. Vincent, and N. M. Le Douarin. 1999. "Anterior Cephalic Neural Crest Is Required for Forebrain Viability." *Development* 126 (1999): 3533–3543.

Fabian, M. R., N. Sonenberg, and W. Filopowicz. 2010. "Regulation of mRNA Translation and Stability by MicroRNAs." *Annual Review of Biochemistry* 79: 351–379.

Faure, G. 1998. *Principles and Applications of Geochemistry: A Comprehensive Textbook for Geology Students.* 2nd ed. Englewood Cliffs, NJ: Prentice Hall.

Felsenstein, J. 2004. *Inferring Phylogenies.* Sunderland, MA: Sinauer Associates.

Feng, W., S. M. Leech, H. Tipney, T. Phang, M. Geraci, R. A. Spritz, L. E. Hunter, and T. Williams. 2009. "Spatial and Temporal Analysis of Gene Expression During Growth and Fusion of the Mouse Facial Prominences." *PLoS One* 4(12): e8066.

Fields, S., and M. Johnston. 2010. *Genetic Twists of Fate.* Cambridge: MIT Press.

Fink, B., K. Grammer, P. Mitteroecker, P. Gunz, K. Schaefer, F. L. Bookstein, and J. T. Manning. 2005. "Second to Fourth Digit Ratio and Face Shape." *Proceedings of the Royal Society of Biological Sciences* 272: 1995–2001.

Finlay, B. L., and R. Uchiyama. 2015. "Developmental Mechanisms Channeling Cortical Evolution." *Trends in Neurosciences* 38: 69–76.

Fish, J. L., C. Dehay, H. Kennedy, and W. B. Huttner. 2008. "Making Bigger Brains—the Evolution of Neural Progenitor-Cell Divisions." *Journal of Cell Science* 121: 2783–2793.

Fisher, R. A. 1915. "The Evolution of Sexual Preference." *Eugenics Review* 7: 184–192.

Fitch, W. T. 2010. *The Evolution of Language.* Cambridge, UK: Cambridge University Press.

Fleagle, J. G. 1999. *Primate Adaptation and Evolution,* 2nd ed. San Diego: Academic Press.

Fondon, J. W. I., and H. R. Garner. 2004. "Molecular Origins of Rapid and Continuous Morphological Evolution." *Proceedings of the National Academy of Sciences USA* 101:18058–18063.

Francis, R. C. 2015. *Domesticated: Evolution in a Man-Made World.* London: W.W. Norton & Co.

Francis-West, P., P. Robson, and D. J. R. Evans. 2003. *Craniofacial Development: The Tissue and Molecular Interactions That Control Development of the Head.* Berlin: Springer-Verlag.

Freiwald, W. A., D. Y. Tsao, and M. S. Livingstone. 2009. "A Face Feature Space in the Macaque Temporal Lobe." *Nature Neuroscience* 12: 1187–1196.

Fundberg, L. 2013. "The Changing Face of America." *National Geographic* 224: 80–91.

Gans, C., and G. R. Northcutt. 1983. "Neural Crest and the Origin of Vertebrates: A New Head." *Science* 220: 268–274.

Garcia, R. R., F. Zamorano, and F. Aboitiz. 2014. "From Imitation to Meaning: Circuit Plasticity and the Acquisition of a Conventionalized Semantics." *Frontiers in Human Neuroscience* 8; doi: 10.3389/fnhum.2014.00605.

Garlake, P. 1995. *The Hunter's Vision: The Prehistoric Art of Zimbabwe.* London: British Museum Press.

Garstang, W. 1929. "The Origin and Evolution of Larval Forms." *British Association for the Advancement of Science* 1929: 77–98.

Gebo, D. L. 2004. "A Shrew-Sized Origin for Primates." *American Journal of Physical Anthropology,* Supp. 39: 40–62.

Ghazanfar, A. A., and D. Y. Takahashi. 2014. "The Evolution of Speech: Vision, Rhythm, Cooperation." *Trends in Cognitive Sciences* 18: 543–533.

Glazko, V., B. Zybaylov, and T. Glazko. 2014. "Domestication and Genome Evolution." *International Journal of Genetics and Genomics* 2: 47–56.

Gould, S. J. 1977. *Ontogeny and Phylogeny.* Cambridge, MA: Belknap Press.

———. 1991. *Wonderful Life: The Burgess Shale and the Nature of History.* London: Penguin Books.

Gould, S. J. 1996. *The Mismeasure of Man.* London: Penguin Books.

Gould, S. J., and N. Eldredge. 1977. "Punctuated Equilibria: The Tempo and Mode of Evolution Reconsidered." *Paleobiology* 3: 115–151.

Graham, S. A., and S. E. Fisher. 2013. "Decoding the Genetics of Speech and Language." *Current Opinions in Neurobiology* 23: 43–51.

Gray, R. T. 2004. *About Face: German Physiognomic Thought from Lavater to Auschwitz.* Detroit: Wayne State University Press.

Green, R. E., J. Krause, and A. W. Briggs. 2010. "A Draft Sequence of the Neanderthal Genome." *Science* 328: 710–722.

Gregory, W. K. 1929. *Our Face from Fish to Man: A Portrait Gallery of Our Ancient Ancestors and Kinsfolk Together with a Concise History of Our Best Features.* New York: G. P. Putnam and Sons.

Gronenberg, W., L. E. Ash, and E. A. Tibbetts. 2008. "Correlation Between Facial Pattern Recognition and Brain Composition in Paper Wasps." *Brain, Behavior, and Evolution* 71: 1–14.

Hahn, S. 2014. "Ellis Englesberg and the Discovery of Positive Control." *Genetics* 198: 455–460.

Hall, B. G. 2001. *Phylogenetic Trees Made Easy: A How-to Manual for Molecular Biologists.* Sunderland, MA: Sinauer Associates.

Hall, B. K. 2000. "The Neural Crest as a Fourth Germ Layer and Vertebrates as Quadroblastic, Not Triploblastic." *Evolution & Development* 2(1): 3–5.

Hanishima, C., L. Shen, S. C. Li, and E. Lai. 2002. "Brain Factor-1 Controls the Proliferation and Differentiation of Neocortical Progenitor Cells Through Independent Mechanisms." *Journal of Neuroscience* 22: 6526–6536.

Hardy, A. S. 1965. *The Living Stream.* New York: Harper & Row.

Hare, B., V. Wobber, and R. Wrangham. 2012. "The Self-Domestication Hypothesis: Evolution of Bonobo Psychology Is Due to Selection Against Aggression." *Animal Behavior* 83: 573–585.

Harvey, P. H., and M. D. Pagel. 1991. *The Comparative Method in Evolutionary Biology.* Oxford, UK: Oxford University Press.

Hauser, M. D., N. Chomsky, and W. T. Fitch. 2002. "The Language Faculty: Who Has It, What Is It, and How Did It Evolve?" *Science* 298: 1569–1579.

Heanue, T. et al. 1998. "Symbiotic Regulation of Vertebrate Muscle Development by Dach2, eya2, and Six1, Homologues of Genes Required for *Drosophila* Eye Formation." *Genes and Development* 13: 3231–3243.

Heard, E., and R. A. Martiennssen. 2014. "Transgenerational Epigenetic Inheritance: Myths and Mechanisms." *Cell* 157: 95–109.

Henn, B. M., L. L. Cavalli Sforza, and M. Feldman. "The great human expansion," *Proceedings of the National Academy of Sciences USA* (2012) 109: 17758–17764.

Hennig, W. 1966. "Phylogenetic Systematics." *Annual Review of Entomology* 10: 97–116.

Hickerson, M. J., B. C. Carstens, J. Cavender-Bares, K. A. Crandall, C. H. Graham, et al. 2010. "Phylogeography's Past, Present, and Future: 10 Years after Avise, 2000." *Molecular Phylogenetics and Evolution* 54: 291–301.

Hill, W. G. 2012. "Quantitative Genetics in the Genomics Era." *Current Genomics* 13: 196–206.

Hobaiter, C., and C. Byrne. 2011. "The Meaning of Chimpanzee Gestures." *Current Biology* 24: 1596–1600.

Hockman, D., et al. 2008. "A Second Wave of Sonic Hedgehog Expression During the Development of the Bat Limb." *Proceedings of the National Academy of Science USA* 105: 16982–16987.

Hof, J., and V. Sommer. 2010. *Apes Like Us: Portraits of a Kinship.* Mannheim: Edition Panorama.

Hofmeyr, J-H. 2007. "The Biochemical Factory That Autonomously Fabricates Itself: A Systems Biological View of the Living Cell." Pp. 217–242 in *Systems Biology: Philosophical Foundations*, edited by F. C. Boogerd, F. J. Bruggerman, and J.-H. Hofmeyr. Amsterdam: Elsevier.

✓ Holloway, R. 1981. "Culture, Symbols, and Human Brain Evolution: A Synthesis." *Dialectical Anthropology* 5: 287–303.

✓ Howells, W. W. 1976. "Explaining Modern Man: Evolutionists versus Migrationists." *Journal of Human Evolution* 5: 477–495.

Hu, D., and R. S. Marcucio. 2009. "Unique Organization of the Frontonasal Ectodermal Zone in Birds and Mammals." *Developmental Biology* 325: 200–210.

✓ Hublin, J. J. 2009. "The Origin of Neanderthals." *Proceedings of the National Academy of Sciences USA* 106: 16022–16027.

✓ Hull, D. L. 1988. *Science as a Process: An Evolutionary Account of the Social and Conceptual Development of Science.* Chicago: University of Chicago Press.

Isbell, L. A. 2006. "Snakes as Agents of Evolutionary Change in Primate Brains." *Journal of Human Evolution* 51: 1–35.

✓ ———. 2009. *The Fruit, the Tree, and the Serpent.* Cambridge, MA: Harvard University Press.

Jablonka, E., and G. Raz. 2009. "Transgenerational Epigenetic Inheritance: Prevalence Mechanisms and Implications for the Study of Heredity and Evolution." *Quarterly Review of Biology* 84: 131–176.

Jablonski, N. G. 2012. *Living Color: The Biological and Social Meaning of Skin Color.* Berkeley: University of California Press.

Jacob, F. 1977. "Evolution and Tinkering." *Science* 196: 1161–1166.

———. 1982. "Molecular Tinkering in Evolution." Pp. 131–144 in *Evolution from Molecules to Men,* edited by D. S. Bendall. Cambridge, UK: Cambridge University Press.

Jacob, F., and J. Monod. 1961. "Genetic Regulatory Mechanisms in the Synthesis of Proteins," *Journal of Molecular Biology* 3: 318–356.

Jandzik, D., A. T. Garnett, T. A. Square, M. V. Cattell, J. K. Yu, and D. M. Medeiros. 2015. "Evolution of the New Vertebrate Head by Co-Option of an Ancient Chordate Skeletal Tissue." *Nature* 518: 534–537.

Janvier, P. 1996. *Early Vertebrates.* Oxford, UK: Clarendon Press.

Jeffrey, W. R., A. G. Strickler, and Y. Yamamoto. 2004. "Migratory Neural Crest-Like Cells Form Body Pigmentation in a Urochordate Embryo." *Nature* 431: 696–699.

Jerison, H. J. 1973. *Evolution of the Brain and Intelligence.* New York: Academic Press.

Judson, H. F. 1979. *The Eighth Day of Creation: Makers of the Revolution in Biology.* London: Jonathan Cape.

✓ Kandel, E. 2012. *The Age of Insight: The Quest to Understand the Unconscious in Art, Mind, and Brain, from Vienna 1900 to the Present.* New York: Random House.

Kanwisher, N., and G. Yovel. 2006. "The Fusiform Face Area: A Cortical Region Specialized for the Perception of Faces." *Philosophical Transactions of the Royal Society B* 361: 2109–2128.

Kawasaki, H., N. Tsuchiya, C. V. Kovatch, K. V. Nourski, H. Oya, et al. 2012. "Processing of Facial Emotion in the Human Fusiform Gyrus," *Journal of Cognitive Neuroscience* 24: 1358–1370.

Kemp C., and G. Kaplan. 2013. "Facial Expressions in Common Marmosets (*Callithrix jacchus*) and Their Use by Conspecifics." *Animal Cognition* 16: 773–788.

Kemp, T. S. 2005. *The Origin and Evolution of Mammals.* New York: Oxford University Press.

Kendrick, K. M. 1991. "How the Sheep's Brain Controls the Visual Recognition of Animals and Humans." *Journal of Animal Science* 69: 5008–5016.

Kerosuo, L., and M. Bronner-Fraser. 2012. "What Is Bad in Cancer Is Good in the Embryo: Importance of EMT in Neural Crest Development." *Seminars in Cell & Develpmental Biology* 23: 320–332.

Kitter, R., M. Kayser, and M. Stoneking. "Molecular evolution of *Pediculus humanus* and the origins of clothing," *Curr. Biol.* 13 (2003): 1414–1417.

Klein, R. G., and B. Edgar. 2002. *The Dawn of Human Culture: A Bold New Theory on What Sparked the "Big Bang" of Human Consciousness.* New York: John Wiley & Sons.

Kohn, L. A. P. 1991. "The Role of Genetics in Craniofacial Morphology and Growth." *Annual Review of Anthropology* 20: 261–278.

Kolodkin, A. L., and M. Tessier-Lavigne. 2010. "Mechanisms and Molecules of Neuronal Wiring: A Primer." *Cold Spring Harbor Perspectives in Biology* 3: a001727.

Kouprina, N., A. Pavlicek, G. H. Mochida, G. Solomon, W. Gersch, et al. 2004. "Accelerated Evolution of the ASPM Gene Controlling Brain Size Begins Prior to Human Brain Expansion." *PLos Biology* 2: 0653–0663.

Kraus, P., and T. Lufkin. 2006. "*Dlx* Homeobox Gene Control of Mammalian Limb and Craniofacial Development" *American Journal of Medical Genetics Part A* 140: 1366–1374.

Krubitzer, L. 2009. "In Search of a Unifying Theory of Complex Brain Evolution." *Annals of the New York Academy of Science* 1156: 44–67.

Krubitzer, L., and D. S. Stolzenberg. 2014. "The Evolutionary Masquerade: Genetic and Epigenetic Contributions to the Neocortex." *Current Opinion in Neurobiology* 24: 157–165.

Kruska, D. C. 2005. "On the Evolutionary Significance of Encephalization in Some Eutherian Mammals: Effects of Adaptive Radiation, Domestication, and Feralization." *Brain, Behavior and Evolution* 65: 73–108

Lack, D. 1947/1983. *Darwin's Finches.* Cambridge, UK: Cambridge University Press.

Laland, K. N. 1994. "Sexual Selection with a Culturally Transmitted Mating Preference." *Theoretical Population Biology* 45: 1–15.

Lane, N., and W. Martin. 2010. "The Energetics of Genome Complexity." *Nature* 467: 929–934.

Larsen, W. K. 2009. *Larsen's Human Embryology,* 4th ed. Amsterdam: Elsevier.

Le Douarin, N., J. M. Brito, and S. Creuzet. 2007. "Role of the Neural Crest in Face and Brain Development." *Brain Research Reviews* 55: 237–247.

Le Douarin, N., and C. Kalcheim. 1999. *The Neural Crest,* 2nd ed. New York: Cambridge University Press.

Leach, H. M. 2003. "Human Domestication Reconsidered." *Current Anthropology* 44: 349–360.

Leakey, R. L. 1994. *The Origin of Humankind*. London: Weidenfeld and Nicholson.

Leopold, D. A., and G. Rhodes (2010). "A Comparative View of Face Perception." *Journal of Comparative Psychology* 124: 233–251.

Lewandoski, M., and S. Machem. 2009. "Limb Development: the Rise and Fall of Retinoic Acid." *Current Biology* 19: R558–561.

Lewis-Williams, D. 2002. *The Mind in the Cave: Consciousness and the Origins of Art*. London: Thames & Hudson.

Lewontin, R. C. 1972. "The Apportionment of Human Diversity." *Evolutionary Biology* 6: 381–398.

———. 2000. *The Triple Helix: Gene, Organism, and Environment*. Cambridge, MA: Harvard University Press.

Li, W.-H. 1997. *Molecular Evolution*. Sunderland, MA: Sinauer Associates.

Lieberman, D. A. 2009. "Speculations About the Selective Basis for Modern Human Cranial Form." *Evolutionary Anthropology* 17: 22–37.

Lieberman, D. A. 2011. *The Evolution of the Human Head*. Cambridge, MA: Harvard University Press.

Lieberman, D. A., B. M. McBratney, and G. Krovitz. 2002. "The Evolution and Development of Cranial Form in *Homo sapiens*." *Proceedings of the National Academy of Sciences USA* 99: 1134–1139.

Liu, F., F. van der Lijn, C. Schurmann, G. Zhu, and M. M. Chakravarty. 2012. "A Genome-Wide Association Study Identifies Five Loci Influencing Facial Morphology in Europeans." *PLOS Genetics* 8: e1002932.

Liu, N., A. C. Barbosa, S. L. Chapman, S. Bezprogvannaya, X. Qi et al. 2009. "DNA Binding-Dependent and -Independent Function of the Hand2 Transcription Factor during Mouse Embryogenesis." *Development* 136: 933–942.

Lovejoy, A. O. (1936). *The Great Chain of Being: A Study of the History of an Idea*. Cambridge, MA: Harvard University Press.

Luo, Z-X. 2007. "Transformation and Diversification in Early Mammal Evolution." *Nature* 450: 1011–1019.

Luo, Z-X., C. X. Yuan, Q. J. Meng, and Q. Ji. 2011. "A Jurassic Eutherian Mammal and Divergence of Marsupials and Placentals." *Nature* 476: 442–445.

MacNeilage, P. F. 1998. "The Frame / Content Theory of Evolution of Speech Production." *Behavioral and Brain Sciences* 21: 499–546.

Maisey, J. G. 1996. *Discovering Fossil Fishes*. New York: Henry Holt & Co.

Mallatt, J. 2008. "The Origin of the Vertebrate Jaw: Neoclassical Ideas versus Newer, Development-Based Ideas." *Zoological Science* 25: 990–998.

Mareckova, K., Z. Weinbrand, M. M. Chakravarty, C. Lawrence, R. Aleong, et al. 2011. "Testosterone-Mediated Sex Differences in Face Shape During Adolescence: Subjective Impressions and Objective Features." *Hormones and Behavior* 60: 681–690.

Margoliash, E. 1963. "Primary Structure and Evolution of Cytochrome C." *Proceedings of the National Academy of Sciences USA* 50(4): 672–679.

Martin, R. D. 1990. *Primate Origins and Evolution: A Phylogenetic Reconstruction*. Princeton, NJ: Princeton University Press.

———. 2006. "New Light on Primate Evolution." The Ernst Mayr Lecture, given in 2003. Berlin: Akademie Verlag.

———. 2010. "Primates." *Current Biology* 22: R785–790.

Martinez-Barbera, J. P., and R. S. Beddington. 2001. "Getting Your Head around Hex and Hesx1: Forebrain Formation in Mouse." *International Journal of Developmental Biology* 45(1): 327–336.

Matsumoto, D. 1992. "American-Japanese Cultural Differences in the Recognition of Universal Facial Expressions." *Journal of Cross-Cultural Psychology* 23: 72–84.

McAvoy, B. P., J. E. Powell, M. E. Goddard, and P. M. Visscher. 2013. "Human Population Dispersal 'Out of Africa' Estimated from Linkage Disequilibrium and Allele Frequencies of SNPs." *Genome Research* 21: 821–829.

McBrearty, S., and A. S. Brooks. 2000. "The Revolution That Wasn't: A New Interpretation of the Origin of Modern Human Behavior." *Journal of Human Evolution* 39: 453–563.

McCann, R. L., M. Stoneking, and A. C. Wilson. 1987. "Mitochondrial DNA and Human Evolution." *Nature* 325: 31–36.

McDougall, I., F. H. Brown, and J. G. Fleagle. 2005. "Stratigraphic Placement and Age of Modern Humans from Kibish, Ethiopia." *Nature* 433: 733–736.

McEvoy, B. P., J. E. Powell, M. E. Goddard, and P. M. Visscher. 2011. "Human Population Dispersal 'Out-of-Africa' Estimated from the Linkage Disequilibrium and Allele Frequencies of SNPs." *Genome Research* 21: 821–829.

McNeill, D. *The Face.* Boston, MA: Little, Brown and Company.

Mellars, P. 2006. "Why Did Modern Human Populations Disperse from Africa ca. 60,000 Years Ago? A New Model." *Proceedings of the National Academy of Sciences USA* 103: 9381–9386.

Mellars, P., K-C. Gori, M. Carr, P. A. Soares, and M. B. Richards. 2013. "Genetic and Archaeological Perspectives on the Initial Modern Human Colonization of Southern Asia." *Proceedings of the National Academy of Sciences USA* 110: 10699–10704.

Meredith, R. W., J. E. Janecka, J. Gatesy, V. A. Ryder, C. A. Fisher, et al. 2011. "Impact of the Cretaceous Terrestrial Revolution and KPg Extinction on Mammal Diversification." *Science* 334: 521–524.

Meulmans, D., and M. Bronner-Fraser. 2007. "Insights from Amphioxus into the Evolution of Vertebrate Cartilage." *PLOS One* 2: e787.

Miller, E. R., G. F. Gunnell, and R. D. Martin. "Deep time and the search for anthropoid origins," *Yearbook of Physical Anthropology* 48 (2005): 60–95.

Miller, G. 2000. *The Mating Mind: How Sexual Choice Shaped the Evolution of Human Nature.* London: Vintage.

Mollison, J. 2005. *James and Other Apes.* London: Chris Boot.

Montague, M. J., G. Li, B. Gandolfi, R. Khan, and B. L. Aken. 2014. "Comparative Analysis of the Domestic Cat Genome Reveals Genetic Signatures Underlying Feline Biology and Domestication." *Proceedings of the National Academy of Sciences USA* 111(48): 17230–17235; doi: 10.1073/pnas.1410083111.

Montgomery, S. H., I. Capellini, C. Venditti, R. A. Barton, and N. I. Mundy. 2010. "Adaptive Evolution of Four Microcephaly Genes and the Evolution of Brain Size in Anthropoid Primates." *Molecular Biology and Evolution* 28(1): 625–638.

Moran, V. A., R. J. Perera, and A. M. Khalil. 2012. "Emerging Functional and Mechanistic Paradigms of Mammalian Non-Coding RNAs." *Nucleic Acids Research* 40(14): 6391–6400.

Morell, V. 1995. *Ancestral Passions: the Leakey Family and the Quest for Humankind's Beginnings.* New York: Simon & Schuster.

Morowitz, H. 2002. *The Emergence of Everything: How the World Became Complex.* Oxford, UK: Oxford University Press.

Morriss-Kay, G. M. 2010. "The Evolution of Human Artistic Creativity." *Journal of Anatomy* 216: 158–176.

———. 2014. "A New Hypothesis on the Creation of the Hohle Fels 'Venus' Figurine." Pp. 1589–1595 in *Pleistocene Art of the World,* edited by J. Clottes. Tarascon-sur-Ariege, France: Actes du Congres.

Nilsson, D. E., and S. Pelger. 1994. "A Pessimistic Estimate of the Time Required for an Eye to Develop." *Proceedings of the Royal Society B: Biological Sciences* 256: 53–58.

Nishihara, H., S. Marayama, and N. Okada. 2009. "Retrotransposon Analysis and Recent Geological Data Suggest Near-Simultaneous Divergence of the Three Superorders of Mammals." *Proceedings of the National Academy of Sciences USA* 106: 5235–5240.

Nobel, D. 2006. *The Music of Life: Biology Beyond Genes.* Oxford, UK: Oxford University Press.

Northcutt, G. R. 2005. "The New Head Hypothesis Revisited." *Journal of Experimental Biology and Zoology (Molecular and Developmental Evolution)* 304B: 274–297.

Northcutt, G. R., and C. Gans. 1983. "The Genesis of Neural Crest and Epidermal Placodes: A Reinterpretation of Vertebrate Origins." *Quarterly Review of Biology* 58: 1–25.

O'Bleness, M. S., C. M. Dickens, L. J. Dumas, H. Kehrer-Sawatzki, G. J. Wycoff, and J. M. Sikela. 2012. "Evolutionary History and Genome Organization of DUF1220 Protein Domains." *G3: Genes, Genomes, Genetics* 2: 977–986.

Oftedal, O. T. 2012. "The Evolution of Milk Secretion and Its Ancient Origins." *Animal* 6: 355–368.

O'Leary, M. A., J. I. Block, J. A. Flynn, J. A. Gaudin, A. Giallombardo, et al. 2013. "The Placental Mammal Ancestor and the Post-K-Pg Radiation of Placentals." *Science* 339: 662–667.

O'Leary, D. D. M., S-J. Chou, and S. Sahara. 2007. "Area Patterning of the Mammalian Cortex." *Neuron* 56: 252–269.

Oppenheimer, S. 2003. *Out of Africa's Eden: The Peopling of the World.* Johannesburg, South Africa: Jonathan Ball Publishers.

———. 2012. "Out-of-Africa, the Peopling of Continents and Islands: Tracing Uniparental Gene Trees Across the Map." *Philosophical Transactions of the Royal Society B* 367: 770–784.

Orr, H. A. 1998. "The Population Genetics of Adaptation: The Distribution of Factors Fixed During Adaptive Evolution." *Evolution* 52: 935–949.

Panksepp, J. 1998. *Affective Neuroscience: The Foundations of Human and Animal Emotions.* New York: Oxford University Press.

Parr, L. A., and B. M. Waller. 2006. "Understanding Chimpanzee Facial Expression: Insights into the Evolution of Communication." *Social Cognitive and Affective Neuroscience* 1: 221–228. doi:10.1093/scan/ns103.

Parr, L. A., B. M. Waller, S. J. Vick, and K. A. Bard. 2007. "Classifying Chimpanzee Facial Expressions Using Muscle Action." *Emotion* 7: 172–181.

Parsons, T. E., E. J. Schmitt, J. C. Boughner, H. A. Jamniczky, R. S. Marcucio, and B. Hallgrimsson. 2011. "Epigenetic Integration of the Developing Brain and Face." *Developmental Dynamics* 240: 2233–2244.

Paternoster, L., A. I. Zhurov, A. M. Toma, J. P. Kemp, and B. St Pourcain. 2012. "Genome-Wide Association Study of Three-Dimensional Facial Morphology Identifies a Variant in PAX3 Associated with Nasion Position." *American Journal of Human Genetics* 90: 478–485.

Perez, S. I., M. F. Tejedor, N. M. Novo, and L. Aristide. 2013. "Divergence Times and the Evolutionary Radiation of New World Monkeys (Platyrrhini, Primates): An Analysis of Fossil and Molecular Dates." *PLOS One* 8: e68029.

Pierani, A., and M. Wassef. 2009. "Cerebral Cortex Development: From Progenitors Patterning to Neocortical Size During Evolution." *Development, Growth, and Differentiation* 51: 325–342.

Phillippe, H., N. Lartillot, and H. Brinkmann. 2005. "Muligene Analyses of Bilaterian Animals Corroborate the Monophyly of Edysozoa, Lophotrochozoa, and Protostomia." *Molecular Biology Evolution* 22: 1246–1253.

✓ Pinker, S. 2011. *The Better Angels of Our Nature*. London: Penguin Books.

Plotnik, J. M., F. B. de Waal, and D. Reiss. 2006. "Self-Recognition in an Asian Elephant." *Proceedings of the National Academy of Sciences USA* 103: 17053–17057.

Pointer, M. A., J. M. Kamilar, V. Warmuth, S. G. Chester, F. Delsuc, et al. 2012. "Runx2 Tandem Repeats and the Evolution of Facial Length in Placental Mammals." *BMC Evolutionary Biology* 12. doi: 10.1186/1471-2148-12-103.

Pringle, H. 2014. "Welcome to Beringia." *Science* 343: 961–963.

Provine, W. 1971. *The Origins of Population Genetics*. Chicago: University of Chicago Press.

Pulquerio, M. J. F., and R. A. Nichols. 2007. "Dates from the Molecular Clock: How Wrong Can We Be?" *Trends in Ecology and Evolution* 22: 180–184.

Pulvers, J. N., J. Bryk, J. L. Fish, M. Wilsch-Braeuniger, Y. Arai, et al. 2010. "Mutations in Mouse Aspm (Abnormal Spindle-Like Microcephaly-Associated) Cause Not Only Microcephaly But Also Major Defects in the Germline." *Proceedings of the National Academy of Sciences USA* 107: 16595–16600.

Quintana-Murci, L., and L. B. Barreiro. 2010. "The Role Played by Natural Selection in Mendelian Traits in Humans." *Annals of the New York Academy of Science* 1214: 1–17.

Rakic, P. 1995. "A Small Step for the Cell, a Giant Leap for Mankind: A Hypothesis of Neocortical Expansion During Evolution." *Trends in Neurobiological Science* 18: 383–388.

Rees, J. L., and R. M. Harding. 2012. "Understanding the Evolution of Human Pigmentation: Recent Contributions from Population Genetics." *Journal of Investigative Dermatology* 132: 846–853.

Regan, B. C., C. Julliot, B. Simmen, F. Vienot, P. Charles-Dominique, and J. D. Mollon. 2001. "Fruits, Foliage, and the Evolution of Primate Colour Vision." *Philosophical Transactions of the Royal Society London B* 356: 229–283.

Reich, D., N. Patterson, M. Kircher, et al. 2010. "Genetic History of an Archaic Hominin Group from Denisova Cave in Siberia." *Nature* 468: 1053–1060.

Reid, J. B., and J. J. Ross. 2011. "Mendel's Genes: Toward a Full Molecular Characterization." *Genetics* 189: 3–10.

Reiss, D. 2001. "Mirror Self-Recognition in the Bottlenose Dolphin: A Case of Cognitive Convergence." *Proceedings of the National Academy of Sciences USA* 98: 5937–5942.

Relethford, J. H. 2008. "Genetic Evidence and the Modern Human Origins Debate." *Heredity* 100: 555–563.

Retaux, S., F. Bourrat, J-S. Joly, and H. Hinaux. 2014. "Perspectives in Evo-Devo of the Vertebrate Brain." Pp. 151–172 in *Advances in Evolutionary Developmental Biology*, edited by J. T. Streelman. Hoboken, NJ: John Wiley & Sons.

Reyes-Centeno, H., S. Ghirotto, F. Detroit, D. Grimaud-Herve, V. Barbujani, and K. Harvati. 2014. "Genomic and Cranial Phenotype Data Support Multiple Modern Human Dispersals from Africa and a Southern Route into Asia." *Proceedings of the National Academy of Sciences USA* 111: 7248–7253.

Richards, M., V. Macaulay, E. Hickey, E. Vega, B. Sykes, et al. 2000. "Tracing European Founder Lineages in the Near Eastern mtDNA pool." *American Journal of Human Genetics* 6: 1251–1276.

Richards, M. R., H.-J. Bandelt, T. Kivisild, and S. Oppenheimer. 2006. "A Model for the Dispersal of Modern Humans out of Africa." Pp. 227–257 in *Human Mitochondrial DNA and the Evolution of* Homo sapiens, edited by H.-J. Bandelt, V. Macaulay, and M. Richards. Hamburg: Springer.

✓ Richerson, P. J., and R. Boyd. 2005. *Not by Genes Alone: How Culture Transformed Human Evolution*. Chicago: University of Chicago Press.

Riddle, R. D., R. L. Johnson, E. Laufer, and C. Tabin. 1993. "Sonic Hedgehog Indicates the Polarizing Activity of the ZPA." *Cell* 75: 1401–1406.

Risch, N., E. Burchard, E. Ziv, and H. Tang. 2002. "Categorization of Humans in Biomedical Research: Genes, Race, and Disease." *Genome Biology* 3: comment2007.1–comment2007.12.

Risen, J., and L. Poitras. 2014. "Banking on Facial Recognition Software, NSA Stores Millions of Images." *International New York Times*, June 2, p. 5.

Roberts, R. B., Y. Hu, R. C. Albertson, and T. D. Kocher. 2011. "Craniofacial Divergence and Ongoing Adaptation via the Hedgehog Pathway." *Proceedings of the National Academy of Sciences USA* 108: 13194–13199.

Roessler, E., E. Belloni, K. Gaudenz, P. Jay, P. Besto, S. W. Scherer, L. C. Tsui, and M. Muenke. 1996. "Mutations in the Human Sonic Hedgehog Gene Cause Holoprosencephaly." *Nature Genetics* 14: 357–360.

Romanski, L. M. 2012. "Integration of Faces and Vocalizations in Ventral Prefrontal Cortex: Implications for the Evolution of Audiovisual Speech." *Proceedings of the National Academy of Sciences USA* 109: 10717–10724.

Roseman, C. C. 2004. "Detecting Inter-Regionally Diversifying Natural Selection on Modern Human Cranial Form by Using Matched Molecular and Morphometric Data." *Proceedings of the National Academy of Sciences USA* 101: 12824–12829.

Rubidge, B. S., and C. A. Sidor. 2001. "Evolutionary Patterns Among the Permo-Triassic Therapsids." *Annual Review of Ecology, Evolution, and Systematics* 32: 449–480.

✓ Russell, J. A. 1994. "Is There Universal Recognition of Emotion from Facial Expression? A Review of the Cross-Cultural Studies." *Psychological Bulletin* 115: 102–141.

Salvini-Plawen, L. V., and E. Mayr. 1977. "On the Evolution of Photoreceptors and Eyes." Pp. 207–263 in *Evolutionary Biology*, 10th ed., edited by M. K. Hecht, W. C. Stene, and B. Wallace. New York: Plenum Press.

Sambasivan, R., S. Kuritani, and S. Tajbakhsh. 2011. "An Eye on the Head: The Development and Evolution of Craniofacial Muscles." *Development* 138: 2401–2415.

✓ Sankararaman, S., S. Mallick, M. Dannemann, K. Prufer, J. Kelso, et al. 2014. "The Genomic Landscape of Neanderthal Ancestry in Present-Day Humans." *Nature* 507: 354–357.

Sankararaman, S., N. Patterson, H. Li, S. Paabo, and D. Reich. 2012. "The Date of In-terbreeding between Neandertals and Modern Humans." *PLOS Genetics* 8(10): e1002947.

Sarnat, H. B., L. Flores-Sarnat. 2005. "Embryology of the Neural Crest: Its Inductive Role in the Neurocutaneous Syndromes." *J. Child Neurol.* 20: 637–643.

Sawyer, G. J., and V. Deak. 2007. *The Last Human: A Guide to the Twenty-Two Species of Extinct Humans.* New Haven, CT: Yale University Press.

Scharff, C., and J. Petri. 2011. "Evo-Devo, Deep Homology, and *FoxP2*: Implications for the Evolution of Speech and Language." *Philosophical Transactions of the Royal Society B* 366; doi: 10.1098/rstb. 2011.0001.

Schlosser, G. 2005. "Evolutionary Origins of Vertebrate Placodes: Insights from Devel-opmental Studies and from Comparisons with Other Deuterostomes." *Journal of Experimental Zoology (Molecular and Developmental Evolution)* 304B: 347–399.

———. 2008. "Do Vertebrate Neural Crest and Cranial Placodes Have a Common Evo-lutionary Origin?" *BioEssays* 30: 659–672.

Schneider, R. A., D. Hu, J. L. Robertson, M. Maden, and J. A. Helms. 2001. "Local Retinoid Signaling Coordinates Forebrain and Facial Morphogenesis by Maintaining Fgf8 and SHH." *Development* 128: 2755–2767.

Schoenebeck, J., and E. Ostrander. 2013. "The Genetics of Canine Skull Shape Varia-tion." *Genetics* 193: 317–325.

Schyueva, D., G. Stampfel, and R. Stark. 2014. "Transcriptional Enhancers: From Prop-erties to Genome-Wide Predictions." *National Review of Genetics* 15: 272–286.

Selemon, L.D. 2013. "A Role for Synaptic Plasticity in the Adolescent Development of Executive Function" *Translational Psychiatry* 3: e238. doi: 10.1038/tp.20137.

Senghas, A., S. Kita, and A. Ozyuerek. 2004. "Children Creating Core Properties of Language: Evidence from an Emerging Sign Language in Nicaragua." *Science* 305: 1780–1782.

Sheehan, M. J., and M. W. Nachmann. 2014. "Morphological and Population Genomic Evidence That Human Faces Have Evolved to Signal Individual Identity." *Nature Communications* 5: 4800–4809.

Sheehan, M. J., and E. A. Tibbetts. 2011. "Specialized Face Learning Is Associated with Individual Recognition in Paper Wasps." *Science* 334: 1272–1275.

Sherwood, C. 2005. "Comparative Anatomy of the Facial Motor Nucleus in Mammals, with an Analysis of Neuron Numbers in Primates." *The Anatomical Record* Part A 287A: 1067–1079.

Shipman, P. 2001. *The Man Who Found the Missing Link: Eugene Dubois and His Life-long Quest to Prove Darwin Right.* New York: Simon & Schuster.

Shu, D-G., H-L. Luo, S. Conway Morris, X-L Zhang, S-X. Hu, et al. 1999. "Lower Cam-brian Vertebrates from South China." *Nature* 402: 42–46.

Shubin, N. 2012. *Your Inner Fish: A Journey into the 3.5 Billion-Year History of the Human Body.* London: Penguin Books.

Simoes-Costa, M., and M. E. Bronner. 2013. "Insights into Neural Crest Development and Evolution from Genomic Analysis." *Genetics Research* 23: 1069–1080.

Simon, H. A. 1992. "A Mechanism for Social Selection and Successful Altruism." *Science* 250: 1665–1668.

Simpson, G. G. 1971. *The Meaning of Evolution: A Study of the History of Life and of Its Significance for Man,* 2nd ed. New York: Bantam Books.

Singer, N. 2014. "An Eye That Picks You Out of Any Crowd." *International New York Times,* May 19, p. 1.

Sliwa, R., J. R. Duhamel, O. Pascalis, and S. Wirth. 2011. "Spontaneous Voice-Face Identity Matching by Rhesus Monkeys for Familiar Conspecifics and Humans." *Proceedings of the National Academy of Sciences USA* 108: 1735–1740.

Smith, T., K. D. Rose, and P. D. Gingerich. "Rapid Asia-Europe-North America geographic dispersal of earliest Eocene primate *Teilhardina* during the Paleocene-Eocene Thermal Maximum," *Proceedings of the National Academy of Sciences USA* 103 (2006): 11223–11227.

Soligo, C. R., and D. Martin. 2006. "Adaptive Origins of Primates Revealed." *Journal of Human Evolution* 50: 414–430.

Steiper, M. E., and E. R. Seiffert. 2012. "Evidence for a Convergent Slowdown in Primate Molecular Rates and Its Implications for the Timing of Early Primate Evolution." *Proceedings of the National Academy of Sciences USA* 109: 6006–6011.

Steiper, M. E., and N. M. Young. 2006. "Primate Molecular Divergence Dates." *Molecular Phylogenetics and Evolution* 41: 384–394.

Stern, D. L., and V. Orgozogo. 2008. "The Loci of Evolution: How Predictable Is Genetic Evolution?" *Evolution* 62(9): 2155–2177.

Stott, R. 2012. *Darwin's Ghosts: In Search of the First Evolutionists.* London: Bloomsbury.

Stow, D. 2012. *Vanished Ocean: How Tethys Reshaped the World.* Oxford, UK: Oxford University Press.

Striedter, G. F. 1994. "The Vocal Control Properties in Budgerigars Differ from Those in Songbirds." *Journal of Comparative Neurology* 343: 35–36.

Stringer, C. B., and P. Andrews. 1988. "Genetic and Fossil Evidence for the Origin of Modern Humans." *Science* 239: 1263–1268.

Stuart, A. J. 1991. "Mammalian Extinctions in the Late Pleistocene of Northern Eurasia and North America." *Biological Reviews of the Cambridge Philosophical Society* 66: 453–562.

Sturm, R. 2009. "Molecular Genetics of Human Pigmentation Diversity." *Human Molecular Genetics* 18: R9–R17.

Sturtevant, A. H. 1965. *A History of Genetics.* New York: Harper & Row.

Sues, H.-D., and N. C. Fraser. 2010. *Triassic Life on Land: The Great Transition.* New York: Columbia University Press.

Sun, X., F. V. Mariani, and G. R. Martin. 2002. "Functions of FGF Signaling from the Apical Ectodermal Ridge in Limb Development." *Nature* 418: 501–508.

Sutrop, U. 2012. "Estonian Traces in the Tree of Life Concept and in the Language Family Tree Theory." *Journal of Estonian and Finno-Ulgric Languages* 3: 297–326.

Takahata, N., Y. Satta, and J. Klein. 1995. "Divergence Time and Population Size in the Lineage Leading to Modern Humans." *Theoretical Population Biology* 48: 198–221.

Tate, A. J., H. Fischer, A. E. Leigh, and K. M. Kendrick. 2006. "Behavioural and Neurophysiological Evidence for Face Identity and Face Emotion Processing in Animals." *Philosophical Transactions of the Royal Society B* 361: 2155–2172.

Tavare, S., C. R. Marshall, O. Will, C. Soligo, and R. D. Martin. 2002. "Using the Fossil Record to Estimate the Age of the Last Common Ancestor of Extant Primates." *Nature* 416: 726–729.

Thewissen, J. G., M. G. Cohn, L. S. Stevens, S. Bajpal, J. Heyning, and W. E. Horton Jr. 2006. "Developmental Basis for Hind-Limb Loss in Dolphins and the Origin of the Cetacean Bodyplan." *Proceedings of the National Academy of Science USA* 103: 8414–8418.

Thompson, D. W. 1917. *On Growth and Form.* Cambridge, UK: Cambridge University Press.

Tibbetts, E. 2002. "Visual Signals of Individual Identity in the Wasp *Polistes fuscatus.*" *Proceedings of the Royal Society London B* 269: 1423–1428.

Tomasello, M. 2008. *Origins of Human Communication.* Cambridge: MIT Press.

Tomasello, M., B. Hare, H. Lehmann, and J. Call. 2007. "Reliance on Head Versus Eyes in the Gaze Following of Great Apes and Human Infants: The Cooperative Eye Hypothesis." *Journal of Human Evolution* 52 (2007): 314–320.

Towers, M., and C. Tickle. 2009. "Generation of Pattern and Form in the Developing Limb." *International Journal of Developmental Biology* 53: 805–812.

True, J. R., and S. B. Carroll. 2002. "Gene Co-option in Physiological and Morphological Evolution." *Annual Reviews of Cell and Developmental Biology* 18: 53–80.

Trut, L., I. Oksina, and A. Kharmalova. 2009. "Animal Evolution During Domestication: The Domesticated Fox as a Model." *BioEssays* 31: 349–360.

Tucker, A. S., A. A. Khamis, C. A. Ferguson, I. Bach, M. G. Rosenfeld, and P. T. Sharpe. 1998. "Conserved Regulation of Mesenchymal Gene Expression by Fgf-8 in Face and Limb Development." *Development* 126: 221–228.

Ueda, S., G. Kumagai, Y. Otaki, S. Yamaguchi, and S. Kohshima. 2014. "A Comparison of Facial Color Pattern and Gazing Behavior in Canid Species Suggests Gaze Communication in Gray Wolves *(Canis lupis).*" *PLOS One* 9(6): e98217.

Vanderhaeghen, P., and H.-J. Cheng. 2010. "Guidance Molecules in Axon Pruning and Cell Death." *Cold Spring Harbor Perspectives in Biology* 2: a001859.

Van Schaik, C. 2004. *Among Orangutans: Red Apes and the Rise of Human Culture.* Cambridge, MA: Harvard University Press.

Vernot, B., and J. M. Akey. 2014. "Resurrecting Surviving Neanderthal Lineages from Modern Human Genomes." *Science* 343: 1017–1021.

Vick, S-J., B. M. Waller, L. A. Parr, M. C. S. Pasqualini, and K. A. Bard. 2007. "A Cross-Species Comparison of Facial Morphology and Movement in Humans and Chimpanzees Using the Facial Action Coding System (FACS)." *Journal of Nonverbal Behavior* 31: 1–20.

Vickaryous, M. K., and B. K. Hall. 2006. "Human Cell Type Diversity, Evolution, Development and Classification with Special Reference to Cells Derived from the Neural Crest." *Biological Reviews* 3: 425–455.

Viskontas, I.V., R.V. Quiroga, and I. Fried. 2009. "Human Medial Temporal Lobe Neurons Respond Preferentially to Personally Relevant Images." *Proceedings of the National Academy of Sciences USA* 106: 21329–21334.

Waddington, C. H. 1957. *Strategy of the Genes.* London: Allen & Unwin.

———. 1961. "Genetic Assimilation." *Advances in Genetics* 10: 257–293.

Wagner, G. P., and V. J. Lynch. 2008. "The Gene Regulatory Logic of Transcription Factor Evolution." *Trends in Ecology and Evolution* 23: 377–385.

Waller, B. M., K. A. Bard, S.-J. Vick, and M. C. S. Pasqualini. 2007. "Perceived Differences between Chimpanzee *(Pan troglodytes)* and Human *(Homo sapiens)* Facial Expressions Are Related to Emotional Interpretation." *Journal of Comparative Psychology* 121: 398–404.

Watson, J. 1970. *The Molecular Biology of the Gene*. New York: W. A. Benjamin.

Weaver, T. D., C. C. Roseman, and C. B. Stringer. 2008. "Close Correspondence Between Quantitative- and Molecular-Genetic Divergence Times for Neanderthals and Modern Humans." *Proceedings of the National Academy of Sciences USA* 105: 4645–4629.

West-Eberhard, M. J. 2003. *Developmental Plasticity and Evolution*. Oxford, UK: Oxford University Press.

Weston, E. M., A. E. Friday, and P. Lio. 2007. "Biometric Evidence that Sexual Selection Has Shaped the Hominin Face." *PLOS One* 8: e710.

Wheeler, B. C., B. J. Bradley, and J. M. Kamiliar. 2011. "Predictors of Orbital Convergence in Primates: A Test of the Snake Detection Hypothesis of Primate Evolution." *Journal of Human Evolution* 61: 233–242.

White, R. 2002. *Prehistoric Art: The Symbolic Journey of Mankind*. New York: Harry N. Abrams.

Wibble, J. R., G. W. Rougier, M. J. Novacek, and R. J. Asher. 2007. "Cretaceous Eutherians and Laurasian Origin for Placental Mammals Near the K/T Boundary." *Nature* 447: 1003–1006.

Wiles, J. F., J. G. Kunkel, and A. C. Wilson. 1983. "Birds, Behavior, and Anatomical Evolution." *Proceedings of the National Academy of Sciences USA* 80: 4394–4397.

Wilkins, A. S. 1986. *Genetic Analysis of Animal Development*. New York: Wiley & Sons.

——.1997. "Canalization: A Molecular Genetic Perspective." *BioEssays* 19: 257–262.

——. 2002. *The Evolution of Developmental Pathways*. Sunderland, MA: Sinauer Associates.

——. 2007a. "Between 'Design' and 'Bricolage': Genetic Networks, Levels of Selection and Adaptive Evolution." *Proceedings of the National Academy of Sciences USA* 104: 8590–8596.

——. 2007b. "Genetic Networks as Transmitting and Amplifying Devices for Natural Genetic Tinkering." Pp. 71–89 in *Tinkering: The Microevolution of Development*, edited by G. Bock and J. Goode. Chichester, UK: John Wiley & Sons.

——. 2008. "Neodarwinism." Pp. 491–516 in *Icons of Evolution*, Vol. 2, edited by B. Regal. Westport, CT: Greenwood Press.

——. 2014. "'The Genetic Tool-Kit': The Life History of an Important Metaphor." Pp. 1–14 in *Advances in Evolutionary Developmental Biology*, edited by J. T. Streelman. Hoboken, NJ: John Wiley & Sons.

Wilkins, A. S., R. W. Wrangham, and W. T. Fitch. 2014. "The 'Domestication Syndrome' in Mammals: A Unified Explanation Based on Neural Crest Cell Behavior and Genetics." *Genetics* 197: 1–14.

Wilkinson, R. D., M. E. Steiper, C. Soligo, R. D. Martin, Z. Yang, and S. Tavare. 2011. "Dating Primate Divergences Through an Integrated Analysis of Paleontological and Molecular Data." *Systematic Biology* 60: 16–31.

Williams, B. A., R. F. Kay, and E. C. Kirk. 2010. "New Perspectives on Anthropoid Origins." *Proceedings of the National Academy of Sciences USA* 107: 4797–4804.

Wilmer, J. B., L. Germine, C. F. Chabris, G. Chatterjee, M. Williams, et al. 2010. "Human Face Recognition Ability Is Specific and Highly Heritable." *Proceedings of the National Academy of Sciences USA* 107: 5238–5241.

Windhajer, S., F. Hutzler, C. C. Carbon, E. Oberzaucher, K. Schaefer, et al. 2010. "Laying Eyes on Headlights: Eye Movements Suggest Facial Features in Cars." *Collegium Antropologicum* 34: 1075–1080.

Windhajer, S., K. Schaefer, and B. Fink. 2011. "Geometric Morphometrics of Male Facial Shape in Relation to Physical Strength and Perceived Attractiveness, Dominance, and Masculinity." *American Journal of Human Biology* 23: 805–814.

Withington, S., R. Beddington, and J. Cooke. 2001. "Foregut Endoderm Is Required at Head Process Stages for Anterior-Most Neural Patterning in the Chick." *Development* 128(3): 309–320.

Wolpert, L. 1969. "Positional Information and the Spatial Pattern of Cellular Differentiation." *Journal of Theoretical Biology* 25: 1–47.

✓ Wolpoof, M. H., X. Wu, and A. G. Thorne. 1984. "Modern *Homo sapiens* Origins, a General Theory of Hominid Evolution Involving the Fossil Evidence from East Asia." Pp. 411–483 in *The Origins of Modern Humans,* edited by F. H. Smith and F. Spencer. London: Alan Liss.

✓ Wood, B., and T. Harrison. 2011. "The Evolutionary Context of the First Hominins." *Nature* 470: 347–351.

Wrangham, R. 2009. *Catching Fire: How Cooking Made Us Human.* London: Profile Books.

Yamamoto, Y., M. S. Byerly, W. R. Jackman, and W. R. Jeffrey. 2009. "Pleiotropic Functions of Embryonic Sonic Hedgehog Expression Link Jaw and Taste Bud Amplification with Eye Loss During Cavefish Evolution." *Developmental Biology* 330: 200–211.

Yanai, I., J. O. Korbel, S. Boue, S. K. McWheeney, P. Bork, and M. J. Lercher. 2006. "Similar Gene Expression Profiles Do Not Imply Similar Gene Function." *Trends in Genetics* 22(3): 132–138.

Yang, M. A., A.-S. Malaspinas, E. Y. Durand, and M. Slatkin. 2012. "Ancient Structure in Africa Unlikely to Explain Neanderthal and Non-African Genetic Similarity." *Molecular Biology and Evolution* 29: 2987–2995.

Young, N. M., H. J. Chong, D. Hu, B. Hallgrimsson, and R. S. Marcucio. 2011. "Quantitative Analyses Link Modulation of Sonic Hedgehog Signaling to Continuous Variation in Facial Growth and Shape." *Development* 137: 3405–3409.

Young, N. M., D. Hu, A. J. Lainoff, F. J. Smith, R. Diaz, et al. 2014. "Embryonic Bauplans and the Developmental Origins of Facial Diversity and Constraint." *Development* 141: 1059–1064.

Yu, J-K. S. 2010. "The Evolutionary Origin of the Vertebrate Neural Crest and Its Developmental Gene Regulatory Networks . . . Insights from Amphioxus." *Zoology* 113: 1–9.

Zelditch, M.L, D. L. Swiderski, H. D. Sheets, and W. L. Fink. 2004. *Geometric Morphometrics for Biologists: A Primer.* San Diego: Elsevier Academic Press.

Zuckerkandl, E., and L. Pauling. 1965. "Evolutionary Divergence and Convergence in Proteins." Pp. 97–166 in *Evolving Genes and Proteins*, edited by V. Bryson and H. J. Vogel. New York: Academic Press.

ACKNOWLEDGMENTS

The creation of this book has had a long history, starting with a lengthy gestation—too long to admit to here—before I began serious writing in 2011. Along the way, many people have expressed interest in the project, read sections of early drafts, discussed the ideas with me, suggested reading, or helped me in numerous other ways. I will try to acknowledge them all but suspect that some individuals have slipped through the cracks of memory. I can only ask those whose names were inadvertently left out to forgive me for the unintended omissions. Furthermore, while I have tried to divide the kinds of assistance into different categories, there were overlaps, of course, that are probably not all noted here.

First, I would like to thank the following for helpful discussions, in person or by e-mail, of aspects of the subject matter of the book. They provided provocative and useful comments or ideas or particular facts. They included Arkhat Abzhanov, Anne Burrows, Robert Cieri, Sophie Creuzet, Michael Depew, Barbara Finlay, Tecumseh Fitch, Luca Giuliani, Benedikt Hallgrimsson, Nick Holland, Eva Jablonska, Daniel Lieberman, David Lewis-Williams, John Maisey, Robert Martin, Daniel Milo, Olaf Oftedal, Maureen O'Leary, Rob Podolski, Joan Richtsmeier, Mark Shriver, Francis Thackeray, the late Philip Tobias, Lyudmilla Trut, and Richard Wrangham.

People who steered me to or directly provided useful source material include Doug Berg, Monique Borgeroff-Mulder, Brian Cusack, Michael Depew, Jill Danzig, Richard Danzig, Barbara Finlay, Andries Gouws, Susan Hamilton, Lewis Held, Louise Holland, Nick Holland, Bertrand Jordan, Daniel Lieberman, Robert Martin, Peter MacNeilage, Jan Plamper, Bruce Rusbridger, Katrin Schaefer, Joern Schmiedel, James Sikela, Michael Tomasello, Brigid Waller, Melinda Wax, Elizabeth von Weiszaecker, and Patricia Wright. In this group, I would especially like to single out Doug Berg, an old friend (of more than fifty years standing!) who was a never-ending source of interesting and valuable references.

I am especially indebted to those individuals who read various chapters and sent me comments on the material. Collectively, they saved me from many errors of both commission and omission. These friends and colleagues are Francisco Aboitiz, Walter Bodmer, Anne Burrows, Richard Danzig, Barbara Finlay, Anthony Grahame, Brian Hall, Nick Holland,

Eva Jablonka, Bertrand Jordan, Natascha Koenigs, Jack Levy, Robert Martin, Gillian Morriss-Kay, Stephen Oppenheimer, Rob Podolski, Constance Putnam, Suzannah Rutherford, Michael Sheehan, Isaac Wilkins, and Richard Wrangham. The usual statement of ownership of flaws, of course, applies: I take full responsibility for all errors of fact, interpretation, or omission that remain.

A special debt of gratitude is owed to my illustrator, Sarah Kennedy, who was able to take my often rather vague suggestions and turn them into clear, informative figures. In addition, the staff at Harvard University Press were tremendously helpful, in particular my former editor, Michael Fisher, and his assistant, Lauren Esdaile. My second editor, Andrew Kinney, coming to the project late, did an excellent job of steering it to publication, ably assisted by Katrina Vassalo. Critical help toward the end was provided by Mary Ribesky, my project manager, and Steven Summerlight, my copyeditor. Early in the project, Peter Nevraumont supplied especially helpful editorial advice.

I am also grateful to various people who sent me pictures for the color plates section of the book. These were Dimitry Bogdanov, Jan Böttger, Dorothy Cheney, Jean Clottes, Thomas Geissman, John Gurche, Thomas Kocher, Maureen O'Leary, Robert Martin, Dieter and Netzin Steklis, Lyudmilla Trut, and Maria von Noordwijk.

Far from least, I would like to express my deep appreciation to two institutions for the support (financial, moral, practical) they provided during the early stages of this project, when I started systematically assembling my knowledge and ideas and began to write. These institutions were the Wissenschaftskolleg zu Berlin (Germany), where I was a fellow from 2009 to 2010, and the Stellenbosch Institute of Advanced Studies (STIAS) (Stellenbosch, South Africa), where I had a fellowship from January to September 2011. The kindness and active assistance of the staff of these two institutions can hardly be sufficiently acknowledged, let alone repaid. A third institution, the Institute of Theoretical Biology (ITB), Humboldt Universität zu Berlin (Germany), provided a much-appreciated base during much of the long writing period from late 2011 to early 2016, and I would like to thank Peter Hammerstein for making possible my connection with the ITB.

Finally, I would like to acknowledge more fully the indirect but important contribution of the three individuals named in the dedication:

my biological parents and my stepfather. To my father, I owe my introduction to and fascination with the world of science, particularly biology and chemistry. My mother, in contrast, imbued me with a deep love of literature and interest in languages. Later, my stepfather opened up the rich world of history and historical research for me. The net result of having three such mentors was the lifelong conviction that our world in all its facets is endlessly interesting, a feeling I experienced repeatedly while constructing this book about the human face and its history.

INDEX

Note: Page numbers in italics refer to illustrations.